Data Mining for Bioinformatics

Data Mining for Bioinformatics

Sumeet Dua
Pradeep Chowriappa

CRC Press
Taylor & Francis Group
Boca Raton London New York

CRC Press is an imprint of the
Taylor & Francis Group, an **informa** business

AN AUERBACH BOOK

CRC Press
Taylor & Francis Group
6000 Broken Sound Parkway NW, Suite 300
Boca Raton, FL 33487-2742

© 2013 by Taylor & Francis Group, LLC
CRC Press is an imprint of Taylor & Francis Group, an Informa business

No claim to original U.S. Government works

Version Date: 20120725

International Standard Book Number: 978-0-8493-2801-5 (Hardback)

This book contains information obtained from authentic and highly regarded sources. Reasonable efforts have been made to publish reliable data and information, but the author and publisher cannot assume responsibility for the validity of all materials or the consequences of their use. The authors and publishers have attempted to trace the copyright holders of all material reproduced in this publication and apologize to copyright holders if permission to publish in this form has not been obtained. If any copyright material has not been acknowledged please write and let us know so we may rectify in any future reprint.

Except as permitted under U.S. Copyright Law, no part of this book may be reprinted, reproduced, transmitted, or utilized in any form by any electronic, mechanical, or other means, now known or hereafter invented, including photocopying, microfilming, and recording, or in any information storage or retrieval system, without written permission from the publishers.

For permission to photocopy or use material electronically from this work, please access www.copyright.com (http://www.copyright.com/) or contact the Copyright Clearance Center, Inc. (CCC), 222 Rosewood Drive, Danvers, MA 01923, 978-750-8400. CCC is a not-for-profit organization that provides licenses and registration for a variety of users. For organizations that have been granted a photocopy license by the CCC, a separate system of payment has been arranged.

Trademark Notice: Product or corporate names may be trademarks or registered trademarks, and are used only for identification and explanation without intent to infringe.

Library of Congress Cataloging-in-Publication Data

Dua, Sumeet, author.
  Data mining for bioinformatics / Sumeet Dua, Pradeep Chowriappa.
  pages cm
  Summary: "Data Mining for Bioinformatics enables researchers to meet the challenge of mining vast amounts of biomolecular data to discover real knowledge. Covering theory, algorithms, and methodologies, as well as data mining technologies, the book presents a thorough discussion of data-intensive computations used in data mining applied to bioinformatics. The book explains data mining design concepts to build applications and systems. It shows how to prepare raw data for the mining process and is filled with heuristics that speed the data mining process"-- Provided by publisher.
  Includes bibliographical references and index.
  ISBN 978-0-8493-2801-5 (hardback)
  1. Bioinformatics. 2. Data mining. I. Chowriappa, Pradeep, author. II. Title.

QH324.27.D83 2013
572'.330285--dc23   2012021082

Visit the Taylor & Francis Web site at
http://www.taylorandfrancis.com

and the CRC Press Web site at
http://www.crcpress.com

Contents

Preface xv
About the Authors xix

SECTION I

1 Introduction to Bioinformatics 3
1.1 Introduction 3
1.2 Transcription and Translation 8
    1.2.1 The Central Dogma of Molecular Biology 9
1.3 The Human Genome Project 11
1.4 Beyond the Human Genome Project 12
    1.4.1 Sequencing Technology 13
        1.4.1.1 Dideoxy Sequencing 14
        1.4.1.2 Cyclic Array Sequencing 15
        1.4.1.3 Sequencing by Hybridization 15
        1.4.1.4 Microelectrophoresis 16
        1.4.1.5 Mass Spectrometry 16
        1.4.1.6 Nanopore Sequencing 16
    1.4.2 Next-Generation Sequencing 17
        1.4.2.1 Challenges of Handling NGS Data 18
    1.4.3 Sequence Variation Studies 20
        1.4.3.1 Kinds of Genomic Variations 21
        1.4.3.2 SNP Characterization 22
    1.4.4 Functional Genomics 24
        1.4.4.1 Splicing and Alternative Splicing 26
        1.4.4.2 Microarray-Based Functional Genomics 30
    1.4.5 Comparative Genomics 32
    1.4.6 Functional Annotation 33
        1.4.6.1 Function Prediction Aspects 33
1.5 Conclusion 37
References 37

2 Biological Databases and Integration 41
- 2.1 Introduction: Scientific Work Flows and Knowledge Discovery 41
- 2.2 Biological Data Storage and Analysis... 44
  - 2.2.1 Challenges of Biological Data.. 44
  - 2.2.2 Classification of Bioscience Databases 48
    - 2.2.2.1 Primary versus Secondary Databases 48
    - 2.2.2.2 Deep versus Broad Databases 48
    - 2.2.2.3 Point Solution versus General Solution Databases 49
  - 2.2.3 Gene Expression Omnibus (GEO) Database 51
  - 2.2.4 The Protein Data Bank (PDB) 53
- 2.3 The Curse of Dimensionality...58
- 2.4 Data Cleaning 59
  - 2.4.1 Problems of Data Cleaning...59
  - 2.4.2 Challenges of Handling Evolving Databases 61
    - 2.4.2.1 Problems Associated with Single-Source Techniques 62
    - 2.4.2.2 Problems Associated with Multisource Integration..62
  - 2.4.3 Data Argumentation: Cleaning at the Schema Level 63
  - 2.4.4 Knowledge-Based Framework: Cleaning at the Instance Level..65
  - 2.4.5 Data Integration 67
    - 2.4.5.1 Ensembl..68
    - 2.4.5.2 Sequence Retrieval System (SRS).........................68
    - 2.4.5.3 IBM's DiscoveryLink..69
    - 2.4.5.4 Wrappers: Customizable Database Software........70
    - 2.4.5.5 Data Warehousing: Data Management with Query Optimization......................................70
    - 2.4.5.6 Data Integration in the PDB 74
- 2.5 Conclusion 76
- References...78

3 Knowledge Discovery in Databases 81
- 3.1 Introduction 81
- 3.2 Analysis of Data Using Large Databases... 84
  - 3.2.1 Distance Metrics 84
  - 3.2.2 Data Cleaning and Data Preprocessing 85
- 3.3 Challenges in Data Cleaning...86
  - 3.3.1 Models of Data Cleaning..89
    - 3.3.1.1 Proximity-Based Techniques................................ 90
    - 3.3.1.2 Parametric Methods 91
    - 3.3.1.3 Nonparametric Methods 93

		3.3.1.4	Semiparametric Methods...............................93
		3.3.1.5	Neural Networks 93
		3.3.1.6	Machine Learning 95
		3.3.1.7	Hybrid Systems...96

3.4 Data Integration 97
    3.4.1 Data Integration and Data Linkage 97
    3.4.2 Schema Integration Issues...98
    3.4.3 Field Matching Techniques 99
        3.4.3.1 Character-Based Similarity Metrics 99
        3.4.3.2 Token-Based Similarity Metrics...................... 101
        3.4.3.3 Data Linkage/Matching Techniques 102
3.5 Data Warehousing..104
    3.5.1 Online Analytical Processing...105
    3.5.2 Differences between OLAP and OLTP 106
    3.5.3 OLAP Tasks 106
    3.5.4 Life Cycle of a Data Warehouse....................................107
3.6 Conclusion 109
References 109

SECTION II

4 Feature Selection and Extraction Strategies in Data Mining..............113
4.1 Introduction 113
4.2 Overfitting 114
4.3 Data Transformation 115
    4.3.1 Data Smoothing by Discretization................................115
        4.3.1.1 Discretization of Continuous Attributes 116
    4.3.2 Normalization and Standardization..............................118
        4.3.2.1 Min-Max Normalization.................................118
        4.3.2.2 z-Score Standardization....................................118
        4.3.2.3 Normalization by Decimal Scaling...................119
4.4 Features and Relevance..119
    4.4.1 Strongly Relevant Features 119
    4.4.2 Weakly Relevant to the Dataset/Distribution 120
    4.4.3 Pearson Correlation Coefficient 120
    4.4.4 Information Theoretic Ranking Criteria 121
4.5 Overview of Feature Selection 121
    4.5.1 Filter Approaches...122
    4.5.2 Wrapper Approaches...123
4.6 Filter Approaches for Feature Selection.....................................124
    4.6.1 FOCUS Algorithm..124
    4.6.2 RELIEF Method—Weight-Based Approach...................126

viii ■ Contents

    4.7   Feature Subset Selection Using Forward Selection 128
              4.7.1   Gram-Schmidt Forward Feature Selection 128
    4.8   Other Nested Subset Selection Methods 130
    4.9   Feature Construction and Extraction 131
              4.9.1   Matrix Factorization 132
                      4.9.1.1   LU Decomposition 132
                      4.9.1.2   QR Factorization to Extract
                                  Orthogonal Features 133
                      4.9.1.3   Eigenvalues and Eigenvectors of a Matrix 133
              4.9.2   Other Properties of a Matrix 134
              4.9.3   A Square Matrix and Matrix Diagonalization 134
                      4.9.3.1   Symmetric Real Matrix: Spectral Theorem 135
                      4.9.3.2   Singular Vector Decomposition (SVD) 135
              4.9.4   Principal Component Analysis (PCA) 136
                      4.9.4.1   Jordan Decomposition of a Matrix 137
                      4.9.4.2   Principal Components 138
              4.9.5   Partial Least-Squares-Based Dimension
                      Reduction (PLS) 138
              4.9.6   Factor Analysis (FA) 139
              4.9.7   Independent Component Analysis (ICA) 140
              4.9.8   Multidimensional Scaling (MDS) 141
    4.10 Conclusion 142
References 143

5 Feature Interpretation for Biological Learning 145
    5.1   Introduction 145
    5.2   Normalization Techniques for Gene Expression Analysis 146
              5.2.1   Normalization and Standardization Techniques 146
                      5.2.1.1   Expression Ratios 148
                      5.2.1.2   Intensity-Based Normalization 148
                      5.2.1.3   Total Intensity Normalization 149
                      5.2.1.4   Intensity-Based Filtering of Array Elements 153
              5.2.2   Identification of Differentially Expressed Genes 155
              5.2.3   Selection Bias of Gene Expression Data 156
    5.3   Data Preprocessing of Mass Spectrometry Data 157
              5.3.1   Data Transformation Techniques 158
                      5.3.1.1   Baseline Subtraction (Smoothing) 158
                      5.3.1.2   Normalization 158
                      5.3.1.3   Binning 159
                      5.3.1.4   Peak Detection 160
                      5.3.1.5   Peak Alignment 160

| | | 5.3.2 | Application of Dimensionality Reduction Techniques for MS Data Analysis | 161 |

		5.3.3	Feature Selection Techniques	162
			5.3.3.1 Univariate Methods	163
			5.3.3.2 Multivariate Methods	164
5.4	Data Preprocessing for Genomic Sequence Data			165
	5.4.1	Feature Selection for Sequence Analysis		166
5.5	Ontologies in Bioinformatics			167
	5.5.1	The Role of Ontologies in Bioinformatics		169
			5.5.1.1 Description Logics	171
			5.5.1.2 Gene Ontology (GO)	171
			5.5.1.3 Open Biomedical Ontologies (OBO)	172
5.6	Conclusion			174
References				176

SECTION III

6 Clustering Techniques in Bioinformatics 181

- 6.1 Introduction 181
- 6.2 Clustering in Bioinformatics 182
- 6.3 Clustering Techniques 183
  - 6.3.1 Distance-Based Clustering and Measures 183
    - 6.3.1.1 Mahalanobis Distance 183
    - 6.3.1.2 Minkowiski Distance 184
    - 6.3.1.3 Pearson Correlation 185
    - 6.3.1.4 Binary Features 185
    - 6.3.1.5 Nominal Features 186
    - 6.3.1.6 Mixed Variables 187
  - 6.3.2 Distance Measure Properties 187
  - 6.3.3 k-Means Algorithm 188
  - 6.3.4 k-Modes Algorithm 190
  - 6.3.5 Genetic Distance Measure (GDM) 190
- 6.4 Applications of Distance-Based Clustering in Bioinformatics 191
  - 6.4.1 New Distance Metric in Gene Expressions for Coexpressed Genes 192
  - 6.4.2 Gene Expression Clustering Using Mutual Information Distance Measure 193
  - 6.4.3 Gene Expression Data Clustering Using a Local Shape-Based Clustering 194
    - 6.4.3.1 Exact Similarity Computation 194
    - 6.4.3.2 Approximate Similarity Computation 194

6.5 Implementation of *k*-Means in WEKA 195
6.6 Hierarchical Clustering 196
    6.6.1 Agglomerative Hierarchical Clustering 196
    6.6.2 Cluster Splitting and Merging 197
    6.6.3 Calculate Distance between Clusters 198
    6.6.4 Applications of Hierarchical Clustering Techniques in Bioinformatics 199
        6.6.4.1 Hierarchical Clustering Based on Partially Overlapping and Irregular Data 200
        6.6.4.2 Cluster Stability Estimation for Microarray Data 201
        6.6.4.3 Comparing Gene Expression Sequences Using Pairwise Average Linking 202
6.7 Implementation of Hierarchical Clustering 202
6.8 Self-Organizing Maps Clustering 203
    6.8.1 SOM Algorithm 203
    6.8.2 Application of SOM in Bioinformatics 206
        6.8.2.1 Identifying Distinct Gene Expression Patterns Using SOM 206
        6.8.2.2 SOTA: Combining SOM and Hierarchical Clustering for Representation of Genes 206
6.9 Fuzzy Clustering 207
    6.9.1 Fuzzy *c*-Means (FCM) 209
    6.9.2 Application of Fuzzy Clustering in Bioinformatics 210
        6.9.2.1 Clustering Genes Using Fuzzy *J*-Means and VNS Methods 210
        6.9.2.2 Fuzzy *k*-Means Clustering on Gene Expression 212
        6.9.2.3 Comparison of Fuzzy Clustering Algorithms 213
6.10 Implementation of Expectation Maximization Algorithm 215
6.11 Conclusion 215
References 216

7 Advanced Clustering Techniques 219
7.1 Graph-Based Clustering 219
    7.1.1 Graph-Based Cluster Properties 219
    7.1.2 Cut in a Graph 221
    7.1.3 Intracluster and Intercluster Density 221
7.2 Measures for Identifying Clusters 222
    7.2.1 Identifying Clusters by Computing Values for the Vertices or Vertex Similarity 222
        7.2.1.1 Distance and Similarity Measure 223
        7.2.1.2 Adjacency-Based Measures 223
        7.2.1.3 Connectivity Measures 224

	7.2.2	Computing the Fitness Measure 224
		7.2.2.1 Density Measure 224
		7.2.2.2 Cut-Based Measures 225
7.3	Determining a Split in the Graph 225
	7.3.1	Cuts 225
	7.3.2	Spectral Methods 225
	7.3.3	Edge-Betweenness 226
7.4	Graph-Based Algorithms 226
	7.4.1	Chameleon Algorithm 226
	7.4.2	CLICK Algorithm 227
7.5	Application of Graph-Based Clustering in Bioinformatics 228
	7.5.1	Analysis of Gene Expression Data Using Shortest Path (SP) 228
	7.5.2	Construction of Genetic Linkage Maps Using Minimum Spanning Tree of a Graph 228
	7.5.3	Finding Isolated Groups in a Random Graph Process 229
	7.5.4	Implementation in Cytoscape 230
		7.5.4.1 Seeding Method 230
7.6	Kernel-Based Clustering 231
	7.6.1	Kernel Functions 232
	7.6.2	Gaussian Function 232
7.7	Application of Kernel Clustering in Bioinformatics 233
	7.7.1	Kernel Clustering 233
	7.7.2	Kernel-Based Support Vector Clustering 234
	7.7.3	Analyzing Gene Expression Data Using SOM and Kernel-Based Clustering 235
7.8	Model-Based Clustering for Gene Expression Data 237
	7.8.1	Gaussian Mixtures 237
	7.8.2	Diagonal Model 237
	7.8.3	Model Selection 238
7.9	Relevant Number of Genes 238
	7.9.1	A Resampling-Based Approach for Identifying Stable and Tight Patterns 238
	7.9.2	Overcoming the Local Minimum Problem in k-Means Clustering 239
	7.9.3	Tight Clustering 239
	7.9.4	Tight Clustering of Gene Expression Time Courses 239
7.10	Higher-Order Mining 240
	7.10.1	Clustering for Association Rule Discovery 240
	7.10.2	Clustering of Association Rules 240
	7.10.3	Clustering Clusters 241
7.11	Conclusion 241
References 241

SECTION IV

8 Classification Techniques in Bioinformatics 247
    8.1   Introduction 247
            8.1.1   Bias-Variance Trade-Off in Supervised Learning.............248
            8.1.2   Linear and Nonlinear Classifiers.......................................248
            8.1.3   Model Complexity and Size of Training Data 251
            8.1.4   Dimensionality of Input Space 253
    8.2   Supervised Learning in Bioinformatics...254
    8.3   Support Vector Machines (SVMs) 257
            8.3.1   Hyperplanes 258
            8.3.2   Large Margin of Separation 259
            8.3.3   Soft Margin of Separation 260
            8.3.4   Kernel Functions 261
            8.3.5   Applications of SVM in Bioinformatics............................263
                    8.3.5.1   Gene Expression Analysis 263
                    8.3.5.2   Remote Protein Homology Detection 265
    8.4   Bayesian Approaches 268
            8.4.1   Bayes' Theorem.. 268
            8.4.2   Naïve Bayes Classification 268
                    8.4.2.1   Handling of Prior Probabilities.........................269
                    8.4.2.2   Handling of Posterior Probability 270
            8.4.3   Bayesian Networks 270
                    8.4.3.1   Methodology...270
                    8.4.3.2   Capturing Data Distributions Using
                                  Bayesian Networks 272
                    8.4.3.3   Equivalence Classes of Bayesian Networks 273
                    8.4.3.4   Learning Bayesian Networks 273
                    8.4.3.5   Bayesian Scoring Metric 273
            8.4.4   Application of Bayesian Classifiers in Bioinformatics........275
                    8.4.4.1   Binary Classification...277
                    8.4.4.2   Multiclass Classification 278
                    8.4.4.3   Computational Challenges for Gene
                                  Expression Analysis 278
    8.5   Decision Trees 279
            8.5.1   Tree Pruning 280
    8.6   Ensemble Approaches 281
            8.6.1   Bagging 283
                    8.6.1.1   Unweighed Voting Methods............................ 284
                    8.6.1.2   Confidence Voting Methods.............................285
                    8.6.1.3   Ranked Voting Methods 286

		8.6.2	Boosting..287
			8.6.2.1 Seeking Prospective Classifiers to Be Part of the Ensemble..288
			8.6.2.2 Choosing an Optimal Set of Classifiers............288
			8.6.2.3 Assigning Weight to the Chosen Classifier.......290
		8.6.3	Random Forest..291
		8.6.4	Application of Ensemble Approaches in Bioinformatics....292
	8.7	Computational Challenges of Supervised Learning........................295
	8.8	Conclusion 295
	References...296
9	**Validation and Benchmarking** **299**
	9.1	Introduction: Performance Evaluation Techniques.......................299
	9.2	Classifier Validation... 300
		9.2.1	Model Selection..301
			9.2.1.1 Challenges Model Selection...............................302
		9.2.2	Performance Estimation Strategies 303
			9.2.2.1 Holdout...303
			9.2.2.2 Three-Way Split... 304
			9.2.2.3 k-Fold Cross-Validation 305
			9.2.2.4 Random Subsampling 306
	9.3	Performance Measures... 306
		9.3.1	Sensitivity and Specificity...307
		9.3.2	Precision, Recall, and f-Measure 308
		9.3.3	ROC Curve 309
	9.4	Cluster Validation Techniques 310
		9.4.1	The Need for Cluster Validation 311
			9.4.1.1 External Measures 312
			9.4.1.2 Internal Measures..313
		9.4.2	Performance Evaluation Using Validity Indices................. 314
			9.4.2.1 Silhouette Index (SI)... 314
			9.4.2.2 Davies-Bouldin and Dunn's Index..................... 315
			9.4.2.3 Calinski Harabasz (CH) Index 315
			9.4.2.4 Rand Index 316
	9.5	Conclusion 316
	References... 316
Index **319**

Preface

The flourishing field of bioinformatics has been the catalyst to transform biological research paradigms to extend beyond traditional scientific boundaries. Fueled by technological advancements in data collection, storage, and analysis technologies in biological sciences, researchers have begun to increasingly rely on applications of computational knowledge discovery techniques to gain novel biological insight from the data. As we forge into the future of next-generation sequencing technologies, bioinformatics practitioners will continue to design, develop, and employ new algorithms that are efficient, accurate, scalable, reliable, and robust to enable knowledge discovery on the projected exponential growth of raw data. To this end, data mining has been and will continue to be vital for analyzing large volumes of heterogeneous, distributed, semistructured, and interrelated data for knowledge discovery.

This book is targeted to readers who are interested in the embodiments of data mining techniques, technologies, and frameworks employed for effective storing, analyzing, and extracting knowledge from large databases specifically encountered in a variety of bioinformatics domains, including, but not limited to, genomics and proteomics. The book is also designed to give a broad, yet in-depth overview of the application domains of data mining for bioinformatics challenges. The sections of the book are designed to enable readers from both biology and computer science backgrounds to gain an enhanced understanding of the cross-disciplinary field. In addition to providing an overview of the area discussed in Section 1, individual chapters of Sections 2, 3, and 4 are dedicated to key concepts of feature extraction, unsupervised learning, and supervised learning techniques prominently used in bioinformatics.

Section 1 of the book contains three chapters and is designed such that readers from the biological and computer sciences can obtain a comprehensive overview of the evolution of the field and its intersection with computational learning. Chapter 1 provides an overview of the breath of bioinformatics and its associated fields. Readers with a computer science background can obtain an overview of the various databases and the challenges these databases pose through the topics elucidated in Chapter 2. Similarly, readers with a biological background can get acquainted with the concepts prominently referred to in computer science and data mining by using the topics covered in Chapter 3. For a course taught at the undergraduate level, Section 1 captures concepts that are vital in data mining and pertain to its applications on biological databases.

Feature extraction and selection techniques are described in Section 2. Chapter 4 contains associated concepts of data mining, and Chapter 5 provides an overview of the concepts discussed in Chapter 4, pertaining to their application on biological data specific to gene expression analysis and protein expression data. These two chapters can be taught at both undergraduate and graduate levels.

Sections 3 and 4 contain intertwining lessons. Section 3 consists of Chapters 6 and 7, which focus on concepts of unsupervised learning, also known as clustering. Chapter 6 provides an overview of unsupervised learning with simpler and more generic clustering techniques and its application on bioinformatics data, and caters to readers at the undergraduate level. Chapter 7 provides a more comprehensive view of advanced clustering techniques applied to large biological databases and caters to readers at the graduate level.

Chapter 8 of Section 4 provides an overview of supervised learning, also known as classification. This chapter is tailored to suit advanced readers and covers a gamut of classification techniques commonly used in bioinformatics. Chapter 9 is the concluding chapter of the book and contains a description of the various validation and benchmarking techniques used for both clustering and classification.

Possible Course Suggestions

As represented in Figure 0.1, a course focusing on clustering techniques in bioinformatics can use Chapters 6, 7, and 9. Similarly, a course that focuses on classifica-

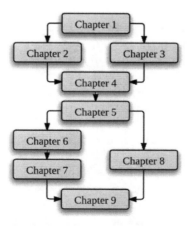

Figure 0.1

tion techniques in bioinformatics can use Chapters 8 and 9. A set of references for additional reading is listed at the end of each chapter.

Organization of the Book

Section 1 of this book is targeted to readers who would be interested in learning the evolution and role of data mining in bioinformatics. It introduces the evolution of bioinformatics and the challenges that can be addressed using data mining techniques.

Simplistically titled "Introduction to Bioinformatics," Chapter 1 provides an introduction and overview of the inception and evolution of bioinformatics, which can serve both as an initial reference and a refresher for readers. It highlights key technological advancements made in the field of biology that have fueled the need for computational techniques to enable automated analysis.

Chapter 2, "Biological Databases and Integration," provides a description of the various biological databases prominently referred to in bioinformatics. This chapter emphasizes the need for data cleaning and cleaning strategies in biological databases that are constantly evolving.

Chapter 3, "Knowledge Discovery in Databases," provides and introduction to the various data mining techniques that can be employed in biological databases. It also emphasizes the various issues and data integration schemes that can be employed for data integration.

Section 2 of this book introduces the role of data mining in analyzing large biological databases. This section is structured such that the reader understands the breath of the various feature selection and feature extraction techniques that data mining has to offer. It also contains application examples of techniques that are prominently used in data-rich fields of proteomics and gene expression data analysis.

Titled "Feature Selection and Extraction Strategies in Data Mining," Chapter 4 focuses on the data mining techniques used to extract and select relevant features from large biological datasets. In this chapter, we touch on topics of normalization, feature selection, and feature extraction that are important for the analysis of large datasets.

It is an important challenge to determine how to interpret the features extracted or selected using the techniques described in Chapter 4. Chapter 5, titled "Feature Interpretation for Biological Learning," therefore focuses on how normalization, feature extraction, and feature selection techniques can be exploited through applications on biological datasets to gain significant insights. This chapter contains descriptions of the application of data mining techniques to areas of mass spectrometry and gene expression analysis that are data rich and introduces the concept of ontologies, abstractions of function for features extracted.

The remaining two sections of the book encapsulate paradigms of both unsupervised and supervised learning in bioinformatics. More specifically, Section 3 focuses on the paradigm of unsupervised learning in data mining, referred to as clustering, and its application to large biological data. The chapters of this section cover important concepts of clustering and provide a gamut of examples of the use of clustering techniques in bioinformatics.

Chapter 6 provides an in-depth description of prominently used clustering techniques and their applications in bioinformatics. Similarly, Chapter 7 contains a comprehensive list of the applications of advanced clustering algorithms used in bioinformatics.

Section 4 gives the reader insight into the challenges of using supervised learning, also known as classification, on biological datasets. This section also addresses the need for validation and benchmarking of inferences derived using either clustering or classification.

"Classification Techniques in Bioinformatics," Chapter 8, contains an overview of classification schemes that are prominently used in bioinformatics. This chapter provides a conceptual view of the challenges encountered during the application of classification on biological databases. The chapter covers systems of both single and ensemble classifiers. Chapter 9 provides the reader insights on model selection and the performance estimation strategies in data mining. The techniques described in this chapter cater to both the validation and benchmarking of clustering and classification techniques.

Acknowledgment

We have been fortunate to have our colleagues and collaborators give us their impressions and contributions toward the contents of this book. We would like to express our gratitude to Mohit Jain for his noteworthy contributions to Chapters 6 and 7, and to Brandy McKnight, who acted as our in-house editorial support. Our gratitude is also due to our current and past collaborators, including Hilary Thompson, Roger Beuerman, James Hill, Brent Christner, and Prerna Dua, for keeping our efforts in perspective and current.

About the Authors

Sumeet Dua is an Upchurch endowed professor of computer science and Interim director of computer science, electrical engineering and electrical engineering technology in the College of Engineering and Science at Louisiana Tech University. He obtained his PhD in computer science from Louisiana State University in 2002. He has coauthored/edited 3 books, has published over 50 research papers in leading journals and conferences, and has advised over 22 graduate thesis and dissertations in the areas of data mining, knowledge discovery, and computational learning in high-dimensional datasets. NIH, NSF, AFRL, AFOSR, NASA, and LA-BOR have supported his research. He frequently serves as a panelist for the NSF and NIH (over 17 panels) and has presented over 25 keynotes, invited talks, and workshops at international conferences and educational institutions. He has also served as the overall program chair for three international conferences and as a chair for multiple conference tracks in the areas of data mining applications and information intelligence. He is a senior member of the IEEE and the ACM. His research interests include information discovery in heterogeneous and distributed datasets, semisupervised learning, content-based feature extraction and modeling, and pattern tracking.

Pradeep Chowriappa is a research assistant professor in the College of Engineering and Science at Louisiana Tech University. His research focuses on the application of data mining algorithms and frameworks on biological and clinical data. Before obtaining his PhD in computer analysis and modeling from Louisiana Tech University in 2008, he pursued a yearlong internship at the Indian Space Research Organization (ISRO), Bangalore, India. He received his masters in computer applications from the University of Madras, Chennai, India, in 2003 and his bachelor's in science and engineering from Loyola Academy, Secunderabad, India, in 2000. His research interests include design and analysis of algorithms for knowledge discovery and modeling in high-dimensional data domains in computational biology, distributed data mining, and domain integration.

BIOINFORMATICS AND KNOWLEDGE DISCOVERY I

Chapter 1

Introduction to Bioinformatics

1.1 Introduction

To understand the functions of the human body, it is first necessary to understand the function of the basic unit of the body—the cell. The human body consists of trillions of cells that perform independent functions and are synchronized to carry out complex bodily functions. Scientists have dug into the functionality of cells, investigating how and why cells perform the tasks that they do. The study of the principles that govern these functions using modeling and computational techniques is the foundation of computational biology.

The human cell possesses hereditary material that is vital for cell replication and duplication and contains several parts, including a plasma membrane and various organelles, which are each designed to render both structure and function for the body (U.S. National Library of Medicine 2011) (Figure 1.1).

Typically, the plasma membrane, also called the lipid bilayer in animal cells, forms an outer lining called the plasma membrane of a cell. This membrane separates the cell from the rest of the environment and selectively allows materials to enter and leave the cell. It is also the characteristic difference between animal and plant cells, as the animal lipid bilayer is characteristically flexible, unlike the rigid plant plasma membrane. The flexibility of the plasma membrane in an animal cell membrane is brought about by its composition of lipid molecules that are characteristically polar, hydrophilic, or hydrophobic in nature. This diversity in composition allows the cell membrane to form various shapes, depending on changes in environmental conditions. The membrane of a cell is coated with

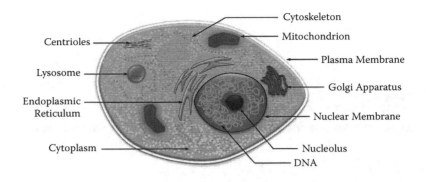

Figure 1.1 A schematic representation of the anatomy of the cell.

surface proteins, such as cell surface receptors, surface antigens, enzymes, and transporters, that bring about the functions of the membrane (Schlessinger and Rost 2005; Tompa 2005). These surface proteins are highly sensitive to the environment, as they are highly hydrophobic or hydrophilic. Research in identifying the structure and function of these membrane proteins has generated interest in recent times (Schlessinger et al. 2006).

The plasma membrane encases the cytoplasm and various organelles of the cell. The bulk of the cell is composed of cytoplasm, which is composed of cytosol (a jelly-like fluid), the nucleus, and other organelle structures. The largest organelle is the cytoskeleton, which is composed of long fibers that spread over the entire cell. Thus, the cytoskeleton provides the vital structure of the cell. Apart from providing the structure and shape of the cell, the cytoskeleton provides several critical functions, including the cell division and movement of the cell.

The endoplasmic reticulum is an organelle of the cell that is a collection of vesicles and tubules held together by the cytoskeleton. Also referred to as the lacey membrane, the endoplasmic reticulum can be one of three types: the rough endoplasmic reticulum (RER), the smooth endoplasmic reticulum (SER), or the sarcoplasmic reticulum (SR). Each of these types of endoplasmic reticulum has specific functions. The RER manufactures proteins through embedded structures known as ribosomes. Ribosomes are organelles that help create proteins by processing genetic instructions coded in the DNA of the nucleus. The ribosomes characteristically attach to the endoplasmic reticulum but, at times, float freely in the cytoplasm. The SER enables the synthesis of lipids and the metabolism of steroids. It is also responsible for regulating the calcium concentration throughout the cell. The SR, which is similar to the SER, functions as a calcium pump. Overall, the endoplasmic reticulum facilitates protein creation, folding, and the transport of the molecules that are in the form of sacs, referred to as the cisternae.

Other organelles in the cell, such as the Golgi apparatus, aid in the packaging of the processed molecules (proteins) from the endoplasmic reticulum for excretion from the cell; this is better known as the recycling center of the cell. Similarly, lysosomes are organelles that break down and digest toxic substances, engulfed bacteria, and viruses in a cell. They also maintain the proper functioning of the cell by recycling worn-out organelles. The organelle responsible for cell function is the mitochondrion, which is responsible for converting food to energy that can be used by the cell. The mitochondrion is a complex organelle that has its own genetic material (deoxyribonucleic acid (DNA)), which is different from the genetic material in the nucleus. This material is known as mitochondrial deoxyribonucleic acid (mtDNA) and enables the mitochondria to self-replicate.

The most important central command center of the cell is the nucleus that houses DNA, the heredity material of the cell. The DNA found in the nucleus is known as the nuclear DNA. Nuclear DNA stores genetic information in the form of a code consisting of four chemical bases, adenine (A), guanine (G), cytosine (C), and thymine (T). Human DNA consists of about 3 billion bases, more than 99% of which are the same in all people. Moreover, nearly every cell in the human body has the same DNA. The nucleus is enveloped by a membrane called the nuclear envelope that protects and separates the DNA from the rest of the cell organelles.

A closer inspection of the DNA sequence shows the existence of an order of the bases in the DNA sequence. This order determines the coded instructions for the cell to grow, mature, divide, or die. In the DNA, the bases A, C, T, and G combine to form base pairs, such as A and T or C and G. A nucleotide consists of an ensemble of these base pairs attached to a sugar molecule and a phosphate molecule (refer to Figure 1.2 for examples of these molecules). The nucleotides in a DNA molecule are arranged in two long strands to form a spiral called the double helix. The structure of DNA is analogous to that of a ladder, where the ladder rungs correspond to the base pairs while the sugar and phosphate molecules correspond to the vertical side pieces of the ladder. This double helix structure of the DNA molecule facilitates replication, and each strand serves as a pattern template for the duplication of sequence bases during cell division, as the resultant child cells should possess the exact copy of the DNA in the parent cell (Figure 1.2).

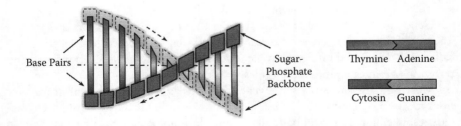

Figure 1.2 Schematic representation of the DNA double helix formed by base pairs attached to a sugar-phosphate backbone. (From http://ghr.nlm.nih.gov/handbook/illustrations/dnastructure.jpg.)

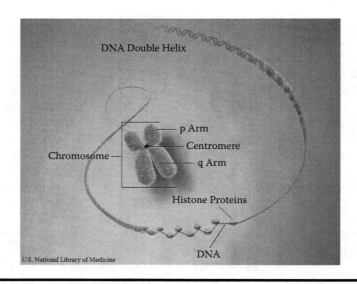

Figure 1.3 DNA and histone proteins are packaged into structures called chromosomes. (From http://ghr.nlm.nih.gov/handbook/illustrations/chromosomestructure.jpg.)

Chromosomes are thread-like structures that contain multiple, tightly packed DNA molecules. These tightly packed units are coiled multiple times around proteins called histones. These histone molecules are believed to provide the necessary structural reinforcement for the chromosome and help in the analysis of the structure of chromosomes. Typically, the structure of a chromosome consists of a central point called the centromere (refer to Figure 1.3), which divides the chromosome into sections called arms. The location of the centromere over the entire chromosome renders the characteristic shape of a chromosome, and acts as the point of reference in locating genes throughout the chromosome. Typically, a chromosome consists of two arms of different lengths. The shorter arm is referred to as the p-arm, and the longer is called the q-arm.

Genes are best known as the basic physical and functional units of heredity. They are found at characteristic locations over the chromosome; these locations are called loci. The coded information (i.e., the DNA) found in genes is translated and transcribed to create protein molecules.

Most humans share the same genes; however, a small number of genes vary from individual to individual. These genes provide individuals their unique characteristics, like hair, eye color, body shape, and skin pigmentation. A particular gene with two or more forms is called an allele. The difference in the gene is exhibited as changes in the DNA bases that contribute to an individual's unique physical features (Figure 1.4).

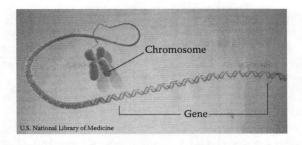

Figure 1.4 Genes are made up of DNA. Each chromosome contains many genes. (From http://ghr.nlm.nih.gov/handbook/illustrations/geneinchromosome.jpg.)

Genes contain codes that are translated into proteins. During translation, the gene codes consisting of trinucleotide units called codons provide the necessary coding for an amino acid. Table 1.1 shows the triplet combinations of nucleotides that result in the creation of 20 known amino acids. The translation is initiated by a START codon (along with nearby initiation factors) and is terminated by a STOP codon. A sequence of amino acids forms a protein, which is a complex molecule that carries out critical functions in the human body. The function of the

Table 1.1 All Amino Acids and Their Corresponding Codons

Amino Acid	Codon	Amino Acid	Codon
Ala/A	GCU, GCC, GCA, GCG	Lys/K	AAA, AAG
Arg/R	CGU, CGC, CGA, CGG, AGA, AGG	Met/M	AUG
Asn/N	AAU, AAC	Phe/F	UUU, UUC
Asp/D	GAU, GAC	Pro/P	CCU, CCC, CCA, CCG
Cys/C	UGU, UGC	Ser/S	UCU, UCC, UCA, UCG, AGU, AGC
Gln/Q	CAA, CAG	Thr/T	ACU, ACC, ACA, ACG
Glu/E	GAA, GAG	Trp/W	UGG
Gly/G	GGU, GGC, GGA, GGG	Tyr/Y	UAU, UAC
His/H	CAU, CAC	Val/V	GUU, GUC, GUA, GUG
Lle/I	AUU, AUC, AUA	START	AUG
Leu/L	UUA, UUG, CUU, CUC, CUA, CUG	STOP	UAA, UGA, UAG

complex protein molecule is determined by its sequence and its three-dimensional (3D) structure, which has direct bearings on the function of the associated gene.

The function of genes is, at times, affected by random changes to naturally occurring sequences. These changes are called mutations. Mutations are random changes in the structure or composition of DNA, which can be caused by mistakes in reproduction or external environmental events, like UV damage. While evolutionary changes in species are caused by beneficial mutations that enable organisms to adapt over time, not all mutations are beneficial. Certain mutations cause diseases such as cancer and could affect the survival of organisms and species over time.

A significant amount of biomedical research has been carried out to determine the functions of protein complexes for medical use. This research has resulted in breakthroughs in drug development.

Section 1.2 contains a description of transcription and translation, closely followed by an introduction to the Human Genome Project (HGP) in Section 1.3, which resulted in an estimate of between 20,000 and 25,000 genes reported in humans.

1.2 Transcription and Translation

The creation of proteins from a gene is complex and consists of two integral steps: transcription and translation. Though most genes contain the information needed to generate proteins, some genes help the cell assemble proteins. Transcription and translation are part of the central dogma of molecular biology, which is the fundamental principle that governs the conversion of information from DNA to RNA to protein (refer to Figure 1.5). The following section provides an overview of the two-stage process of transcription and translation.

The first step of transcription occurs in the nucleus of the cell where the information stored in the DNA (of a gene) is transferred to the mRNA (messenger ribonucleic acid). Typically, both RNA and DNA are composed of nucleotide base chains; however, they differ in properties and chemical composition. The mRNA is a type of RNA that holds the chemical blueprint of the protein product. The resultant protein product carries the encoded information from the DNA within the nucleus to the DNA within the cytoplasm of the cell for the production of the protein complex.

The second step of translation occurs outside the walls of the nucleus, in which the ribosomes present on the rough endoplasmic reticulum read the encoded information from the mRNA to produce the protein. The mRNA sequence consists of a string of codons, three bases that represent independent amino acids. The assembly of amino acids into the corresponding protein sequence is brought about by the transfer RNA (tRNA) one amino acid at a time. This process of assembly continues until the stop codon in the mRNA is encountered. This two-step process is called the central dogma of molecular biology (refer to Figure 1.5).

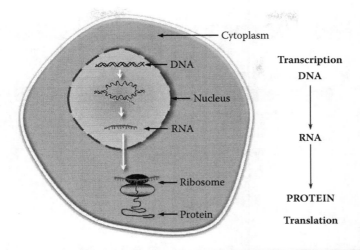

Figure 1.5 The central dogma of molecular biology. The processes of transcription and translation of information from genes are used to make proteins. (From http://ghr.nlm.nih.gov/handbook/illustrations/proteinsyn.jpg.)

1.2.1 The Central Dogma of Molecular Biology

As described previously, each gene contains the genetic makeup of an individual and the coded information required to manufacture both noncoding RNA and proteins. The expression of a gene is carried out by the two-stage process of translation and transcription (refer to Figure 1.6).

The first step in this process is called transcription, which involves the replication of gene content by copying the content of the DNA to an equivalent RNA molecule also known as the primary transcript. The primary transcript is essentially the same sequence as the gene, except that it is complementary in its base pair content. This similarity enables the sequence to convert from DNA and RNA and vice versa, in the presence of certain enzymes. The resultant RNA sequence reflecting the transcribed DNA is called a transcription unit encoding one gene. The nucleotide composition of the resultant RNA includes uracil (U) in place of thymine (T) in the DNA complement. DNA transcription is regulated and directed by regulatory sequences. The DNA sequence before the coding sequence is called the five prime untranslated region (5'UTR); similarly, the sequence following the coding sequence is called the three prime untranslated region (3'UTR). The direction of transcription moves from the 5' to the 3'. Each gene is further divided into intermediate regions called exons and introns. The exons carry information required for protein synthesis. As shown in Figure 1.6, the messenger RNA (mRNA) contains information from the exons. The process of splicing filters out the intron sequence from the primary transcripts.

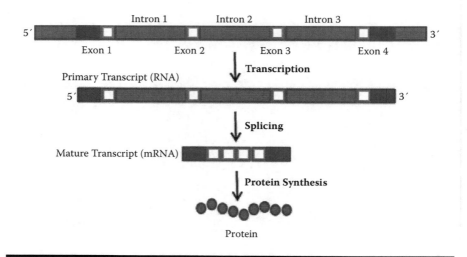

Figure 1.6 An overview of the transcription to translation. The gene is first transcribed to yield a primary transcript, which is processed to remove the introns. The mature transcript (mRNA) is then translated into a sequence of amino acids, which defines the protein. (From http://genome.wellcome.ac.uk/assets/GEN10000676.jpg.)

The second step is translation, also known as protein synthesis. In this step, the resultant mRNA from transcription is translated to the resultant protein complex with the help of ribosomes. Translation occurs in the cytoplasm of the cell, outside the nuclear wall. The decoding of mRNA is initiated when the ribosome binds to the mRNA with the help of tRNAs, which transfer specific amino acids from the cytoplasm to the ribosome. The ribosome helps build the protein complex as it reads the information encoded in the mRNA.

The process of translation begins when the ribosome binds to the 5' end of the mRNA. The codons of the mRNA specify which amino acid needs to be appended to create the polypeptide chain. This process is terminated when the ribosome encounters the 3' (stop codon) of the mRNA. The resultant chain of amino acids folds to form the structure of the protein. This process is called translation, as there is no direct correspondence between the nucleotide sequence of the DNA and the resultant protein complex.

Transcription and translation is a regulated process that enables the controlled expression of genes. With evolution and differences in species, it is known that all genes are not expressed in the same way. With the exception of the housekeeping genes, genes that are always expressed in all cells (performing the basic functions) are expressed differently during different phases of development. Proteins known as transcription factors (TFs) regulate genes. These proteins bind to DNA sequences, preventing them from being transcribed and translated, and thereby switching them on or off as desired. Thus, the gene expression can be a controlled process based on the activity of transcription factors.

Transcription factors, being proteins themselves, require genes to produce them. This requirement opens a conundrum in which one gene expression affects the expression of the other genes. In this manner, genes and proteins are linked in a regulatory hierarchy. This process of turning genes on and off is called gene regulation. Gene regulation is an important part of normal development; however, a number of human diseases are the result of the absence or malfunction of transcription factors and the resultant disruption of gene expression. Considering the importance of gene regulation, a significant amount of research should be performed to understand how genes regulate each other (Figure 1.6) (Baumbach et al. 2008; Cao and Zhao 2008).

1.3 The Human Genome Project

The Human Genome Project (HGP) was initiated as a joint endeavor and sponsored by the Office of Biological and Environmental Research at the Department of Energy (DOE) and the National Human Genome Research Institute at the National Institutes of Health (NIH), with the goal of sequencing the human genome within 15 years (Collins 1998). More than 2,000 scientists from over 20 institutions in 6 countries collaborated to produce the first working draft of the human genome, a landmark in scientific research. The final phase of the HGP (1993–2003) has fulfilled its promise as the single most important project in biology and the biomedical sciences. Although the initial sequence had ~150,000 gaps, and the order and orientation of many of the smaller segments had yet to be established, the finished sequence contained 2.85 billion nucleotide base pairs (bp) and just 341 gaps (Figure 1.7).

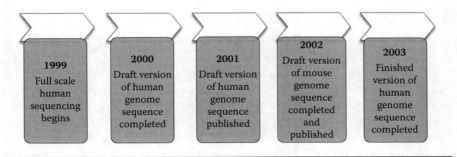

Figure 1.7 Key milestones achieved in the last 5 years of the HGP (1999–2003) (Constructed based on information from http://www.genome.gov/Images/press_photos/highres/38-300.jpg.)

The comprehensive human genome sequence made available through this project has increased our ability to analyze genomes, and has aided research in areas such as large-scale biology, biomedical research, biotechnology, and health care. Though researchers involved with the project have proclaimed it to be complete, certain aspects of the project have yet to be fully implemented. The methods and outcomes of this project are constantly evolving and can lead to a better understanding of gene environment interactions, structures, and functions, thereby eventually leading to the creation of accurate DNA-based medical diagnostics and therapeutics that would be important to the biomedical research community (Collins 1998).

Genetic sequence variation is necessary for the study of evolution. The HGP provides a comprehensive availability of the human genome sequence, thereby presenting unique scientific and research avenues for collaborative research. Apart from providing a means to understand numerous medically important and genetically complex human diseases, the HGP is also focused on delivering (1) genetic tests, (2) a better understanding of inherited diseases, and (3) patient-specific therapies.

Bioinformatics and computational biology are important components of making these goals a reality. The HGP (along with the other genome projects) has provided us with a description of the complete sequences of all the genes in more than a dozen organisms, and continuously provides more complete genome sequences as research continues. With technological innovations, the data generated have been growing at an exponential rate and are stored in distributed databases across the world. These databases provide challenges and opportunities for the analysis and exploitation of genes and protein sequences. In order to reap the intellectual and commercial benefits of this genetic information, researchers must be able to find the function of individual gene products. In the following section, we highlight the goals laid by the HGP and the corresponding strides made thereof in achieving the goals.

1.4 Beyond the Human Genome Project

With the completion of the sequencing of the human genome, the HGP focus switched to making the sequence publicly available to its mapping. The extraction of 3 billion base pairs was in itself a humongous task, and the analysis of this magnitude of data presented its own set of challenges and opportunities requiring a huge number of resources. Researchers from around the world realized the importance and the significant scientific contributions that could be made in the areas of human health and participated in the global endeavor to map the entire human genome (Figure 1.8).

The following sections describe the technological strides made thus far in five key areas: (1) sequencing technologies, (2) sequence variation studies, (3) functional genomics, (4) comparative genomics, and (5) functional annotation.

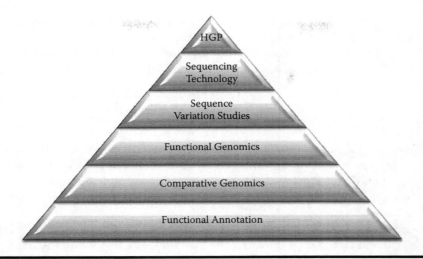

Figure 1.8 The five key areas that have been formed since the completion of the human genome project (HGP).

1.4.1 Sequencing Technology

With technological innovations, DNA sequencing technology continues to improve dramatically. Since the HGP began, the growth in data generated from sequencing projects has been exponential. This growth is caused by the emphasis given to sequencing technologies, due to:

1. Reduced costs and increased throughput of current sequencing technology
2. Support for novel technologies that can significantly improve sequencing technologies
3. Newly developed effective methods that introduce new sequencing technologies

The consequent technological innovations in the recent past have brought about a decline in the per-base cost of DNA sequencing at an exponential rate. These innovations are attributed to the improvement in the read length and accuracy of sequencing traces and have resulted in the consequent exponential growth of the genome databases (Shendure et al. 2008). The introduction of instruments capable of producing millions of DNA sequences read in a single run provides the ability to answer questions with unimaginable speed. These technologies are aimed at providing inexpensive, genome-wide sequence readouts as endpoints to applications.

There are six distinct techniques for DNA sequencing: (1) dideoxy sequencing, (2) cyclic array sequencing, (3) sequencing by hybridization, (4) microelectrophoresis,

(5) mass spectrometry, and (6) nanopore sequencing. The primary objective of these sequencing technologies is to identify the primary nucleotides, such as adenine (A), guanine (G), cytosine (C), and thymine (T), in the content of the DNA strands. The following sections provide an overview of these various sequencing strategies used.

1.4.1.1 Dideoxy Sequencing

Dideoxy sequencing was initially proposed by the Sanger Institute. The process proceeds by primer-initiated, polymerase-driven synthesis of DNA strands complementary to the template with the determined sequence. Numerous identical copies of the sequencing template undergo the primer extension reaction within a single microliter-scale volume.

Generating sufficient quantities of a template for a sequencing reaction is typically achieved by either (1) miniprep of a plasmid vector into which the fragment of interest has been cloned, or (2) polymerase chain reaction (PCR) followed by a cleanup step.

In the sequencing reaction, both the natural deoxynucleotides (dNTPs) and the chain-terminating dideoxynucleotides (ddNTPs) are present at a specific ratio. The ratio determines the relative probability of incorporation of dNTPs and ddNTPs during the primer extension. Incorporation of a ddNTP instead of a dNTP results in the termination of a given strand. Therefore, for any given template molecule, or strand, elongation will begin at the 3' end of the primer and will terminate upon the incorporation of a ddNTP. In older protocols for dideoxy sequencing, four separate primer extension reactions are carried out, each containing only one of the four possible ddNTP species (ddATP, ddGTP, ddCTP, or ddTTP), along with template, polymerase, dNTPs, and a radioactively labeled primer. The result is a collection of many terminated strands of different lengths within each reaction. As each reaction contains only one ddNTP species, fragments with only a subset of possible lengths will be generated, corresponding to the positions of that nucleotide in the template sequence. The four reactions are then electrophoresed in four lanes of a denaturing polyacrylamide gel to yield size separation with single nucleotide resolution. The pattern of bands (with each band consisting of terminated fragments of a single length) across the four lanes allows researchers to directly interpret the primary sequence of the template under analysis.

Current implementations of dideoxy sequencing differ in several key ways from the protocol described above. Only a single primer extension reaction is performed. This reaction includes all four species of ddNTP, which are labeled with fluorescent dyes that have the same excitation wavelength but different emission spectra, allowing for identification by fluorescent energy resonance transfer (FRET).

To minimize the required amount of template DNA, a cycle sequencing reaction is performed, in which multiple cycles of denaturation, primer annealing, and primer extension are performed to linearly increase the number of terminated strands.

1.4.1.2 Cyclic Array Sequencing

All of the recently released or soon-to-be-released non-Sanger commercial sequencing platforms, including systems from 454/Roche, Solexa/Illumina, Agencourt/Applied Biosystems, and Helicos BioSystems, fall under the rubric of a single paradigm, called cyclic array sequencing. Cyclic array platforms are cheap because they simultaneously decode a 2D array bearing millions (potentially billions) of distinct sequencing features. The sequencing features are "clonal," in that each resolvable unit contains only one species of DNA (as a single molecule or in multiple copies) physically immobilized on the array. The features may be arranged in an ordered fashion or randomly dispersed. Each DNA feature generally includes an unknown sequence of interest (distinct from the unknown sequence of other DNA features on the array) flanked by universal adaptor sequences. A key point in this approach is that the features are not necessarily separated into individual wells. Rather, because they are immobilized on a single surface, a single reagent volume is applied to simultaneously access and manipulate all features in parallel. The sequencing process is cyclic because in each cycle an enzymatic process is applied to interrogate the identity of a single base position for all features in parallel. The enzymatic process is coupled to either the production of light or the incorporation of a fluorescent group. At the conclusion of each cycle, data are acquired by charge-coupled device (CCD)-based imaging of the array. Subsequent cycles are aimed at interrogating different base positions within the template. After multiple cycles of enzymatic manipulation, position-specific interrogation, and array imaging, a contiguous sequence for each feature can be derived from an analysis of the full series of imaging data covering its position.

1.4.1.3 Sequencing by Hybridization

The principle of sequencing by hybridization (SBH) is that the differential hybridization of target DNA to an array of oligonucleotide probes can be used to decode the target's primary DNA sequence. The most successful implementations of this approach rely on probe sequences based on the reference of a genome sequence of a given species, such that genomic DNA derived from individuals of that species can be hybridized to reveal differences relative to the reference genome (i.e., resequencing, rather than *de novo* sequencing). The difference between SBH and other genotyping array platforms that use similar methods is that SBH attempts to query all bases, rather than only bases at which common polymorphisms have been defined. In resequencing arrays developed by Affymetrix and Perlegen, each feature consists of a 25 bp oligonucleotide of a defined sequence. For each base pair to be resequenced, there are four features on the chip that differ only at their central position (dA, dG, dC, or dT), while the flanking sequence is constant and is based on the reference genome. After hybridization of the labeled target DNA to the chip and the imaging of the array, the relative intensities at each set of four features targeting a given position can be used to infer the target DNA's identity.

1.4.1.4 Microelectrophoresis

As mentioned above, conventional dideoxy sequencing is performed with microliter-scale reagent volumes, with most instruments running 96 or 384 reactions simultaneously in separate reaction vessels. The goal of microelectrophoretic methods is to make use of microfabrication techniques developed in the semiconductor industry to enable significant miniaturization of conventional dideoxy sequencing. A key advantage of this approach is the retention of the dideoxy biochemistry, which has proven robustness for >1,011 bases of sequencing. Until alternative methods achieve significantly longer read lengths than they can today, there will continue to be an important role for Sanger sequencing. Microelectrophoretic methods may prove critical to continue to reduce costs for this well-proven chemical process. There may also be a key role for lab-on-a-chip integrated sequencing devices that will provide cost-effective, clinical point-of-care molecular diagnostics.

1.4.1.5 Mass Spectrometry

Mass spectrometry (MS) has established itself as the key data acquisition platform for the emerging field of proteomics. There are also applications for MS in genomics, including methods for genotyping, quantitative DNA analysis, gene expression analysis, analysis of indels and DNA methylation, and DNA/RNA sequencing.

Matrix-assisted laser desorption/ionization time-of-flight mass spectrometry (MALDI-TOF-MS) is an MS sequencing technique that relies on the precise measurement of the masses of DNA fragments present within a mixture of nucleic acids. *De novo* sequencing using MALDI-TOF-MS read lengths are limited to <100 bp. Applications of MS sequencing include:

1. Deciphering sequences that appear as compression zones by gel electrophoresis
2. Direct sequencing of RNA (including for identification of posttranslational modifications of ribosomal RNA)
3. Robust discovery of heterozygous frameshift and substitution mutations within PCR products in resequencing projects
4. DNA methylation analysis

1.4.1.6 Nanopore Sequencing

Nanopore sequencing is an approach for single-molecule sequencing that involves passing single-stranded DNA through a nanopore. The nanopore is a biological membrane protein or a synthetic solid-state device. As individual nucleotides are expected to obstruct the pore to varying degrees in a base-specific manner, the resulting fluctuations in electrical conductance through the pore can, in principle, be measured and used to infer the primary DNA sequence. Published examples of the nanopore-based characterization of single nucleic acid molecules include:

1. The measurement of duplex stem length, base pair mismatches, and loop length within DNA hairpins (Vercoutere et al. 2001)
2. The classification of the terminal base pair of a DNA hairpin, with approximately 60 to 90% accuracy with a single observation, and >99% accuracy with 15 observations of the same species (Winters-Hilt et al. 2003)
3. Reasonably accurate (93 to 98%) discrimination of deoxynucleotide monophosphates from one another with an engineered protein nanopore sensor (Astier et al. 2006)

Significant pore engineering and technology development may be necessary to accurately decode a complex mixture of DNA polymers with single-base pair resolution and useful read lengths. Provided these challenges can be met, nanopore sequencing has the potential to enable rapid and cost-effective sequencing of populations of DNA molecules with comparatively simple sample preparation.

1.4.2 Next-Generation Sequencing

With the advancements made in sequencing technologies, there has also been recent advancement in the form of a new generation of sequencing instruments. These instruments cost less than the techniques described in the previous section and promise faster sequence readings, as they require only a few iterations to complete an experiment. These faster reads foster the potential to add to the exponential increase of sequence data. The expected increase of data is also attributed to the next-generation sequence technology's ability to process millions of reads in parallel, rather than the traditional 96 reads. Thus, with the introduction of next-generation sequencing technology, large-scale production gene sequence data may require specialized use of robotics and high-tech instruments, computer databases for storage of the huge data, and bioinformatics software for analysis.

An added advantage of the proposed next-generation sequence reads is that they are generated from fragment libraries that have not been subjected to conventional vector-based cloning and *Escherichia coli*-based amplification stages used in capillary sequencing rendering the sequences of any prevalent biases caused by cloning.

Three commercially used and commonly cited next-generation sequencing platforms include the Roche (454) GS FLX Sequencer, the Illumina Genome Analyzer, and the Applied Biosystems SOLiD Sequencer (refer to Table 1.2 for a detailed comparison). The generic work flow for creating a next-generation sequence library is simple. Fragments of DNA are prepared for sequencing by ligating specific adaptor oligos to both ends of each DNA fragment. Typically, only a few micrograms of DNA are needed to produce a library. Each of these platforms applies a unique or

Table 1.2 Comparison of Metrics and Performance of Next-Generation DNA Sequencers

	Platform		
	Roche (454)	Illumina	AB SOLiD
Sequencing chemistry	Pyrosequencing	Polymerase-based sequencing by synthesis	Ligation-based sequencing
Amplification approach	Emulsion PCR	Bridge amplification	Emulsion PCR
Mb/run	100 Mb	1,300 Mb	3,000 Mb
Time/run (paired ends)	7 h	4 days	5 days
Read length	250 bp	32–40 bp	35 bp

Source: Mardis, E.R., *Trends Genet* 24, no. 3 (2008): 133–141.

modified approach to sequence the paired ends of a fragment, the scope of which is not covered in this book. For details refer to Mardis (2008).

Since next-generation sequencing technology is relatively new, there is little insight on the accuracy of the reads, and the quality of the results obtained have yet to be understood. When compared to the more traditional capillary sequencers, next-generation sequencers produce shorter reads, ranging from 35 to 250 base pairs (bp), than the traditional 650 to 800 bp created by other methods. The length of the reads could impact the utilization of the generated data. Efforts are being pursued currently to benchmark the reads with the traditional capillary electrophoresis.

Although next-generation sequence technology provides many advantages over traditional methods, it also poses several computational challenges. Many storage and data management systems cannot handle the amount of data generated. The data storage must be scalable, dense, and inexpensive to handle the exponential growth. Various centers of bioinformatics around the globe are investing heavily in high-performance disk systems and data pipelines to overcome the challenge of handling the large number of files that are expected to be accessed when the demand arises.

Software pipelines are also required to provide the necessary analysis and visualization of the data generated. More importantly, software has to be in place to provide annotations of the sequences generated.

1.4.2.1 Challenges of Handling NGS Data

The challenges of handling the deluge of NGS data stem from two key concepts that are used to analyze the sequence reads. These concepts focus on *de novo* assembly and alignment. The following sections describe the computational algorithms used to handle the massive amounts of Illumina sequencing data for both *de novo* assembly and alignment of reads (Paszkiewicz and Studholme 2010).

1.4.2.1.1 *De Novo* Assembly

De novo sequence assembly is the process whereby we merge individual sequence reads to form long contigs (continuous sequences) that share the same nucleotide sequence as the original template DNA from which the sequence reads were derived.

Two algorithms are prominently used to assemble sequence reads: (1) algorithms based on the overlap-layout-consensus (OLC) approach (Huang and Madan 1999) and (2) algorithms based on a de Bruijn graph (Simpson et al. 2009). These have been well-reviewed techniques and have been implemented in effective genome-assembly software packages. However, these genome sequence assembly programs are not well suited to short sequence reads generated by Illumina and AB SOLiD platforms (Paszkiewicz and Studholme 2010).

1.4.2.1.2 Alignment

Once the assembly is performed, the contigs are subject to alignment algorithms (Li and Homer 2010), which focus on the creation of auxiliary data structures called indices for the sequence reads and the reference sequence. We can categorize these structures into three algorithms: (1) hash table-based algorithms, (2) suffix tree-based algorithms and (3) algorithms based on merge sorting.

1.4.2.1.2.1 Hash Table-Based Algorithms — These algorithms create a hash table index that can be used to trace back to specific basic local alignment search tool (BLAST) matches as they rely on a seed-and-extend paradigm. In the first phase of the algorithm, BLAST maintains the position of each k-mer subsequence of the query in a hash table with the k-mer sequence being the key, and scans the database sequences for k-mer exact matches called seeds. Once this phase is complete, BLAST extends and joins the seeds without gaps. Further refinements are carried out using Smith-Waterman alignment to refine the seeds, which achieves statistically significant results. The tools that are prominently using the hash table-based algorithms are MAQ, the SOAP family of alignment tools, viz., SOAP, SOAP2, and SOAP3/GPU, and Abyss (Simpson et al. 2009).

1.4.2.1.2.2 Suffix-Based Trees — With the short sequence reads it is a challenge to obtain the exact matches of the reads using BLAST. Thus researchers tend to favor inexact matches of sequence for alignments. The suffix-based approaches aim to essentially reduce the inexact matching problem to the exact matching problem using two steps: (1) identifying exact matches and (2) building inexact alignments supported by exact matches. To find exact matches, these algorithms use a certain representation of suffix trees. The advantage of using suffix trees is that alignment to multiple identical copies of a substring in the reference is only needed once because these identical copies collapse on a single path in the tree, whereas with a typical hash table index, an alignment must be performed for each copy. The tools that prominently use the suffix-based trees for alignment of sequences are MUMmer and REPuter (Paszkiewicz and Studholme 2010).

1.4.3 Sequence Variation Studies

Nature retains diversity in a population of organisms living in varied environmental conditions. This diversity is the result of genetic variations: traits that vary and are coded in the genes of the population. Since the inception of the HGP, several studies have been conducted to understand the effect of genetic variations between individuals.

Natural sequence variation is the fundamental property of all genomes. It is believed that any two haploids exhibit multiple kinds of genetic variations and polymorphisms (see Figure 1.9). There are three basic forms of genetic variations: mutations, gene flow, and sex. Not all of these genetic variations have functional implications. Sequence polymorphisms also include duplications, rearrangement, insertions, and deletions. The most common polymorphism in the human genome

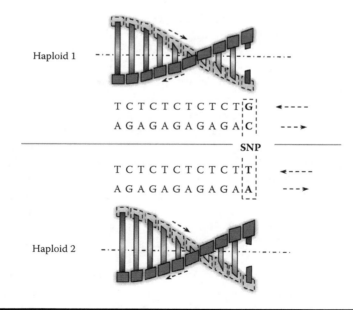

Figure 1.9 A schematic representation of a single nucleotide polymorphism between two haploids.

is the single-base pair difference, better known as a single nucleotide polymorphism (SNP). When two haploid human genomes are compared, it is observed that SNPs occur at every kilobase of the gene sequence. SNPs are abundant, stable, and widely distributed across the genome. Because of these properties, SNPs can be used for the mapping of complex traits such as cancer, diabetes, and mental illness. However, the occurrences of these variations across the entire genome are rare, making it a challenge to challenge to identify and understand these variations (Figure 1.9).

Keeping this challenge in mind, the objective of sequence variation studies is to provide dense maps of SNPs that will make genome-wide association studies possible. These maps are powerful means for identifying genes that contribute to disease risk. They will also permit the prediction of individual differences in drug responses. When the maps are made available to the public, maps of a large number of SNPs distributed across the entire genome come together with technology for rapid, large-scale identification. The scoring of SNPs must be developed to facilitate this research. The HGP envisioned the following goals concerning genetic variation analysis. First, the goal is to develop technologies for rapid, large-scale identification or scoring, or both, of SNPs and other DNA sequence variants. In order to achieve this goal, the following objectives had to be met:

1. The creation of an SNP map of at least 100,000 markers
2. The development of concepts and methods to study multigene traits and map DNA sequence variations to phenotypic variations such as complex disease
3. The creation of public resources containing DNA samples and cell lines to enable SNP discovery using the public resources

To this end, large bodies of works have been conducted through primary data sources that contain SNP data, including the dbSNP (current build 134) containing approximately 6,961,883 human reference SNP clusters, the Human Gene Mutation Database (HGMD) containing 113,247 entries (professional release 2011.2), and the disease-specific Online Mendelian Inheritance in Man (OMIM) (September 2011) that contains approximately 2,648 genes with disease-causing mutations. Several tools are available for the analysis of SNPs, of which SNPper is prominently used. Furthermore, BioPerl provides an API for the analysis of SNPs and Genewindow provides visualization technology. Other online resources that enable effective visualization of SNP data include the UCSC Genome Browser (see Figure 1.10) and the Ensembl Genome Browser (Table 1.3).

1.4.3.1 Kinds of Genomic Variations

HGP focuses on the creation of a repository of all known SNPs derived from a diverse population across the United States and the creation of appropriate tools to

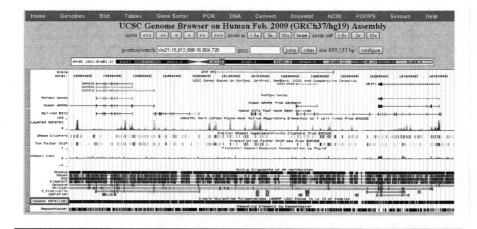

Figure 1.10 A screenshot of the UCSC Genome Browser, a tool to visualize SNP data.

analyze SNPs. The HGP suggests that approximately 95% of the discovered SNPs belong to the noncoding regions of the genome. Furthermore, it is still an open challenge to determine the functional aspect of the SNPs found near or in genes. However, it is still believed that based on their location on the genome, SNPs can potentially alter the functions of DNA, RNA, and proteins alike. A general categorization of SNPs based on their location is shown in Table 1.4 (Mooney 2005; Rebbeck et al. 2004).

Generally, nonsynonymous SNPs (nsSNPs) cause a change in the amino acid sequence of the resultant protein sequence, either by substituting amino acids or introducing a nonsense/truncation mutation (Ng and Henikoff, 2006). Table 1.4 shows variants that affect the expression of a gene translation by interrupting a regulatory region known as a regulatory SNP. Similarly, those variants that interfere with normal splicing and mRNA functions are categorized as intronic SNPs or synonymous SNPs. Due to increasing research efforts, the molecular effects of variations are becoming better understood, which allows us to shed more light on genetic diseases.

1.4.3.2 SNP Characterization

To understand the patterns of sequence variations in coding regions of genes, bioinformatics strategies have been focused on analyzing disease-associated mutations that focus precisely on where diseased alleles occur with respect to their corresponding protein structures. It is important to understand the underpinnings of these mutations and what properties guide such mutations.

Table 1.3 SNP Resources Widely Used

	Description
Genome Resources	
dbSNP	The primary repository for SNP data
Ensembl	Genome database
GoldenPath	Genome database
HapMap Consortium	Haplotype block information
JSNP	Japanese SNP database
Mutation Repositories	
HGVBase	Public genotype-phenotype database
HGMD	Mutation database with many annotations
Swiss-Prot	Protein database with extensive variant annotations
List of Locus-Specific Databases	
CGAP-GAI	Cancer Gene Anatomy Project at the National Cancer Institute
Other databases and tools	Tools for SNP analysis and gene characterization
Tools	
SNPper	Novel software for SNP analysis
BioPerl	A programming application program interface (API) for bioinformatics analysis
Genewindow	Interactive tool for visualization of variants

Table 1.4 Existing SNP Categorization

Coding SNPs	cSNP	Positions that fall within the coding regions of genes
Regulatory SNPs	rSNP	Positions that fall in regulatory regions of genes
Synonymous SNPs	sSNP	Positions in exons that do not change the codon to substitute an amino acid
Nonsynonymous SNPs	nsSNP	Positions that incur an amino acid substitution
Intronic SNPs	iSNP	Positions that fall within introns

It is hypothesized that mutations on the gene sequence (position specific) are conserved through evolution and are reflected to the protein structure (Ng and Henikoff 2002; Krishnan and Westhead 2003).

One of the tasks of SNP analysis is to gauge the impact of each nsSNP on protein function. Due to the size of the SNP data, this task is experimentally infeasible. Thus, researchers have looked into computational methods to predict changes in protein function if an amino acid changes. This technique, also known as amino acid substitution (AAS), focuses on disease-causing mutations that are likely to occur at positions that are conserved through evolution. It is further believed that disease-causing AASs affect the structural characteristics of the resulting protein, suggesting that protein structural information can be used to analyze these mutations (Table 1.5).

1.4.4 Functional Genomics

With the entire human genome sequence publicly available, a new approach to address biological challenges has taken form. This approach, called functional genomics, entails the functional understanding of the human DNA on a genome scale. Functional genomics is viewed as an intermediate step that brings biological research to being applied in medicine (from bench-side to bedside). Based on successes of previous studies of sequences within organisms, it is inferred that the function of genes and other functional elements of the genome can be inferred more accurately only when the genome is studied in its entirety.

Table 1.5 Strategies That Have Been Used for Analysis of AAS

Method	Algorithm
SIFT (Ng and Henikoff 2002)	Sequence homology and position-specific scoring matrices
PolyPhen (Stitziel et al. 2004)	Sequence conservation, structural information modeling
SNPs3D (Yue and Moult 2005)	Structure-based support vector machines (SVMs) and sequence conservation-based SVMs
PANTHER PSEC (Thomas et al. 2003)	Sequence homology and scores obtained from PANTHER hidden Markov models of protein families
TopoSNP (Stitziel et al. 2004)	Characterization of residues based on topological information such as buried, on-surface, or pocket information

At the end of the HGP, knowledge about a gene's structure and other elements was only the tip of the iceberg. Further insights about the function of a gene can be derived from its interaction with the environment.

Existing methods for analyzing DNA function at a genomic scale include the comparison and analysis of sequence patterns, large-scale analysis of mRNA, various approaches of gene distribution, and the analysis of protein complexes (for gene products). Despite these methods, there is still a need for novel strategies to elucidate the function of genes. Thus, functional genomics focuses on the development of technology that can be used for the large-scale analysis of the human genome in its entirety rather than in parts. In functional genomics, emphasis is given to gene transcripts and their protein products, including the identification and sequencing of full-length cDNAs that represent the entire human genome. Thus, the following were the objectives of functional genomics:

1. Extend support for the creation of global approaches, improved technologies, and the creation of relevant libraries for the comparative and computational analysis of noncoding sequences: It is imperative to understand these sequences, as they are noncoding and carry out other functions, such as RNA splicing, sequences that are responsible for the formation of chromatin domains, sequences that maintain chromosome structure, sequences that are responsible for recombination and replication, and sequences that specify numerous functional untranslated RNAs.
2. Enable and support the creation of technology for the comprehensive analysis of gene expression so that it is possible to analyze spatial and temporal patterns of gene expression in both human and model organisms, thereby providing a means to understand the expression of genes: To make this analysis possible, cost-effective and efficient technology that measures the parameters of gene expression in a reliable manner and can be easily reproduced must be developed. In addition to the required technological innovations, complementary DNA (cDNA) sequences and validated sets of clones with unique identifiers are also needed to analyze gene expression data. Other required developments include novel methods to quantify, represent, analyze, and archive the resulting gene expression data.
3. Investigate alternate means of studying functions, like methods for genome-wide mutagenesis: This step includes the creation of mutations that cause loss or alteration in gene functions. Associated technologies for large-scale *in vivo* and *in vitro* are also required to generate and find mutations in each gene and phenotype.
4. Understand protein functions on a genome-wide scale to develop technology for global protein analysis to provide a comprehensive understanding of genome functions: The development of computational and experimental models to analyze both spatial and temporal patterns of protein expression, protein-ligand interactions, and protein modification is required.

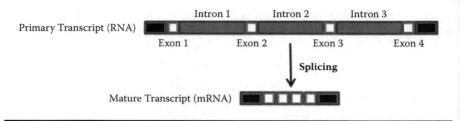

Figure 1.11 The process of splicing, in which the introns are removed from the primary transcript (RNA) and the exons are combined to form the mature transcript (mRNA).

1.4.4.1 Splicing and Alternative Splicing

Splicing, the first step to understanding the functions of genes and the roles they play in an organism, is the alteration of the primary transcript RNA after transcription. In this process, introns are removed, and the remaining exons are joined (see Figure 1.11). It is necessary for the mature transcript (of the mRNA) to be subject to splicing, as it enables the production of the correct protein during translation. However, it is commonly observed that a set of unique proteins can be created by varying the exon composition of the mRNA through the process of splicing. This process is referred to as alternative splicing. Alternative splicing can occur in many ways using different combinations of exon units. Moreover, exons can be skipped, or introns can be retained, creating a complex system requiring the need for computational modeling and interpretation.

The sequencing of the human genome has raised the importance of alternative splicing as an RNA regulatory mechanism. Furthermore, alternative splicing has provided a means for researchers to explain why there is such a large repertoire of proteins. It has also potentially helped identify and explain defects that occur in the splicing mechanism and that result in complex diseases such as cancer.

Bioinformatics has played a key role in cataloguing splice variations in humans and other eukaryotic genomes (Modrek et al. 2001). Tools and algorithms have also been developed to characterize splice regulatory elements that control the expressions of genes (Florea 2006). Instead of focusing on an organism's total number of genes to explain its functional and behavioral complexity, researchers are now interested in determining how each gene can be "reused" to create multiple functions and new modes of regulation. To this end, studies on both human and mouse sequence data have resulted in algorithms that have clustered genes and samples based on their alternative splicing patterns, indicating the importance of alternative splicing to differentiate between genes (Lee and Wang 2005).

1.4.4.1.1 Types of Alternative Splices

Alternative splicing of pre-mRNA is an important regulatory mechanism to modulate genes and their corresponding protein complexes within a cell. It is believed

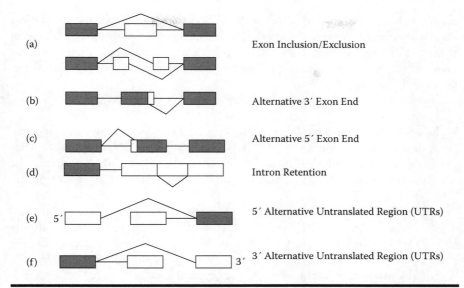

Figure 1.12 Schematic representation of the types of alternative splicing events. Alternatively, spliced elements (exons or portions of exons) are shown in red, and those constitutively spliced are shown in blue. The exons are represented as boxes, and the introns by straight lines connecting the exons. (From Florea, L., *Briefings Bioinformatics* 7, no. 1 (2006): 55–69. With permission.)

that the proteins obtained from alternative splices can be used to regulate a gene expression within a cell. It is therefore necessary to understand and catalog all possible combination of exons obtained from a gene.

With the perspective of gene structure, alternative splicing is categorized into four types of events (see Figure 1.12). It should be noted that due to the data's intrinsic property of being noisy, the identification of gene boundaries is difficult. Therefore, it is an open challenge to identify and characterize the 5' and 3' alternative untranslated regions (UTRs), as shown in Figure 1.12e–f.

1.4.4.1.2 Alternative Splicing for Gene Annotation

The role of bioinformatics in alternate splicing is prevalent in areas of gene annotation and splice regulation (Lee and Wang 2005). Traditional gene discovery, better known as gene prediction (Birney et al. 2004), has been performed through a combination of *ab initio* and comparative methods for the identification of linear exon-intron models of genes. With the completion of the HGP and the resultant large-scale annotation projects such as the Ensembl (Hubbard et al. 2002) and UCSC Genome Browser database (Fujita et al. 2010) with different data, dependent models came into existence. These models are based on different prediction methods that create the "evidence" of the existence of a gene and use a combiner algorithm to associate the collected evidences into a unified representative model of a gene. With the inclusion of alternative spliced transcripts or alternative splicing events as part of the annotation process through manual curation, these databases improve the quality of their datasets.

There are four prominent approaches used in gene prediction:

1. *Ab initio* **programs:** These programs do not require any prior or additional information to predict a gene for a given DNA sequence. They rely on the hidden Markov model (HMM) framework to provide the parameterization and decoding of a probabilistic model of gene structure (Zhang 2002).
2. **Evidence-based techniques:** There are two classes of evidence-based techniques for gene prediction. The first class uses the well-known pairwise HMM methods. The second class uses external evidence to score potential exons (Parra et al. 2003; Birney et al. 1996; Alexandersson et al. 2003).
3. **Informant approach:** This technique predicts a gene based on information of exons derived from two or more sample genomes (Pedersen and Hein 2003).
4. **Feature-based approaches:** These approaches do not rely on a probabilistic model or prior knowledge from the underpinning DNA. However, the framework facilitates the integration of multiple component features derived from the DNA sequence (Howe et al. 2002).

1.4.4.1.3 Regulation of Alternative Splicing

To regulate splicing, it is important to identify what causes or controls the variation in splicing. The control of alternative splicing affects the abundance, structure, and function of transcripts and encoded proteins from a gene through the modification of their properties, such as its binding affinity, intracellular localization, stability, and enzymatic activity (Stamm et al. 2005). Furthermore, exon selection in alternative splicing is tissue specific, and is determined based on the developmental stage, or disease specific (Florea 2006). Thus, the regulation of alternative splicing is more specific and case driven than transcriptional regulation.

Though little is known about splicing regulation through regulatory proteins, there is an alternative form of regulation that focuses on splice regulation that is not part of the basal spliceosome function. The basal spliceosome function is regulated by families of splicing regulatory proteins. These proteins bind to the RNA in the surrounding regions of exons, thereby catalyzing the exon's inclusion or exclusion by activating or inhibiting the function of the splice site. Little is known about the characteristics of regulatory proteins and the corresponding RNA binding sites, and these issues are being actively investigated.

1.4.4.1.4 Splice Variants

Gene annotation using alternative splicing and the regulation of alternative splicing form the crux of research that relies on computational methods. The resulting bioinformatics techniques focus mainly on cataloging the various splice variants.

Despite the tendency of genomes to remain the same for different tissues or cell types in an organism, their transcriptomes (set of all RNAs of the tissues/cell) can be significantly different.

The motivation for using splice annotation is to identify and catalog all mRNA transcripts of a cell at different stages, using both spatial and temporal expression, along with functional information of the splices. This objective is difficult, if not impossible to achieve, considering the incomplete and fragmented nature of the data along with insufficient experimental characterization.

Several computational approaches have been suggested to overcome these limitations and identify splice variants. These techniques rely on characterizing splicing patterns obtained from partial cDNA or protein sequences, or exon-level alternative splicing events to analyze and characterize transcriptomes. The varying splice patterns can be applied in the design of diagnostic markers that can be validated using either *in vitro* microarray and proteomic experiments or *in silico* via the identification and annotation of splice forms.

Bioinformatics techniques used for the annotation of full-length alternative spliced transcripts include:

1. **Gene indices:** Gene indices refer to gene- or transcript-oriented collections of express sequence tags (ESTs) and micro-RNA (mRNA) sequences grouped by sequence similarity (Lee et al. 2005; Liang et al. 2000). This method employs a pairwise sequence similarity for comparison between two sequences. Here, all the EST and mRNA sequences are subject to a one-to-one comparison to identify overlaps between each other. These sequences are then grouped and assembled into disjointed clusters (a consensus sequence) based on a threshold of overlaps.

    The creation of a gene index is complex and suffers from two drawbacks: (a) overclustering, in which different paralogs (similar sequences belonging to different genes) are put into the same cluster creating a false overlap, and (b) underclustering, in which several clusters are produced for a single gene.

2. **Genome-based methods for clustering spliced alignments:** In this approach, unlike gene indices, the spliced alignments of cDNA or protein sequences are clustered at a point of reference along the reference genome sequence (loci) (Florea et al. 2005). Splice graphs are one such technique that is prominently used in alternative splicing annotation for capturing splice variants in a gene (Kim 2005). Using the concept of directed acyclic graphs, with each node representing an exon and edge that connects two exons representing an intron, a splice variant corresponds to the paths obtained

through the graph traversal from a predetermined source vertex (vertex with no incoming edges) to a sink vertex (vertex with no outgoing edges). The advantage of this technique is that it results in all possible combinations of exon-intron combinations. However, not all of the combinations are biologically significant. Several filtering strategies prioritize the combinations and rank splice variants that are biologically significantly higher.

1.4.4.2 Microarray-Based Functional Genomics

Microarray technology has been an important contribution to functional genomics as it provides a means to analyze the expressions of hundreds of thousands of genes that belong to an organism for a specific reaction at a given instance, simultaneously. This technology has facilitated an understanding of the fundamental aspects of growth and development. Moreover, it has aided in the exploration of the genetic causes of complex genetic diseases such as cancer. Typically, microarray data are classified into three categories, based on the types of the samples used to construct the microarrays (see Table 1.6, Figure 1.13).

Gene regulatory network analysis (Huang et al. 2007) is an analytic technique that is used to extract gene regulatory features (i.e., activation and inhibition) from gene expression patterns. Changes of gene expression levels across samples provide information that allows reverse engineering techniques to construct the network of regulatory relations among those genes (Lockwood et al., 2006).

For instance, the expression of a gene is regulated by a transcriptional control mediated by a complex cis-regulatory system. Transcriptional factors activate or repress gene expression by binding to their respective binding sites: comparatively short sequences (several hundred to several thousand base pairs, depending on the species) upstream, downstream, or far away from the transcriptional start sites. Specific sites within such regions, which are generally composed of dense clusters, are recognized by the regulatory proteins (transcription factors (TFs)) that control the rate of gene transcription.

Table 1.6 The Categorization of Microarrays and Their Associated Applications

Microarray Type	Application
CGH	Tumor classification, risk assessment, and prognosis prediction
Expression analysis	Drug development, drug response, and therapy development
Mutation/polymorphism analysis	Drug development, therapy development, and tracking disease prognosis

Source: NCBI, *NCBI: A Science Primer,* July 27, 2007, http://www.ncbi.nlm.nih.gov/About/primer/microarrays.html#ref1 (accessed September 13, 2011).

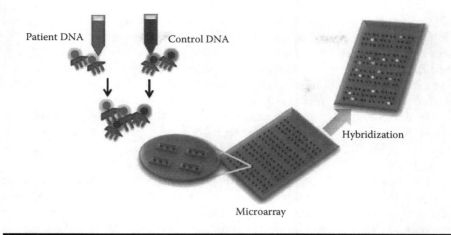

Figure 1.13 Schematic representation of the microarray-based comparative genomic hybridization (CGH) process.

1.4.4.2.1 Types of Regulatory Regions

Regulatory regions of higher eukaryotes can be subdivided into proximal regulatory units—promoters—which are located close to the 5' end of the gene, and distal transcription regulatory units called enhancers or cis-regulatory modules (CRMs). CRMs may be located far upstream or downstream of the target gene, and are much more difficult to recognize because they lack proximal specific transcriptional signals, such as position relative to coding sequence, the TATA box, the CAAT box, the transcription start site consensus, etc. Therefore, recognition of CRMs is even more difficult than recognition of promoters (Abnizova and Gilks 2006).

1.4.4.2.2 Experimental Determination of Regulatory Region Function

Biochemical characteristics can identify binding sites precisely and are the only way to determine whether consensus sequences differ among species. There are several methods available for producing DNA-protein interaction data. Nitrocellulose binding assays, electrophoretic mobility shift assay (EMSA), enzyme-linked immunosorbent assay (ELISA), DNase footprinting assays, DNA-protein cross-linking (DPC), and reported conducts are examples of *in vitro* techniques that are used to determine DNA binding sites and analyze the difference in binding specificity for different protein-DNA complexes. The major disadvantage of these methods is that they are not suited to high-throughput experiments.

A microarray-based assay called chromatin immunoprecipitation (ChIP) was developed for genome-wide determination of protein binding sites on DNA.

Other types of experiments are systemic evolution of ligands by exponential enrichment (SELEX) and phage display (PD), which offer a high-throughput possibility to select high-affinity binders, DNA and protein targets, respectively. Both SELEX and PD suffer the same disadvantage: most sequences obtained from these experiments are good binders, but it is hard to say anything about their relative affinities. It is assumed that the best binders occur more frequently.

In dsDNA microarrays are presented for exploring sequence-specific protein-DNA binding. The major advantage over the aforementioned methods is that it is a high-throughput method resulting in data with associated relative binding affinities.

Finally, x-ray crystallographic and NMR spectroscopic data provide a base for studying the structural details of protein-DNA interactions. Protein-DNA complexes have successfully been co-crystallized, and the data have been deposited into the protein data bank and nucleic acid database (NDB). However, these experiments are time-consuming.

Unfortunately, for technical reasons, the numbers experimentally verified, binding sites are nearly always underestimated, and the physical length of regulatory regions is rarely well defined.

1.4.5 Comparative Genomics

Due to evolution, all organisms are believed to be related. Thus, the study of one species could lead to valuable information about another species. Molecular genetics enables researchers to understand the genes of one species based on the genetic makeup of related genes in other species. To this end, several experiments provide insights into the universality of biological mechanisms, through comparisons between genomes. Thus, valuable insights relating to the gene structure and function of closely related species are brought to the forefront using comparative genomics.

The comparative analysis of the human genome with a variety of modeled organisms is advantageous and is an important field of research. The underpinning rationale that governs cross-species sequence comparative genomics, as stated above and in Pennacchio and Rubin (2003), is based on the observation that sequences and functions are conserved across evolutionary distant species. This conservation enables researchers to identify and distinguish between functional and nonfunctional genetic sequences in both gene sequence data and protein sequence data. This rationale lays the impetus for gene expression, regulation, and control experiments.

It has also been shown that the inverse also holds true in orthologous genomic sequences from different vertebrates. Thus, the comparative analysis of evolutionary conserved sequences is a viable strategy to identify biologically active regions over the human genome.

Various genomic visualization, annotation tools, and databases are available to the biomedical research community and are publicly available. These tools have been successfully used to identify biologically important genes and sequences involved in gene regulation.

1.4.6 Functional Annotation

Supporting the above genomic research is one of the keystones of the HGP. This support includes the effective recoding, distribution, and analysis of all results and discoveries. Bioinformatics and computational biology are core components that are targeted toward satisfying this goal. Thus, the services that bioinformatics offers can be categorized into two areas: (1) databases and (2) analytical tools.

This section is devoted to the effective collection, analysis, annotation, and storage of sequence data that is exponentially growing. For effective use of the data generated in the public domain, it is important to provide effective mapping of all gene sequence data to expression data and protein sequence data. User-friendly interfaces and user-friendly databases are imperative to the success of the genome project. Additionally, a range of computational algorithms that allow researchers to extract, view, and annotate gene and protein sequences effectively will benefit the research community. Such algorithms address the following objectives:

1. Improve the content and utility of databases.
2. Develop better tools for data generation, capturing, and annotation.
3. Develop and improve tools and databases for comprehensive functional studies.
4. Develop and improve tools for representing and analyzing sequence similarity and variation.
5. Create mechanisms to support effective approaches for explorative and robust software that can be widely used in different applications.

The successes of these objectives have been documented primarily in the creation and maintenance of large databases such as the PDB, Ensembl, and SwissProt. However, bioinformatics and computational biology has been actively pursued as an area of research for the creation of better analysis techniques, algorithms, and tools in fields like gene sequence analysis, microarray analysis, protein sequence and structural analysis, and functional annotation.

1.4.6.1 Function Prediction Aspects

One of the problems arising from the completion of the HGP was the functional annotation of generated sequences. Biologists were then and are now faced with the challenge of analyzing the functional significance of genes with traditional statistical techniques. Not only is the volume of sequence and structure data growing, but the diversity of the sources that generate the data also poses significant challenges that require computational expertise and has led to a disproportionate growth in the number of uncharacterized gene sequences.

The established and traditionally used method for gene and protein annotation is based on homology modeling in which new sequences are assigned functions based on the similarity they share with sequences of known annotations. However, homology modeling amplifies existing erroneous annotations. Because of this problem, the efficacy of this method is questionable considering the constant growth of sequence information. Thus, there is a need for standardized, large-scale sequence annotation tools that use machine learning and are free of manual curation. This automated function prediction of sequences could be incorporated into larger work flows. This section explains some computational protein function prediction techniques (Friedberg 2006).

The definition of biological function is ambiguous, and the exact meaning of the term varies based on the context in which it is used. Further, there is a multiperspective view of protein function that is categorized into three classes:

1. **The biochemical aspect:** In this class, the protein function is derived from the specific substrate information. This definition requires only a disembodied protein performing alone *in vitro*.
2. **The physiological aspect:** In this class, function is defined in respect to the function of a protein within an organism from the subcellular level to the whole organism. Here, sequences could derive functional information from the signal pathways that the protein is a part of or from their interacting partners.
3. **The phenotypic (medical) aspect:** In this class, the functional information is derived from the mutations that occur in the sequence of the protein.

Keeping these aspects in mind, there are several methods proposed in the automated functional annotation of sequences, and the following section enumerates them.

1.4.6.1.1 Computational Functional Annotation

The basic challenge faced in the computational annotation of sequences is determining what constitutes functional information and how that function should be described in a computationally interpretable manner. Two forms of information can be adopted to define protein function. From a data mining perspective, these forms of information include protein sequence information and protein structure information that can be included as features of interest in the algorithm.

Protein sequences are represented as character strings that are used in an array of tasks: pairwise and multiple sequence alignments and motifs, all of which can easily be included as features for analysis using computational algorithms. Protein structural information, on the other hand, is more complex. Here, the PDB files (.pdb) have vast amounts of information, in the form of 3D coordinates, which can be exploited to find similarities between two proteins.

Apart from features of interest, there is also a need for controlled vocabulary, or keywords that can be used to annotate functionally significant regions of a protein, and well-defined relationships in describing functions. The Enzyme Commission Classification (EC) (Webb 1992) is one such annotation system that classifies reactions based on a four-level hierarchy (represented using a four-position identifier) that moves from a general to a specific categorization.

For example, the hierarchy starts with a generalized lyase (4.-.-.-), in the first position and moves through a more specific nitrogen lyase (4.3.-.-) or to ammonia lyases (4.3.1.-) to the more specific histidine-ammonia lyase (4.3.1.3) in the fourth position.

Several other such annotation schemes provide a controlled vocabulary to annotate sequences; the most prominently used annotation scheme is that of Gene Ontology (GO). The adopted controlled vocabulary is based on three aspects of gene product function: molecular function, biological process, and cellular location.

The primary purpose of the GO Annotation (GOA) project is to annotate genomes and their by-products using GO terms. When GO terms are assigned to a gene product, an evidence code stating how the annotation was obtained is assigned as well, so that the source of the annotation is noted. Thus, GO provides a standard means for programs to describe their functional predictions.

1.4.6.1.1.1 Sequence Homology-Based Functional Annotation

Traditional means of predicting the function of sequences rely on homology. These techniques are also known as the homology transfer technique, as they traverse databases of sequences to find matches between sequences and the query sequence. From the reported matches, a transfer of relevant functional information takes place in the query sequence.

A commonly adopted tool, basic local alignment search tool (BLAST), matches significant sequence similarity to other sequences in a database of experimentally annotated sequences. The biological rationale for using homology transfer is that if two sequences have a high degree of similarity, then they have evolved from a common ancestor and hence have similar, if not identical, functions.

However, this rational does not seem to hold with growing databases and fails in three conditions:

1. High sequence similarity does not guarantee accurate annotation transfer: When two sequences share functional similarity, it is observed that only certain regions of the sequences (subregions) contribute to the functional characterization of sequences. Thus, if two sequences share a higher degree of similarity, it does not imply that the subregions contribute to the exact matches or are being conserved. Moreover, it has been shown that enzymes that are supposedly analogous due to undetectable sequence similarity are in fact similar. It is believed that 35% sequence identity and 60% aligned enzymes share four EC numbers.

    Domain shuffling also contributes to the failure of homology transfer by adding, deleting, or redistributing domains of the sequence between homologous sequences.

2. Growing databases exhibit greater diversity in sequences that affect sequence-based tools to discover similarity between proteins: Here, with the evolution of databases, categorization of sequences is constantly changing. These changes make homology transfer more challenging, as the number of clustered similar proteins for which there is no reference sequence is also growing at the same rate.
3. Chances of propagating erroneous annotations throughout the database: As more sequences enter the database, errors in annotation are often propagated and amplified based on a single erroneous annotation.

The Pfam database is the most commonly used database for protein sequence analysis. A slew of other databases, such as InterPro, SMART, CDD, and PRODOM, use the annotations at the domain level derived from Pfam and provide the user multiple alignments of protein domains. Users of these programs need to take into consideration that Pfam does not address domain shuffling, and thus the results obtained could not be as accurate as anticipated.

1.4.6.1.1.2 Structure-Based Functional Annotation — Protein structural information is represented by a collection of 3D coordinates that correspond to the amino acids that make up the protein. This representation is computationally expensive; thus, algorithms have been designed to find ways of reducing this 3D representation while preserving the spatial and physicochemical information.

The functional annotation of proteins using the 3D structural information of proteins is built on the pretext that more information can be extracted from the structure than just the sequence information. That is, knowing the structure can yield better insight into the biochemical mechanism of how proteins function. The underlying hypothesis in structural methods is that if the 3D structure is of a known fold, then that protein may possess the function of proteins processing the corresponding fold. Moreover, structure is better conserved than sequence; thus, proteins with little or no sequence similarity still have structural similarity.

Traditional structural methods are dependent on structural alignment, which entails aligning a novel protein with other proteins from its fold. Functional transfer is performed by verifying whether the aligned proteins share the same catalytic sites that are believed to be conserved by amino acid content and side-chain orientations.

With proteins of unknown structural folds and low similarity to any known fold, functional annotation is still possible by analyzing structural patterns of the protein. Here, just like sequence-based patterns, the program looks for shared structural patterns between a novel protein and a protein of a known function.

Structural patterns are best described as 3D shapes completely dissociated from the amino acids or a string of characters representing amino acids and their physical environment. For example, one can look for 3D motifs to describe the function of a protein. Here, an algorithm creates a library of 3D motifs with associated functions. A search algorithm scans the library, attempting to match extracted 3D motifs from the protein molecule. The result is a map of potential functional sites for a given protein to a library of existing function sites.

1.5 Conclusion

In this chapter we highlight the accomplishments made after the completion of the HGP that have led to the formation of key areas of research. Though bioinformatics and computational biology has created a niche for itself, its applications can be felt in other areas, such as comparative genomics, functional genomics, and sequence variation analysis. With new technological innovations being made in these areas, there has been a volume of data that require analysis. To this end, this book is dedicated to understanding the principles of data mining and its applications in the area of bioinformatics.

References

Abnizova, I., and W.R. Gilks. Studying statistical properties of regulatory DNA sequences, and their use in predicting regulatory regions in the eukaryotic genomes. *Briefings Bioinformatics* 7, no. 1(2006): 48–54.
Alexandersson, M., S. Cawley, and L. Pachter. SLAM: Cross-species gene finding and alignment with a generalized pair hidden Markov model. *Genome Res* 13 (2003): 496–502.
Astier, Y., O. Braha, and H. Bayley. Toward single molecule DNA sequencing: Direct identification of ribonucleoside 5'-monophosphate by using an engineered protein nanopore equipped with a molecular adapter. *J Am Chem Soc* 128 (2006): 1705–1710.
Baumbach, J., A. Tauch, and S. Rahmann. Towards the integrated analysis, visualization and reconstruction of microbial gene regulatory networks. *Briefings Bioinformatics* 10, no. 1 (2008): 75–83.
Birney, E., M. Clamp, and R. Dirbin. GeneWise and Genomewise. *Genome Res* 14 (2004): 988–995.
Birney, E., J.D. Thompson, and T.J. Gibson. Pairwise and searchwise: Comparison of a protein profile to all three translation frames simultaneously. *Nucl Acids Res* 24 (1996): 2730–2739.
Cao, J., and H. Zhao. Estimating dynamic models for gene regulation networks. *Bioinformatics* 20, no. 14 (2008): 1619–1624.
Collins, F. S., A. Patrinos, E. Jordan, A. Chakravarti, R. Gesteland, and L. Walters. New goals for the U.S. Human Genome Project: 1998–2003. *Science* 282 (1998): 682–689.
Florea, L. Bioinformatics of alternative splicing and its regulation. *Briefings Bioinformatics* 7, no. 1 (2006): 55–69.
Florea, L., et al. Gene and alternative splicing annotation with AIR. *Genome Res* 15, no. 1 (2005): 54–66.
Friedberg, I. Automated protein function prediction—The genomic challenge. *Briefings Bioinformatics* 7, no. 3 (2006): 225–242.

Fujita, P.A., et al. The UCSC Genome Browser database: Update 2011. *Nucl Acids Res* 39 (2010): D876–D882.

Howe, K.L., T. Chothia, and R. Durbin. GAZE: A generic framework for the integration of gene-prediction data by dynamic programming. *Genome Res* 12 (2002): 1418–1427.

Huang, X., and A. Madan. CAP3: A DNA sequence assembly program. *Genome Res* 9, no. 9 (1999): 868–877.

Huang, Z., J. Li, H. Su, G.S. Watts, and H. Chen. Large-scale regulatory network analysis from microarray data: Modified Bayesian network learning and association rule mining. *Decision Support Syst* 43 (2007): 1207–1225.

Hubbard, T., et al. The Ensembl genome database project. *Nucl Acids Res* 30, no. 1 (2002): 38–41.

Kim, N., S. Shin, and S. Lee. ECgene: Genome-based EST clustering and gene modeling for alternative splicing. *Genome Res* 15, no. 4 (2005): 566–576.

Krishnan, V.G., and D.R. Westhead. A comparative study of machine-learning methods to predict the effects of single nucleotide polymorphisms on protein function. *Bioinformatics* 19, no. 17 (2003): 2199–2209.

Lee, C., and Q. Wang. Bioinformatics analysis of alternative splicing. *Briefings Bioinformatics* 6, no. 1 (2005): 23–33.

Lee, Y., et al. The TIGR Gene Indices: Clustering and assembling EST and known genes and integration with eukaryotic genomes. *Nucl Acids Res* 3 (2005): D71–D74.

Li, H., and N. Homer. A survey of sequence alignment algorithms for next-generation sequencing. *Briefings Bioinformatics* 11, no. 5 (2010): 473–483.

Liang, F., I. Holt, G. Pertea, S. Karamycheva, S.L. Salzberg, and J. Quackenbush. An optimized protocol for analysis of EST sequences. *Nucl Acids Res* 28, no. 18 (2000): 3657–3665.

Lockwood, W.W., R. Chari, B. Chi, and W.L. Lam. Recent advances in array comparative genomic hybridization technologies and their applications in human genetics. *Eur J Hum Genet* 14 (2006): 139–148.

Mardis, E.R. The impact of next-generation sequencing technology on genetics. *Trends Genet* 24, no. 3 (2008): 133–141.

Modrek, B., A. Resch, C. Grasso, and C. Lee. Genome-wide detection of alternative splicing in expressed sequences of human genes. *Nucl Acids Res* 29, no. 13 (2001): 2850–2859.

Mooney, S. Bioinformatics approaches and resources for single nucleotide polymorphism functional analysis. *Briefings Bioinformatics* 6, no. 1 (2005): 44–56.

NCBI. *NCBI: A science primer.* July 27, 2007. http://www.ncbi.nlm.nih.gov/About/primer/microarrays.html#ref1 (accessed September 13, 2011).

Ng, P.C., and S. Henikoff. Accounting for human polymorphisms predicted to affect protein function. *Genome Res* 12, no. 3 (2002): 436–446.

Ng, P.C., and S. Henikoff. Predicting the effects of amino acid substitutions on protein function. *Annu Rev Genom Human Genet* 7 (2006): 61–80.

Parra, G., P. Agarwal, J.F. Abril, T. Wiehe, J.W. Fickett, and R. Guigo. Comparative gene prediction in human and mouse. *Genome Res* 13 (2003): 108–117.

Paszkiewicz, K., and D.J. Studholme. De novo assembly of short sequence reads. *Briefings Bioinformatics* 11, no. 5 (2010): 457–472.

Pedersen, J.S., and J. Hein. Gene finding with a hidden Markov model of genome structure and evolution. *Bioinformatics* 19 (2003): 219–227.

Pennacchio, L.A., and E.M. Rubin. Comparative genomic tools and databases: Providing insights into the human genome. *J Clin Invest* 111 (2003): 1099–1106.

Rebbeck, T.R., M. Spitz, and X. Wu. Assessing the function of genetic variants in candidate gene association studies. *Genetics* 5, no. 8(2004): 589–597.

Schlessinger, A., and B. Rost. Protein flexibility and rigidity predicted from sequence. *Proteins* 61 (2005): 115–126.

Schlessinger, A., G. Yachdav, and B. Rost. PROFbval: Predict flexible and rigid residues in proteins. *Bioinformatics* 22, no. 7 (2006): 891–893.

Shendure, J.A., G.J. Porreca, and G.M. Church. Overview of DNA sequencing strategies. *Curr Protoc Mol Biol* 81, no. 7.1.1–7.1.11 (2008): 1–11.

Simpson, J.T., K. Wong, S.D. Jackman, J.E. Schein, S.J.M. Jones, and I. Birol. ABySS: A parallel assembler for short read sequence. *Genome Res* 19, no. 6 (2009): 1117–1123.

Stamm, S., et al. Function of alternative splicing. *Gene* 344 (2005): 1–20.

Stitziel, N.O., T.A. Binkowski, Y.Y. Tseng, S. Kasif, and J. Liang. topoSNP: A topographic database of non-synonymous single nucleotide polymorphisms with and without known disease associations. *Nucl Acids* 32 (2004): D520–522.

Thomas, P.D., M.J. Campbell, A. Kejariwal, H. Mi, and B. Karlak. PANTHER: A library of protein families and subfamilies indexed by function. *Genome Res* 13 (2003): 2129–2141.

Tompa, P. The interplay between structure and function in intrinsically unstructured proteins. *FEBS Lett* 579 (2005): 3346–3354.

U.S. National Library of Medicine. *Genetics home reference: Your guide to understanding genetic conditions.* September 5, 2011. http://ghr.nlm.nih.gov/ (accessed September 13, 2011).

Vercoutere, W., S. Winter-Hilt, H. Olsen, D. Deamer, D. Haussler, and M. Akeson. Rapid discrimination among individual DNA hairpin molecules at single-nucleotide resolution using an ion channel. *Nat Biotechnol* 19, no. 3 (2001): 248–252.

Webb, E.C. *Enzyme nomenclature 1992: Recommendations of the Nomenclature Committee of the International Union of Biochemistry and Molecular Biology on the nomenclature and classification of enzymes.* San Diego: Academic Press, 1992.

Winters-Hilt, S., W. Vercoutere, V.S. DeGuzman, D. Deamer, M. Akeson, and D. Haussler. Highly accurate classification of Watson-Crick basepair on termini of single DNA molecules. *J Biophys* 84, no. 2 (2003): 967–976.

Yue, P., and J. Moult. Identification and analysis of deleterious human SNPs. *J Mol Biol* 356 (2005): 1263–1274.

Zhang, M.Q. Computational prediction of eukaryotic protein-coding genes. *Nat Rev Genet* 3 (2002): 698–709.

Chapter 2

Biological Databases and Integration

Since the beginning of the Human Genome Project (HGP), as described in Chapter 1, the numbers of published results on bioinformatics experiments have grown substantially, and datasets and refined computational models have been created to solve critical biological problems. However, these models and results seldom reach the depth and breadth of the biomedical community and are seldom interpreted correctly. This challenge is even more apparent in integrative approaches, in which data inflows from disparate sources and several models are used to analyze a single problem (Reich et al. 2006). In this chapter we wish to familiarize the readers with prominent databases and BioMarts used in bioinformatics.

2.1 Introduction: Scientific Work Flows and Knowledge Discovery

Scientific work flows, formal descriptions of a process or processes, aimed at addressing this challenge have been applied (Deelman et al. 2009). Advances in research and technologies have resulted in an explosion of information and knowledge. The ability to characterize and understand diseases is growing exponentially based on information obtained from genetic and proteomic studies, clinical studies, and other research endeavors. The depth and breadth of information already available in the research community at large presents an enormous opportunity for individual care. Because our knowledge of this domain is still rudimentary, investigations are now moving away from hypothesis-driven research and are moving toward data-driven research, in which an analysis is based on a search for biologically significant patterns (Potamias et al. 2007; Ng and Wong 2004).

By definition, knowledge discovery is "the non-trivial process of identifying valid, novel, potentially useful, and ultimately understandable patterns in data" (Fayyad 1996). It is important for the user to know that data mining is an interactive and iterative process. It is due to this interactive and iterative nature that data mining finds its place as an experimental approach and that researchers are able to try various possibilities before discovering a single solution.

With the advent of high-throughput technologies such as microarrays and next gene sequencing, one predicted application lies in the areas of genome-wide association studies (GWASs). Bioinformatics is seen to be vitally important in the storage, analysis, and distribution of the data generated from such analysis.

Thus, the challenges of analyzing biological data differ significantly from the challenges of analyzing traditional data. Therefore, it is common in bioinformatics that work flows work on smaller datasets. It is equally challenging to validate tests by domain experts, and make the discovered knowledge known to a wide audience. Thus, any work flow associated with knowledge discovery should adequately satisfy the following constraints.

1. **Share knowledge about the semantics of the data:** It is well known that, in data mining, finding an optimal representation of data is critical for obtaining good results. That is, care must be taken during preprocessing techniques, e.g., feature selection and construction.
2. **The plausibility of results:** When there is not enough statistical information about the validity of a hypothesis, one can look for external evidence for or against this hypothesis in scientific literature, which usually contains much more knowledge than what is encoded in the specific dataset. To make use of this knowledge, the interpretability of the models must be ensured.

Data integration, through work flows, can only be adequately performed if the user knows what services exist and where to find those services. With the large number of existing services in bioinformatics and the operations they perform, it is a challenge to integrate data using work flows. Moreover, this challenge is exacerbated by the arbitrary nomenclature followed and the lack of documentation available.

To effectively integrate data using knowledge discovery in databases (KDD), the following points of contrast are required:

1. **Data centric:** In typical work flows, the functions associated with analysis are treated as primary and the data used for analysis are treated as secondary. That is, data are treated as a variable, and the functions associated with the analysis of the data are important. The KDD process, to the contrary, treats the data as primary or central to the analysis, and the methods associated with analysis are considered to be secondary. This perspective renders

various functions to be applied to the data to solve research questions. Here, a researcher would typically execute myriad functions on the same dataset and readily rule out any function that fails to answer the research question.
2. **Iterative and interactive:** The KDD process as described previously is an iterative and interactive process. As such, a user is given the flexibility to choose appropriate functions on a trial-and-error basis based on how the data analysis is handled. The KDD process is structured so that the results obtained at every step enable decision making to proceed or restart at any step of the process.
3. **Dependencies between discovered knowledge:** Typically, bioinformatics involves the analysis of data from multiple datasets and their associated transformations. It is imperative that the researcher keep track of the transformations applied and results obtained. This procedure is a challenge at times due to the magnitude of data handled and hypothesis-driven work style used by researchers in bioinformatics.
4. **Handling of data types:** Several challenges arise in the handling of data types from disparate sources. It is thus imperative that researchers map the results to the metadata and their associated descriptions, especially while handling data from disparate sources (Figure 2.1).

In this chapter, we describe the intricacies involved in handling prevalent databases used in bioinformatics. Descriptions of the tools used in data mining for bioinformatics are detailed in the following sections.

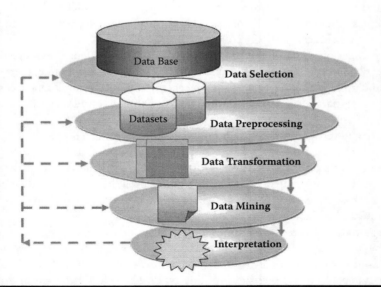

Figure 2.1 The steps involved in the KDD process.

2.2 Biological Data Storage and Analysis

The areas of data quality, cleaning, and integration are discussed below. Data quality refers to challenges pertaining to the characteristics of data stored in large bioinformatics databases and their associated schemas. Data quality usually addresses whether or not the records within the database are accurate, timely, complete, and consistent. Three methods for managing data quality are data cleaning, data reconciliation, and data integration.

Before we understand the three methods of data quality, we first enlist the characteristics of biological databases that are frequently referred to in the bioinformatics literature (Li 2006).

2.2.1 Challenges of Biological Data

Most, if not all, biological databases are created by biologists who have limited knowledge of how to effectively store data. As a result, data stored in these biological databases are often arranged in a hierarchical fashion. This hierarchical fashion of storing data mimics the evolutionary relationships between organisms. Moreover, it is also observed that the data types are tightly coupled to the specific technologies used for data acquisition. These factors attribute to the inconsistencies that plague many of these databases. These inconsistencies have far-reaching effects, as these databases have a large scope of applicability. For example, biological data pertaining to the human species in the hierarchy encompasses organisms from the highest level to the lowest level in the hierarchy, for example, organs, tissues, cells, organelles, and pathways or networks. The applications of these data include genomics, proteomics, phenomics, localizomics, ORFeomics, pharmacogenomics, and pharmacogenetics clinical trials (Li 2006).

The evolutionary nature of the biological data renders unique characteristics that are described as highly heterogeneous, large in data volume, dynamic, hierarchical, not standardized, lacking database management applications and data access tools for biological databases, and data integration and annotation (Table 2.1).

Highly heterogeneous: Brought about by the inherent complexity of biology and the array of technologies associated with the generation of data. The resultant databases are diverse in the associated data types and data schemas that are closely coupled with bioinformatics. Examples include genome databases, gene expression databases, protein databases, and protein-protein interaction databases.

Large data volume: With the unique data types and data accumulation witnessed over the past decade, data volume is expected to grow further. Considering the number of genes in the human body (20,000 to 25,000), the completed gene expression profiling of all genes, for all organs and tissues, along with cell types across development stages and timelines, will result in

Table 2.1 Examples of Molecular Databases by Categories in Biosciences

Category	Names	Database Contents	Types
Genes and transcript	EMBL	DNA sequences and derived protein sequences hosted by European Bioinformatics Institute (EBI)	Flat file
	UniGene	Clustering of human, mouse, and rat DNA and Expressed sequence tag (EST) sequences into gene-oriented, nonredundant clusters hosted by NCBI	ASN.1
Gene expression	GEO	A gene expression and hybridization array repository system hosted by NCBI	Relational models
	ArrayExpress	A public microarray data repository system hosted by EBI	Relational Models
Genome	MGD	Mouse genome data hosted by Jackson Laboratory	Flat file
	UCSC	Human, mouse, rat genomic sequences hosted by UCSC	Relational models
Protein	SwissProt	Protein sequences and annotation hosted by EBI	Flat file
Protein domains and motifs	InterPro	Protein families, domains, and functional sites in which identifiable features are found in known proteins hosted by EBI	Relational models
	Pfam	Protein sequence alignments and domain profiles hosted by Sanger Institute	Relational models
PPI	BIND	Biomolecular interaction database hosted by Blueprint Institute of Mount Sanai Hospital, Canada	Object

(Continued)

Table 2.1 Examples of Molecular Databases by Categories in Biosciences (Continued)

Category	Names	Database Contents	Types
	DIP	Protein-protein interaction database hosted by Harward Hughes Medical Institute	Relational models
Protein structures	MMDB	Curated protein structures, related sequences, and literature hosted by NCBI	Flat file
	PDB	Experimentally determined 3D structures of biological macromolecules hosted by RCSB	Relational models
Pathways	TransPath	Signal transduction pathways and reactions hosted by BioBase	Object

Source: Li, A., Eng Lett 13, no. 3 (2006): EL_13_3_13. With permission.

large amounts of data. Similarly, while considering the sequences of DNA and proteins, the volume of biological data has and will continue to expand in the number of sequences and in related graphics, images, and two-dimensional (2D) gel experiments.

The dynamic nature of sources of bioscience: To capture the complexity of the DNA and proteins, new technologies are rapidly increasing the number of dimensions (ways to analyze a problem) in biosciences. To keep up with the changes, new databases are created, and existing databases are constantly being updated with new data structures and features at every release.

The hierarchical structure of biological data: Though common, the hierarchical characteristics create a bottleneck for modeling and querying in traditional data models, such as relational or object-oriented models. For example, the DNA contained within the nucleus of a cell contains coded fragments (an integral part of the chromosome) called genes. Each gene will encode one or more proteins through one or more mRNAs. Each protein, in turn, will function in one or more pathways of various tissues. Modeling this flow of information is highly complex.

Lack of standardization in data formats and in controlled vocabularies in scientific domains: The vocabularies used to describe many biological objects are ambiguous. This ambiguity has been attributed to the fact that these databases vary in origin and history, resulting in widely used synonyms and homonyms. Another important aspect to consider while handling the databases in bioinformatics is the different formats used to represent the data. This diversity in data formats makes it difficult to use standard querying software in these databases. Moreover, it is observed that these databases lack explicit database schema, in which data are stored in relational tables consisting of a well-defined set of attributes that describe the data stored. Thus, it is also a challenge to index stored data. For example, the data formats and types adhered to for gene expression profiling using Affymetrix oligonucleotide arrays will be different than those of cDNA arrays.

Lack of database management applications and data access tools for biological databases: The lack of standardization in both data formats and data types inhibits the development of application tools in biological database management systems that are comprehensive and usable to a large community. The effects of the lack of standardization are also felt in retrieval efficiency, which is complicated, and heterogeneous applications need to be developed to handle information extraction and analysis.

Data integration and annotation: The advances made in web technology and the use of hypertext have enabled data integration of diverse domains. Thus, hypertext constitutes a part of the database contents and provides added annotation or meaning to biological entities. Nonetheless, hypertext does not provide the required standardization among databases, as it is vulnerable to the ambiguity in the identifiers or terminology system.

2.2.2 Classification of Bioscience Databases

According to Li (2006), databases can be classified based on two criteria, the goals with which they were designed and built, and their content. This classification of databases provides both computer scientists and biologists with an idea of what functions the database has to offer. Moreover, it also provides an abstraction of the application tools and database management systems these databases provide.

The major classifications of molecular databases are primary versus secondary databases, deep versus broad databases, and point solution versus general solution databases, each of which is described below.

2.2.2.1 Primary versus Secondary Databases

This distinction between databases is based on the original goals that were laid out during the inception of the database. As proposed by 3rd Millennium, Inc. (2002), primary databases are considered to be mainly data repositories and serve as data archives. Their functionality is defined by the two basic operations of storage and retrieval with limited or no complexity. These databases, apart from storage of primary data, allow a limited degree of freedom in the form of additional annotation information. GenBank is an example of such a database. The GenBank database primarily stores nucleotide sequences and their corresponding functional information pertaining to associated experimental labs and projects. The standardization enforced by GenBank on its input information and taxonomy enables effective internal interpretation. Secondary databases, on the other hand, store data from several publically available sources. The Pfam data are an example of such a secondary database, in which information regarding protein sequences is extracted from related primary databases or archives. The extracted information in the Pfam database is performed both manually (PfamA) and automatically (PfamB) and provides for a bifurcation of the holistic database.

2.2.2.2 Deep versus Broad Databases

In this classification, the databases are categorized into deep databases and broad databases based on the scope of the data contained in them. As proposed by Cornell et al. (2003), the scope of the databases is defined by the key features of the databases, the source of the databases, and the formats by which the data are defined. For example, the SwissProt database is a protein sequence database that contains protein sequences from all known species; thus it is considered a broad database. In contrast, the deep databases contain information specific to species. For example, the *Saccharomyces* Genome Database (SGD) contains all known information pertaining to the *Saccharomyces* genome. The primary purpose of these databases is to provide browsing and visualization for discovered data, along with complex query processing through limited data integration.

2.2.2.3 Point Solution versus General Solution Databases

Proposed by Wong (2002), this categorization aims to differentiate biological databases into two categories, systems point solution and general solution databases. As the name suggests, the goals of a system point solution database are specific to a predefined biological problem or question. Hence, these databases are small and have limited scalability. In comparison, a general solution database has neither predefined data sources nor questions that are addressed during its design. Thus, a general solution database can be flexibly extended by incorporating additional data sources to answering general queries during its design. The applications of such databases are described in Li (2006) and are provided in Table 2.2.

The following section encapsulates the aforementioned characteristics of databases and the issues entailed in determining quality data for mining in light of commonly used databases for data mining. The following databases are described below: the Gene Expression Omnibus (GEO) (Edgar et al. 2002) database and the Worldwide Protein Data Bank (PDB) (Berman et al. 2003).

The following sections highlight some commonly used databases and their related types.

Table 2.2 All Databases: Classification of Molecular Databases

	Name	Pri.	Sec.	D	B	PS	GS	Rep.	Bro.	Vis.	Query	Ana.
Genes and transc.	GenBank	*			*	*		*	*	*	*	
	EMBL	*			*	*		*	*	*	*	
	DDBJ	*			*	*		*	*	*	*	
	RefSeq		*		*	*			*	*	*	
	UniGene	*	*		*	*			*	*	*	
	dbEST	*			*	*		*	*	*	*	
Genomes	Flybase	*		*		*		*	*	*	*	
	MGD	*		*			*	*	*	*	*	
	SGD	*		*			*	*	*	*	*	
	UCSC		*	*			*		*	*	*	
Expr.	GEO	*		*			*	*	*	*	*	
	ArrayExpress	*		*			*	*	*	*	*	*
	REMBRANDT	*			*	*		*	*	*	*	*
	GXD	*			*	*		*	*	*	*	

(Continued)

Table 2.2 All Databases: Classification of Molecular Databases (Continued)

	Name	Pri.	Sec.	D	B	PS	GS	Rep.	Bro.	Vis.	Query	Ana.
Proteins	SwissProt	*			*	*		*	*	*	*	*
	Trembl	*			*	*		*	*	*	*	
	Enzyme	*			*	*		*	*	*	*	
	UniProt	*			*	*		*	*	*	*	
	MIPS	*			*	*		*	*	*	*	
	BRENDA	*			*	*			*	*	*	
Nome.	HUGO	*		*		*			*	*	*	
	Enzyme nome.	*			*	*		*	*	*	*	
P seq. and motifs	InterPro		*		*	*			*	*	*	*
	Prosite	*			*	*			*	*	*	
	ProDom	*			*	*			*	*	*	
	SMARTS	*			*	*			*	*	*	
	PRINTS	*			*	*			*	*	*	
	BLOCKS	*			*	*			*	*	*	
	PFAM								*	*	*	
Protein structure	PDB	*		*		*		*	*	*	*	
	MMDB		*		*	*		*	*	*	*	
	FSSP/Dali		*		*	*			*	*	*	
	SCOP		*		*	*			*	*	*	
	CATH		*		*	*			*	*	*	
	HSSP								*	*	*	
PPI Pathways	YPD		*	*		*		*	*	*	*	
	BIND	*			*	*		*	*	*	*	
	DIP	*			*	*			*	*	*	
	MINT	*		*		*			*	*	*	
	PathDB	*			*	*		*	*	*	*	*
	TransPath	*			*	*			*	*	*	*
	Kegg		*		*	*		*	*	*	*	
	MPW	*			*	*		*	*	*	*	
	UM-BBD	*			*	*		*	*	*	*	

Table 2.2 All Databases: Classification of Molecular Databases (Continued)

	Name	Pri.	Sec.	D	B	PS	GS	Rep.	Bro.	Vis.	Query	Ana.
Annot.	Gene Ontology	*			*		*	*	*	*	*	
	Taxonomy		*		*	*		*	*	*	*	
	OMIM	*			*	*		*	*	*	*	

Note: The classifications are based on design goals and the contents of the databases, as well as the applications on the databases.

Abbreviations: Pri., primary database; sec., secondary database; D, deep database; B, broad database; PS, point solution; GS, general solution; rep., repository; bro., browser; vis., visualizing; ana., analysis; transc., transcripts; expr., expression; nome., nomenclature; PPI, protein-protein interaction; and annot., annotation.

2.2.3 Gene Expression Omnibus (GEO) Database

In this section, we cover the important characteristics that gene expression databases possess (Do et al. 2003). Some characteristics of the data may be omitted or may be only partially included in the database. Importance is given to specific characteristics, while other characteristics are derived (implied) from the specific characteristics. Thus, due consideration needs to be given to analyze and segregate the characteristics of data.

In gene expression data, raw data are obtained in the form of microarray chip images, a product of the microarray experiment. Typically, a record in a gene expression database consists of three parts, image data, expression data, and annotation data. Image data are a scanned image of the microarray chip. Expression data are the normalized version of the image scanned. It is a sequence of numbers that represents the expression of a gene for a given sample. This information constitutes the core of the gene expression database and is accessed frequently. Taking into consideration the high volume of data and the frequency of references made to it in the database, it is desirable to apply effective indexing and store schemas for quicker and more effective access of these data. The third component to a record is the annotation data. Annotation data are the metadata that are appended to the microarray data. These data add additional information to the record and consist of textual descriptors that help interpret the detected gene expression levels or keywords that describe the associated gene function. The annotation information can be further categorized as follows.

Gene annotation: Annotation information pertaining to the gene sequence's place on the microarray is categorized in this section. Annotations pertaining to the gene name, its known functions, and location over the chromosome are found here. These annotations are collected over time and are publically accessed from different databases.

Sample annotations: Similar to the gene annotation, annotation pertaining to the sample studied is stored in this section. Information pertinent to the hybridization used to extract the targets, the corresponding biological descriptions pertaining to source and sample characteristics, like information that describes whether the sample is normal or diseased, and information that describes if there are any *in vitro* or *in vivo* treatments that have been applied are found here.

Experiment annotations: Experiment annotations contain the information regarding the protocols followed during the experiment and parameter settings used by the associated tools and software during hybridization.

The data stored in the databases have associated descriptors that add value to the data. These annotations are manually entered or derived from external databases. Thus, it is imperative to organize annotation data in a uniform manner to improve its effectiveness for analyzing gene expression data.

The current standard used to capture annotation data renders two challenges that must be addressed. The first challenge is that of standardization. As annotation information is entered manually through free text, different sources have adhered to different vocabularies. The discrepancies that arise due to varied vocabularies affect the integration and matching of records. The second challenge stems from the lack of standards in the use of vocabularies. Many terms may be used to describe the same things, making the querying of these databases a challenge.

As a solution to the above challenges, the use of free text to describe annotations should be avoided. The advent of ontologies to this end, and more specifically, the Gene Ontology, created by the GeneOntology (GO) Consortium, is a specialized hierarchy of categories that provides the basis for standardizing annotation vocabulary in gene expression data storage.

Initiated by the need of a public repository for high-throughput gene expression data, the Gene Expression Omnibus (GEO) project (Edgar et al. 2002) was designed to provide a flexible and open design to store, retrieve, and insert data from high-throughput gene expression and genomic hybridization experiments. It is intended to act as a central data distribution hub of gene expression data derived from coherent datasets.

As seen in Figure 2.2, GEO segregates data into three principal components, platform, sample, and series stored and accessed in a relational database model. Here, the data are not fully granulated within the database. Instead, a tab-delimited ASCII table is stored for each platform and each sample. The resultant tables of the GEO database are shown in Figure 2.3. These tables consist of multiple columns with accompanying column header names. The data within this table are partially extracted for indexing, but may be further extracted for more extensive search and retrieval.

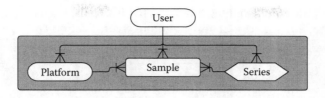

Figure 2.2 The entity relationship diagram of the GEO database. (From Edgar, R., et al., *Nucl Acids Res* 30, no. 1 (2002): 207–210. With permission.)

Similar to the GEO database, there are several other publically available databases that provide the necessary information regarding genes and their expressions. Table 2.3 enumerates a few popularly referenced software tools, packages and databases in this area.

2.2.4 The Protein Data Bank (PDB)

The PDB is one of the largest repositories of known protein structures in the world. It contains information of all experimentally determined structures of proteins, nucleic acids, and complex assemblies and their corresponding 3D coordinates. As of March 2010, the database contained an estimated 63,956 known structures, publically accessed over the Internet. The growth of this database has been exponential, and the number of known structures doubled between 2005 and 2010, as shown in Figure 2.4 and Table 2.4.

Formerly referred to as the Brookhaven Protein Database, this steady and substantial growth in the number of protein structures is because data are pooled to

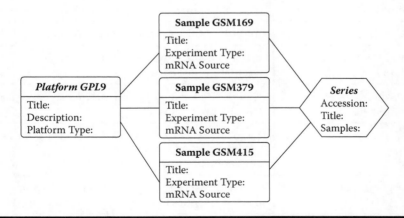

Figure 2.3 An example of three samples referencing one platform and contained in a single series.

Table 2.3 Other Publically Available Software Tools, Packages, and High-throughput Gene Expression and Genomic Hybridization Data Resources

Software Tools	Description
Agile Protein Interaction Data Analyzer (APID)	Provides exploratory analysis of protein-protein interactions
Database of Interacting Proteins (DIP)	Provides data integration from various sources to create single, consistent set of protein-protein interactions
GeneXPress	Provides a visualization and analysis tool for gene expression data, integrating clustering, and gene annotation
Gapasi	This is a software package for modeling biochemical systems
GOstat	A tool used to identify statistically overrepresented GO terms within a group of genes
Data Resources	**Institution**
ExpressDB	Harvard-Lipper Center for Computational Genetics
Global Gene Expression Group	Science Park-Research Division, University of Texas M.D. Anderson Cancer Center
MAExplorer	National Cancer Institute, NIH
Microarray Center	Public Expression Profiling Resource
Microarray Project	National Human Genome Research Institute, NIH
SAGENET	Johns Hopkins University School of Medicine
Yeast Microarray Global Viewer	Laboratoire de genetique moleculaire, Ecole Normale Superieure
RNA Abundance Database (RAD)	Computational Biology and Informatics Laboratory, University of Pennsylvania
Gene Expression Omnibus	National Center for Biotechnology information, NIH
Code	**Environment**
MetageneCreator	MATLAB© package used to identify overlapping clusters of genes in arbitrarily large datasets.
Deal	R package used to create Bayesian networks with both continuous and/or discrete variables

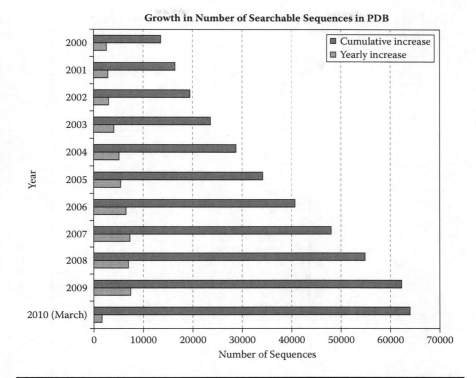

Figure 2.4 Exponential growth of the PDB (as of March 15, 2010).

the database from various organizations. These organizations act as deposition, data processing, and distribution centers for PDB data. The organizations constitute what is now known as the Worldwide Protein Data Bank (wwPDB), consisting of Research collaboratory for structural Bioinformatics (RCSB) PDB in the United States, PDBe in Europe, and PDBj in Japan. The Biological Magnetic Resonance Data Bank (BMRB) group from the United States joined the wwPDB in 2006. The mission of the wwPDB is to maintain a centralized Protein Data Bank Archive of macromolecular structural data that are freely and publicly available to the global community.

With the substantial growth in the number of proteins, efforts of the PDB are focused on data cleaning and data integration, and eliminating data inconsistencies. To make this data integrating and cleaning possible, it is important to understand the relation database model of the PDB.

A relational database is, in essence, a set of related tables (entities), each of which is uniquely identified by a primary key. One table may contain a field/attribute that is a primary key in another table. Records may not be added to a table unless there is a corresponding record in the related table. This dependency between tables is known as referential integrity and ensures that changes made to one table are

Table 2.4 The Growth in the Number of Sequences in the PDB from 2000 to 2010

Year	Yearly Increase in Number of Sequences	Cumulative Increase in Number of Sequences
2010 (March)	1,612	63,956
2009	7,439	62,344
2008	7,004	54,905
2007	7,232	47,901
2006	6,492	40,669
2005	5,372	34,177
2004	5,192	28,805
2003	4,172	23,613
2002	3,004	19,441
2001	2,832	16,437
2000	2,628	13,605

reflected in the other table. As all tables should be related to at least one other table, there should be no stand-alone tables in a relation model.

Relational databases are created through a process of normalization, during which redundant data are removed and data consistency and integrity is enforced. Formal methods for staging normalization are called normal forms.

The PDB has been recently reengineered to become a rational database based on the macromolecular Crystallographic Information File (mmCIF) schema. The adhered mmCIF dictionary, to this end, has been viewed as an ontology that details key concepts and relationships in functional genomics experiments.

By definition, an ontology is a representation of a preexisting domain of reality that (1) reflects the properties of the objects within its domain in such a way that it obtains a systematic correlation between reality and representation and (2) is intelligible to a domain expert (3) if formalized in a way that allows it to support automatic information processing. Generally speaking, an ontology consists of four components: classes, a hierarchical structure (is-a relations), relations (other than is-a relations), and axioms. Unfortunately, the mmCIF does not meet this definition. The failure to follow ontology standards has resulted in many poor design failures in mmCIF, and this in turn has resulted in a poor PDB relational database design.

The mmCIF dictionary of the PDB is written in the Self-Defining Text Archive and Retrieval (STAR) language, which consists of a set of data names with associated data values. Multiple values for one data name are allowed. Identifying the data names that may appear in a loop construct is important for the first stage of normalization. These are the fields that would be a repeating group and would then be removed at this stage. The International Union of Crystallography (IUCr) states that only items that need to be repeated should appear in a looped list and gives guidelines to the mmCIF categories that are normally represented in this format.

One of the most difficult aspects of basing a relational database on mmCIF is the lack of consistency in the recording of experiments. The majority of the mmCIF data items for the individual protein structures are omitted. mmCIF labels these items as optional, and as a result, sometimes only the minimum amount of experimental information has been provided. Other problems include the amount of data repetition in mmCIF and data redundancy; for example, protein entry ID assesses most mmCIF categories but is not indicative of a dependent repeating group. Thus, care has to be taken when normalizing.

The simplest way to familiarize oneself with the associate relational database is by studying its associated schema diagram. As the PDB is both complex and large, its associated schema is also large. A centralized table containing a reference to the PDB entry and all the other tables relates to it directly, rendering a donut shape to the schema. It is also observed that there are no table-to-table relationships, which thereby exhibit a hierarchical model to the data stored.

The following are characteristics of tables in the PDB:

1. The primary key for every table is the same, and the key name changes when it is involved in a relationship. This change in primary key name adds confusion when tracking relationships.
2. Several stand-alone tables are derived and are not part of the database.
3. Inconsistencies in field size show that the fields of any record in a table of the PDB could have varied lengths.

Every category in the mmCIF is allocated a table in the relational database, and no normalization is applied. This allocation attributes to the unusual shape of the PDB schema. Though the schema is equivalent to the database STAR schema, the PDB has been intentionally denormalized, and therefore does not meet the requirements of a data warehouse.

Data repetition among the mmCIF categories is not resolved, which implies that every relationship in the database is the same, and interdependencies between other tables have been ignored. Thus, the PDB does not satisfy the requirements of a successful relational model.

On the other hand, the Pfam database (Bateman et al. 2002) follows all the requirements of a relational database.

2.3 The Curse of Dimensionality

With the advent of massive storage and rapid-throughput technologies to generate data, recent decades have witnessed data analysis transform to a realm that is beyond the scope of traditional statistical approaches. However, the belief that these developments in information technology will solve any structural problems for data analysis is not true. Over the last 30 years data mining in particular has been formalized in the form of software packages, and has been the key in transforming the paradigm of hypothesis-driven research into a data-driven paradigm. This data-driven paradigm has been brought about by addressing the fundamental problems that are omnipresent and require additional support of data analysis, to convert raw data into information for effective decision making (Donoho 2000).

More specifically, the inherent large number of dimensions, called the curse of dimensionality, has ubiquitous effects throughout the sciences, specifically in bioinformatics. The curse of dimensionality refers to the large number of features p (dimensions) that describe each record n in the database, that is, large n and small p. Hence, the curse of dimensionality is also referred to as the small n big p problem. Standard statistical approaches do not hold true in such scenarios. They are based on the assumption that $p < N$ and $N \to \infty$. Many of the methods used in statistical data analysis are derived from linear algebra and group theory to develop close to exact distribution results. These results all fail when $p > N$. They are also based on the assumption that $N \to \infty$ with fixed p, which does not always hold true in reality; on the contrary, p could tend to ∞ and N being fixed, as in the case of many genes describing relatively few samples of genetic diseases.

The effect of large dimensions on modeling data in high-dimensional space is best captured when we take into consideration data points in a 10-dimensional space. The distance between independent data points increases with the inclusion of more dimensions. The density or distribution of the points becomes sparse, making it difficult to apply traditional approaches to fit a model to these points in 10 dimensions. The application of traditional approaches is especially difficult when we consider $p > N$, where the number of points are smaller than the dimensions analyzed.

Though statistically challenging, the curse of dimensionality has opened up many avenues to help researchers understand the role of features in describing the data. It was observed that many identical dimensions, dimensions that represent redundant information, existed. This redundant information laid the foundation for numerous feature selection and feature extraction techniques. A detailed description of these techniques is given in Chapter 4.

The section below includes descriptions of dimensions and their roles in data integration and data cleaning. Many biological databases cross-reference data that are derived from external databases. This cross-referencing renders challenges for effectively representing data. Thus, the following section is dedicated to addressing these issues.

2.4 Data Cleaning

Biological data are rich with issues, such as data inconsistencies and data duplications that can be addressed with data cleaning and integration methodologies. Data cleaning in biological data is an important function necessary for the analysis of biological data. This step can standardize the data for further computation and improve the quality of the data for quicker search and retrievals. The primary purpose of most biological databases is to create repositories that integrate work from numerous scientists. This use requires sophisticated data cleaning strategies. Chapter 3 provides the various data cleaning strategies that encompass data from single sources/databases. However, in this section, we provide the description of data cleaning strategies designed to overcome traditional problems that can be avoided using data mining techniques.

As discussed in previous sections, biological data are evolutionary in nature. Most of the well-known databases mimic this inherent property by storing the data in a hierarchical fashion (as a phylogenetic tree). This hierarchical fashion of data storage possesses the following problems:

1. As emphasized in the previous sections, nomenclature and vocabulary used in data annotation do not adhere to a set of standards.
2. It is frequently observed that the data from biological databases lack a consistent format, especially when performing operations on data from phylogenetic systems.
3. Data from legacy phylogenetic systems require cleaning and extensive modification.
4. It is a challenge to find duplicates within the structural data (trees) and recodes within the dataset.
5. It is difficult to remove duplicates when required.
6. Finding clusters similar to structural data (trees) and records, merging similar records, and finding anomalous structural data (trees) and data are also difficult.

Data cleaning, also called data cleansing or scrubbing, is the process of detecting and removing errors and inconsistencies from data in order to improve data quality. The above-mentioned data quality problems are present in most biological data collections, such as files and databases, e.g., in data warehouses, federated database systems, or global web-based information systems traditionally used in bioinformatics. Table 2.5 contains a list of popular data cleaning methodologies applied on biological databases.

2.4.1 Problems of Data Cleaning

The quality of data is gauged by the number of errors, discrepancies, redundancies, ambiguities, and the degree of incompleteness therein that diminishes the quality

Table 2.5 Table of Popular Data Cleaning Methodologies

Methodology	Example System
ETL	Talend—an open source data integration tool
Multi-pass sorted neighborhood	Merge/Purge (Hernandez and Stolfo 1995)
Disambiguation methods	ConQuer (Fuxman et al., 2005)
Knowledge-based technique	Intelliclean™ (Low et al., 2001)

results obtained from data analysis or data mining. Thus, data cleaning is the process of detecting and removing the above-mentioned factors to improve the overall quality of the data for mining purposes. The problems associated with data cleaning tend to fall into two categories. The first category is the detection of erroneous data. Problems from erroneous data usually stem from, but are not limited to, errors caused by user inputs such as inconsistency in input, missing values, misspelling, improper generation of data, and differences between input data and legacy data. The second category is the detection of duplicate records. In the past, duplicate detection has been applied to large databases where duplication control is not very strong. The associated algorithms were used to detect similarity between strings for file-based systems and similarity between records in rational databases. However, in large databases with complex schemas, the feasibility of the same logic failed to detect duplicate records and files. This problem reduces the quality of data in large databases with complex schemas. These problems are magnified in databases that evolve with time, as is the case with biological databases.

In addition, these problems are more prevalent in biological databases. Most biological databases are fueled by the data generated by experiments from around the world. The sources of these data include large submissions by high-throughput sequence and gene expression experiments. Based on the global scale of bioinformatics, it has been a challenge to ensure adequate quality control of the submission process.

Moreover, according to the 2008 annual review of databases, the number of molecular biology databases increased by 95 in 2008 (McLeod et al. 2009). Most of these databases have their own data formats, nomenclatures, and schemas. This disparity in database characteristics requires standardization. Some of these databases derive or replicate their content from well-known archives such as GenBank. This replication has its own negative implications, as it fosters propagation of resubmission of the same sequence if not monitored or regulated.

In addition, the sequences in the databases are manually curated. For example, the SwissProt section of the UniProtKB/SwissProt database is manually curated by experts from the Swiss Institute of Bioinformatics. It is known that errors do seep in despite stringent quality control mechanisms. These errors are further magnified when automated systems use the erroneous annotations caused unintentionally.

2.4.2 Challenges of Handling Evolving Databases

The challenges in handling evolving databases can be divided into two categories, as shown in Figure 2.5 (Rahm and Do 2000). These problems can be solved using data cleaning and data transformations techniques. Data transformations techniques include changing the data types and various summarization schemes, and are used to enforce changes in the structure and representation of data content. Data transformations help map the data from their given formats into the format expected by the application (Muller and Freytag 2003). These transformations are important for handling evolving databases, especially those that help in the migration of a legacy system to a new information system or those that integrate multiple data sources.

Data quality problems (Rahm and Do 2000) consist of two categories, single-source and multisource problems. These two categories are further divided into schema and instance-related problems. Instance-related problems refer to errors and inconsistencies in the actual data contents that are not visible at the schema level. They are the primary focus of data cleaning. Schema-level problems are the problems found at the schema level. These errors are also reflected on data instances. They can be addressed by incorporating changes into the schema design, i.e., evolving the schema by performing schema translation and schema integration operations.

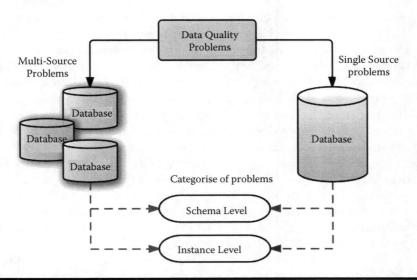

Figure 2.5 Categorization of data quality problems in data sources.

2.4.2.1 Problems Associated with Single-Source Techniques

In single-source techniques, problems of data quality pertain to the integrity constraints that are applied to the data during the schema design. The integrity constraints control the fields and their associated data types that are allowed to be entered into the database. Since biological databases consist of databases that are based on file systems and loosely defined relational schemes, the sources without schema, such as file systems, have few restrictions on what data can be entered and stored. This lack of restrictions gives rise to inherent inconsistencies and results in a high probability of errors. Sources that are based on relational models specify data constraints during schema design or inception. The constraints enforce attribute restrictions that prevent values that do not confine to a specific range or format. Similarly, uniqueness constraints enforce uniqueness of values entered in a field of the record, if desired. Most of these constraints are application specific. Data quality violations that are associated with the data schema are categorized as schema-related issues. They occur because of poor schema design or lack of proper constraints during schema inception. On the other hand, problems that are associated with errors and inconsistencies that cannot be prevented at the schema level, such as typographic errors, missing values, duplicate records, and misspellings, are defined as instance-specific problems.

2.4.2.2 Problems Associated with Multisource Integration

Typically, bioinformatics databases require data from two or more data sources. The inherent problems of data cleaning databases of single sources are magnified when data from two or more sources are integrated. In multisource integration, the problems faced are derivatives of the problems of each independent source. These problems stem from the fact that data from different sources can be represented differently, overlap, or contradict each other because the databases are tailored to suit specific applications. The differences in the ways that the databases are deployed and maintained results in heterogeneity in data models, schema designs, and data management systems.

The differences at the schema level are addressed by schema translation and schema integration. The specific problems at the schema level are the naming and structural conflicts in the databases. When an attribute or feature is assigned the same name in different databases to represent the same object (synonyms), or when different names are used to represent different objects (homonyms), the errors that are associated with this conflict are known as naming conflicts. Conflicts that arise due to variations in representation of the same object in different sources are called structural conflicts. These conflicts can occur with different component structures and different data types.

Conflicts at the instance level stem from differences in the representation of data in different sources. These conflicts typically result in duplicate and contradicting records. Moreover, the attributes or features with the same name and data types could follow different standards and interpretations. Time is also an important factor when considering data from different sources. In such a case, care should be taken, as data could refer to different points of time.

Thus, the main challenge of cleaning data from different sources is identifying overlapping data. These overlapping data enable effective matching of records between sources. The problem is also referred to as the object identify problem, the merge/purge problem, or the duplicate elimination problem (Rahm and Do 2000). In an ideal scenario, the data from different sources may complement each other and add information about the entity. To make this happen, it is important to filter out duplicate information and retain complementing information by merging them to existing information, thereby providing a consistent view of the real-world entities. Figure 2.6 provides a categorization of the errors that are typically found in biological databases. Again, the categorization is described for the attribute, record, single-database, and multiple-database problems.

2.4.3 Data Argumentation: Cleaning at the Schema Level

By definition, data reconciliation is the process of comparing data from multiple sources for creating consistency in the data. As mentioned in the previous section, the number of databases for molecular biology has grown. In addition, changes have been made to 68 previously existing databases (McLeod et al. 2009). These data sources are riddled with feature inconsistencies and incomplete information. Moreover, the data, at times, contain conflicting information. With the abundance of Internet-based tools to analyze the data, newer inferences are being derived from these databases on a regular basis. Any inconsistencies in the data exacerbate misleading conclusions, emphasizing the need for better quality data. To this

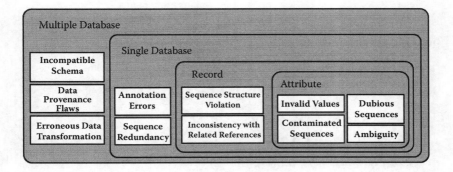

Figure 2.6 The classification of errors in biological databases provided by Judice (2007).

end, there is a need for techniques that evaluate the datum before it is used. Data reconciliation plays a vital role in removing conflicts between data sources.

The method of *argumentation* for data reconciliation has been proposed by McLeod et al. (2009). An argument is a reason to believe something is true; it is used to support or attack a conclusion. Arguments can also attack and defeat each other. Once defeated, an argument can be reinstated if the argument that defeats it is defeated. When presented with the arguments for/against a conclusion, the user can evaluate the evidence and make a decision as to whether or not to believe it. As time passes, new information becomes available, and new arguments can be created. These new arguments may defeat existing arguments, thus reinstating other arguments. When presented to users, these changes may alter their perspective and so alter their opinion of the conclusion.

Argumentation was implemented by McLeod and Burger (2008) over two gene *in situ* expression databases. An *in situ* database consists of 3D images of organisms that highlight the areas where a particular gene is expressed. In the analysis of *in situ* gene expression data, two images of samples (mouse or zebrafish) are compared to obtain a spatial processing of where genes are expressed. Figure 2.7 provides a brief conceptual view of the process of argumentation followed in this study.

Commonly used databases for the mouse genes are GXD (Smith et al. 2007) and EMAGE (Venkataraman et al. 2008). These databases are complementary, as they publish the same information and are based on the same ontology— Edinburgh Mouse Atlas Project (EMAP). When these databases are queried independently for a specific gene, the results vary in regard to the displayed records. To resolve this issue and obtain a more accurate result, biologists typically treat the results from the databases as mutually exclusive and based on laborious related research in published paper surveys, and decide whether the results obtained are conclusive. This issue is prevalent in most biological analyses and studies that involve biological databases.

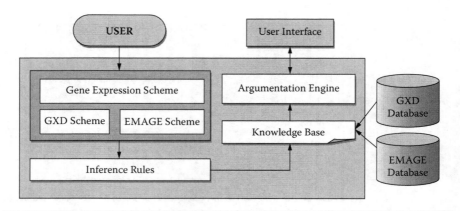

Figure 2.7 Conceptual schematic of the process of argumentation.

A closer analysis of the databases GXD and EMAGE reveals that some of the experiments present in GXD are not present in EMAGE and vice versa. Furthermore, EMAGE maps some of its embryo 2D images onto its 3D embryo model of EMAP. This mapping entails both textual annotation and spatial transformation. GXD contains results that are mapped to the EMAP ontology, laying the basis for applying argumentation to resolve the inconsistencies between the two biological databases.

The process of argumentation involves the use of an argumentation engine (Fox et al. 2007). Using domain information and expert knowledge in the form of inference rules, the argumentation engine interprets these rules to create arguments by backward chaining through the rules in response to a query from the user. Expert knowledge is provided by a domain expert. All information from the domain expert is recorded using a natural language. Argument schemes are then employed to act as an interface between the domain expert and the argument engine (Verheij 2003).

Another important aspect of the arguments generated by the argument engine is to resolve conflicts between arguments. This conflict resolution is brought about using a ranking scheme that allows the domain expert to further provide weights to different schemes based on the order of importance. These scores are then propagated back to the rules that generated them, and thus establish an order of importance to the rules.

When a query needs to be processed, for example, the user specifies a specific gene and a corresponding structure through a specialized client interface. The client first pulls up all relevant data, and then transforms it to a format that can be used by the argumentation engine. Simultaneously, both domain data and expert knowledge is loaded into the argumentation engine knowledge base. Once this network is set up, the query (Is the gene expressed in the structure?) is sent to the argumentation engine, and the results are displayed to the user.

2.4.4 Knowledge-Based Framework: Cleaning at the Instance Level

Data cleaning that uses domain knowledge to duplicate record identification and for de-duplication is a necessary component of data preprocessing. This method, in contrast to the previous method of argumentation, uses data from a single source.

As the title of this section suggests, data cleaning at the instance level uses domain knowledge as the key ingredient for cleaning. Thus, the proposed framework employed by the system IntelliClean™ (Lee et al. 2000; Low et al. 2001) provides a viable representation and utilization of domain knowledge for data cleaning. The framework, as seen in Figure 2.8, also supports effective record standardization, duplicate elimination, anomaly detection, and removal of unclean data.

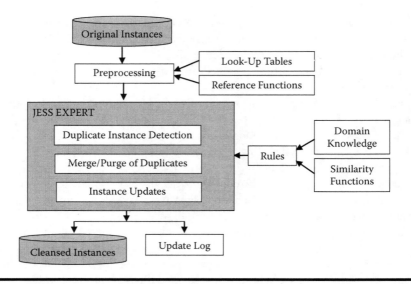

Figure 2.8 Schematic representation of the IntelliClean framework.

The framework is composed of three stages, preprocessing, processing, and verification and validation.

The preprocessing stage: In this stage, data from a single source are subjected to various standardization and format checks to free the data from inconsistencies like variations in abbreviations and formats. This process is called record scrubbing. Record scrubbing is brought about through the use of reference functions and lookup tables. The lookup tables are used to compare all the abbreviations used in the applications with their occurrence in the records. If differences are observed, the equivalent from the lookup table is used to replace the incorrect abbreviation in the record. The result of this stage is a set of conditioned instances that is subjected to the next stage.

The processing stage: In this stage, the conditioned instances are compared to a set of rules that enable the effective identification of inconsistencies between instances. Rules in this system are generated using the Rete algorithm (Rete 1982) that is implemented using the Java° Expert System Shell (JESS) (Friedman-Hill 1999). The rules used for this algorithm are further classified as follows:

1. **Rules for duplicate identification:** These rules are specifically used to identify duplicate instances in the conditioned data.
2. **Rule to merge/purge:** Once the rules have been identified or detected, these rules are used to delete the duplicate records. For example, a rule could specify the deletion of one of the records, depending on the degree

of match or prevalence of the rule. The duplicate instances are deleted throughout the database.
3. **Rules for updates:** These rules help modify or update an instance, but may not be required.
4. **Rules to generate alerts:** If there are instances that violate certain constraints of functional dependence or integrity constraints, then these rules are useful to generate the corresponding alerts.

Generating and comparing the rules to instances is a time-consuming process. There are variations in the implementation of the system to make it time-effective.

The validation and verification stage: The validation and verification stage requires human intervention to manipulate duplicate instances for which the merge/purge rules do not work or have not been defined. The entire system is log based, meaning that there is record of all corrections or updates to the data maintained in a log file. This file allows the users or domain experts to verify whether the corrections carried out in the preprocessing and processing stages are accurate and enable these experts to undo any incorrect actions. It also helps gauge the correctness of the rules. Rules that perform incorrect updates and duplicate detection could be removed from the system.

2.4.5 Data Integration

Let us consider a typical challenge faced by biologists attempting to collect data from multiple databases. Typically, when searching for evidence linking phenotypes to genes, data are gathered based on phenotype differences and allelic variants between the strains, genotypes, and pathways in which these genes belong. This process involves the gathering of data from multiple sources. With myriad databases available, it becomes a challenge for anyone to identify the corresponding databases and what services they offer.

The laborious steps in this process allow the users to learn how to utilize these databases based on what each one has to offer. Typically, such data collection includes copying the data from these databases into Microsoft Word® or Excel® files for further analysis. This process is error-prone and leads to computational bottlenecks, as the method does not scale up to the magnitude of the data.

These bottlenecks call for data integration approaches. Data integration has been a constant endeavor since the early 1990s and the inception of the HGP, in which data were generated on a large scale by sources being geographically distributed across the globe. With the geographically distributed sources, it was important then to find a way of integrating this data. To this end, Kleisli (Wong 2000), a powerful query system, was developed by the University of Pennsylvania to solve this predicament. This system consists of a nested relational data model, a high-level query language, and a powerful query optimizer. It can handle multiple source databases and can withstand the issue of heterogeneity among these source databases. This heterogeneity allows the user to create structured query language (SQL)-like queries that are independent of the location of the data, its format (relational/flat file), and the disparate access protocols implemented at the sources. Moreover, it can store, update, and manage complex nested data through application program interfaces (APIs) available in Java and Perl*.

Several other approaches have been proposed since Kleisli. These include Ensembl (Hubbard et al. 2002), SRS (Etzold and Argos 1996), and DiscoveryLink (Haas et al. 2001).

2.4.5.1 Ensembl

The Ensembl software technology is the outcome of a joint venture between the European Bioinformatics Institute and the Sanger Institute (http://www.ensembl.org). Ensembl, though not an ideal example of data integration, provides a feel of the benefits of integrating data from different sources. This tool provides for query processing of all eukaryotic genomic sequence data. It gathers and assembles sequences from various data sources to their corresponding locations on the genome. Based on these derived sequence assemblies and using GenScan, the tool automatically predicts genes in these data. These predictions are then made publically available. Moreover, Ensembl also has tailor-made functions that enable complex operations of annotation of these sequences.

2.4.5.2 Sequence Retrieval System (SRS)

The Sequence Retrieval System (SRS) tool is marketed by LION biosciences and is known for its query and navigation system (http://srs.ebi.ac.uk). It is one of the most widely used data traversal systems in life science, as it provides access to several biological databases that include sequence databases, metabolic pathway databases, and literature abstracts. SRS is built using a programming language called Interpreter of Commands and Recursive Syntax (Icarus) (Wong 2002). SRS allows its users the facility to add their own databases to be traversed and compared if desired. The addition of new databases into SRS requires the submission of both the new database in the flat file format and its corresponding schema or structure; both must be available as an Icarus script that acts as a wrapper to the data submitted. These Icarus script wrappers constitute a wrapper programming language of SRS, which is responsible for creating indexes for each of the parsed flat files that are described by the Icarus script. By doing so, a biologist can access the data using keywords and constraints in the SRS query language.

The SRS query language is an information retrieval language, which means that the results obtained after query execution are simple data aggregates that match the specific constraints. SRS processes limited data joining and restructuring capabilities. The SRS frontend offers users accessibility to the multiple data sources independently without the hassles of handling these sources independently. Thus, SRS, though popularly perceived as a data integration tool, is an interface integrating tool.

2.4.5.3 IBM's DiscoveryLink

Proposed by IBM and based on IBM's DB2 database management system, DiscoveryLink (http://www.ibm.com/discoverylink) stands out due to its explicit relational data model that acts as an intermediary between the data sources and the end user. This intermediate data model enables the user to query it when required. This feature is in line with the SRS system. However, the data model supports most SQL queries if it follows the relational model, allowing the user to process complex queries in contrast to SRS limited *join* and *retrieval* operations. However, DiscoveryLink suffers from the complexities that biological databases possess with respect to complex nested data. For example, it is not straightforward to add new databases to DiscoveryLink, taking into consideration the laborious task of making legacy biological databases into relational data models. According to Wong (2002), if SwissProt were subjected to the third normal form, each record would be split into 30 pieces. This large number of possible splits exposes how infeasible it is to use DiscoveryLink considering the join operations required to process a query. However, considering the flexibility of SQL queries processes, it does seem feasible to utilize DiscoveryLink as a data integration solution.

With the evolution of database stands XML has become the *de facto* standard for data formatting and exchange over the Internet. Though not a solution to the data integration problem, it is appreciated for its flexibility in formatting hierarchically nested documents and its uncanny data modeling using tag definitions. These features make it ideal to model the complex and evolving nature of biological data with the flexibility it offers. Moreover, it has fueled the creation of semistructured data processing languages such as XQL and XQuery. These languages help query across multiple data sources and transform the results into a form that supports further processing. It is thus evident that the biomedical community is adhering to the standards of XML, as in the case of databases such as PIR and Entrez.

With the advances in data integration, the different data integration strategies are categorized as wrappers and warehouses. Both of these categories of integration schemes are prevalent in the bioinformatics community. The wrapper strategy is considered to support both flexible and loosely coupled models. In this strategy, different resources are combined dynamically, and generic features of data are modeled and queried using query-based logic in the form of API abstractions.

The warehousing strategy is fixated at creating a centralized architecture to store data from distributed sources in a locally stored data warehouse. Thus, in the warehousing strategy, data from different sources are moved to a centralized data model.

This method requires considerable effort, as not all sources can be altered. It is also challenging to keep this warehouse up-to-date considering that biological databases constantly evolve, and data structures vary simultaneously at the sources.

We describe these two categories with the following two well-known approaches in the biomedical community.

2.4.5.4 Wrappers: Customizable Database Software

As an example of the wrapper category of integration strategies, MOLGENIS (Swertz et al. 2004) is an open-source package that is dedicated to providing backend storage solutions, graphical frontends, and a programmable environment for users to tailor multiple data sources. Intended to design and generate database software for new research projects, this process has become increasingly useful in easy access to known databases. It also facilitates the storage, navigation, and location of data across multiple databases, and has an API that can integrate software services, processing tools, and web services that are written in R*, Java, or HTTP. It is controlled by a domain-specific language (DSL) that helps map data types to their outputted form to create user-defined software.

All these functions are provided through a graphical programming interface to help users to use this tool with ease. Moreover, the DSL provides an abstraction to the actual work that needs to be carried out. For example, a single change in the source DSL helps control the multiple changes across the software code. Thus, making it more user-friendly only enhances its usability within the biomedical community.

2.4.5.5 Data Warehousing: Data Management with Query Optimization

Data warehousing has been proven successful when used with commercial databases. However, due to the descriptive nature of biological databases, it is a challenge to apply data warehousing in biological databases. The integration of information from disparate biological data sources and reconciling frequently conflicting data in an efficient, yet scalable manner have proven to be major bottlenecks for the application of data warehousing in the biomedical community (Aberer and Hemm 1996). The majority of biological databases are designed to facilitate the unambiguous storage and update of large amounts of data, and therefore have complex, normalized schemas that are specific for a given type of data. Consequently, large-scale querying of the stored data is computationally expensive, must be designed specifically for a given database, and requires domain-specific software solutions. However, efforts are being pursued to make data warehousing a reality for the entire biological community. One such effort is known as BioMarts (Smedley et al.).

BioMarts was initially called EnsMart (Kasprzyk et al. 2004). EnsMart was capable of organizing data from individual databases into one query-optimized system using a data warehousing technique specifically designed for descriptive biological data. The impetus of creating a data warehousing technique was to provide an integration mechanism to integrate data from disparate sources, along with an effective querying mechanism that is unified yet domain independent. The key features of the provided solution were used to increase the scalability in large datasets and provide rapid and flexible data access and support for easy integration with third-party data and programs and intuitive user interfaces.

EnsMart provides a consistent genome annotation across a variety of metazoan genomes using an automated pipeline system to predict genes and to carry out cross-species analysis. EnsMart uses the data derived from the numerous databases that constitute the Ensembl genome database (relating predominantly to genes and single nucleotide polymorphisms (SNPs)), functional annotation, and expression. Table 2.6 contains the list of datasets that constitute the EnsMart.

A web-based tool known as MartView helps to query EnsMart. A query is executed in MartView in three stages: the start, filter, and output stages.

In the start stage, the data are selected based on the species and focus of the query. The start stage is followed by the filtering stage, in which the user is provided with the flexibility to narrow his search to a subset with characteristics of interest. The tool feature for region filtering allows a search to be carried out on the full genome, on a single chromosome (as determined by markers, bands, or base pairs). The availability of other filter options depends on the data content for a particular species and focus.

Finally, we have the output stage. In this stage, the data that satisfy the filter criteria are organized into a number of topics reflecting the kinds of data that are most likely to be required in different types of analyses. Again, the topics

Table 2.6 Datasets of EnsMart

Species	Category	Dataset	Primary Source
Homo sapiens	Genomic	Ensembl genes	Ensembl
		EST genes	Ensembl
		Vega genes	VEGA
		SNP	dbSNP/HGVbase
		Markers	UCSC
	Disease	OMIM morbid map	OMIM
	Expression	eVOC	SANBI
		GNF	Novartis

(*Continued*)

Table 2.6   Datasets of EnsMart (Continued)

Species	Category	Dataset	Primary Source
		EST	dbEST
	Protein annotation	InterPro	Ensembl
		Pfam	Ensembl
		Prosite	Ensembl
		PRINTS	Ensembl
		PROFILE	Ensembl
		FAMILY clusters	Ensembl
Mus musculus	Genomic	Ensembl genes	Ensembl
		EST genes	Ensembl
		SNP	dbSNP
		Markers	MGI
	Protein annotation	As for *Homo sapiens*	Ensembl
Rattus norvegicus	Genomic	Ensembl genes	Ensembl
		EST genes	Ensembl
		SNP	MDC
		Markers	RMR/WTCHG
	Disease	QTL	RGD
	Protein annotation	As for *Homo sapiens*	Ensembl
Caenorhabditis elegans	Genomic	WormBase genes	AceDB
	Protein annotation	As for *Homo sapiens*	Ensembl
Caenorhabditis briggsae	Genomic	Ensembl genes	Ensembl
	Protein annotation	As for *Homo sapiens*	Ensembl
Danio rerio	Genomic	Ensembl genes	Ensembl
		Markers	EMBL STS
	Protein annotation	As for *Homo sapiens*	Ensembl

Table 2.6 Datasets of EnsMart (Continued)

Species	Category	Dataset	Primary Source
Fugu rubripes	Genomic	Ensembl genes	IMCB
	Protein annotation	As for *Homo sapiens*	Ensembl
Anopheles gambiae	Genomic	Ensembl genes	Ensembl
		SNP	Ensembl
		Markers	Anobase
	Protein annotation	As for *Homo sapiens*	Ensembl
Drosophila melanogaster	Genomic	FlyBase genes	FlyBase
	Protein annotation	As for *Homo sapiens*	Ensembl

available will depend on the species and focus. A variety of output formats are supported.

Built on the success of EnsMart, BioMart (http://www.ebi.ac.uk/biomart) is an open-source data management system that comes with a range of query interfaces that allow users to group and refine data based on many criteria. In addition, the software features a built-in query optimizer for fast data retrieval. BioMart installation can provide domain-specific querying of a single data source or function as a one-stop shop (web portal) to a wide range of BioMarts, as the central portal does. All BioMarts have the same look and feel, which has obvious advantages to users moving between resources. However, the power of the system comes from integrated querying of BioMarts. If any datasets share common identifiers (such as Ensembl gene IDs or UniProt IDs), or even mappings to a common genome assembly, these can be used to link BioMarts in integrated queries. Additionally, these datasets do not have to be located on the same server or even at the same geographical location. This distributed solution has many advantages, not least of which is the fact that each site can utilize its own domain expertise to deploy its own BioMart.

1. BioMart enables scientists to perform advanced queries on biological data sources through a single web interface.
2. It performs integrated querying of data sources regardless of their geographical locations.
3. BioMart capabilities are extended by integration with several widely used software packages, such as BioConductor (Gentleman et al. 2004), DAS (Dowell et al. 2001), Galaxy (http://galaxy.psu.edu/), Cytoscape*, and Taverna*.

4. BioMart is now an integral part of large data resources such as Ensembl (Flicek et al. 2008), UniProt (UniProt Consortium 2010), and HapMap (International HapMap Consortium 2007), to name a few.

Biologists need to ask complex queries of these data to test and drive their research hypotheses. Typically, each data source provides an advanced query interface on its site. However, each site has its own solution, and subsequently, the user must overcome a learning curve before he or she can start interacting with data (Table 2.7).

2.4.5.6 Data Integration in the PDB

In integrating information for the proteins in the PDB, information pertaining to structure, biological function, cellular location, and associated disease is integrated and presented to the user. This information for each protein molecule is derived from a wide spectrum of sources and presented to the user. Thereby, the RCSB PDB fully exposes the features of each protein. This process is achieved through weekly updates of integrated information from sources such as the Gene Ontology (GO), Enzyme Commission (EC), KEGG pathways, and National Center for Biotechnology Information (NCBI) resources that include sources such as the OMIM, SNP, and BookShelf.

Table 2.7 All Publicly Accessible BioMarts to Date

Name of BioMart	Description of Contents	Location of BioMart
Ensembl genes	Automated annotation of over 40 eukaryotic genomes	EMBL-EBI, UK
Ensembl homology	Ensembl Compara orthologs and paralogs	EMBL-EBI, UK
Ensembl variation	Ensembl variation data from dbSNP and other sources	EMBL-EBI, UK
Ensembl genomic features	Ensembl markers, clones, and contigs data	EMBL-EBI, UK
Vega	Manually curated human, mouse, and zebrafish genes	EMBL-EBI, UK
HTGT	High-throughput gene targeting/trapping to produce mouse knockouts	Sanger, UK
Gramene	Comparative grass genomics	CSHL, United States

Table 2.7 All Publicly Accessible BioMarts to Date (Continued)

Name of BioMart	Description of Contents	Location of BioMart
Reactome	Curated database of biological pathways	CSHL, United States
Wormbase	*C. elegans* and *C. briggsae* genome database	CSHL, United States
Dictybase	*Dictyostelium discoideum* genome database	Northwestern University, United States
RGD	Rat model organism database	Medical College of Wisconsin, United States
PRIDE	Proteomic data repository	EMBL-EBI, UK
EURATMart	Rat tissue expression compendium	EMBL-EBI, UK
MSD	Protein structures	EMBL-EBI, UK
UniProt	Protein sequence and function repository	EMBL-EBI, UK
Pancreatic Expression Database	Pancreatic cancer expression database	Barts and the London School of Medicine, UK
PepSeeker	Peptide mass spectrometer data for proteomics	University of Manchester, UK
ArrayExpress	Microarray data repository	EMBL-EBI, UK
GermOnLine	Cross-species knowledge base of genes relevant for sexual reproduction	Biozentrum/SIB, Switzerland
DroSpeGe	Annotation of 12 *Drosophila* genomes	Indiana University, United States
HapMap	Catalog of common human variations in a range of populations	CSHL, United States
VectorBase	Invertebrate vectors of human pathogens	University of Notre Dame, United States

(*Continued*)

Table 2.7 All Publicly Accessible BioMarts to Date (Continued)

Name of BioMart	Description of Contents	Location of BioMart
Paramecium DB	*Paramecium tetraurelia* model organism database	CNRS, France
Eurexpress	Mouse *in situ* expression data	MRC Edinburgh, UK
Europhenome	Mouse phenotype data from high-throughput standardized screens	MRC Harwell, UK

PDB structure mapping is performed by enabling accurate assignment of references (identifiers) to external databases; these identifiers include those from the GenBank (Benton et al. 2009), PubMed, EC (Webb 1992), and SwissProt, now referred to as UniProt (UniProt Consortium 2010) databases, along with the taxonomy of the source organism (see Figure 2.9).* All structural information for the sequence is obtained from the Structural Classification of Proteins (SCOP) database. However, information relating to structure often exhibits a one-to-many relationship, as structure consists of one or more components, such as multiple polypeptide chains. This representation of structures as a number of constituent components, each with external data assignments, is an ongoing process at the RCSB PDB (Deshpande et al. 2005).

Relevant information from external databases is retrieved by parsing related files to identify related information and is stored in the database. For example, KEGG pathways associated with a given EC number are retrieved by issuing a web service call to the KEGG database at query runtime. Under an agreement with the U.S. National Library of Medicine, PubMed identifiers for the primary citation associated with a structure are used to load PubMed abstracts into the RCSB PDB database. These abstracts can then be searched by keyword(s) as an alternative means to find structures of interest.

This structure results in the creation of a single consolidated report for every protein and is presented in Table 2.8.

2.5 Conclusion

With the exponential growth of biological data, this chapter is aimed at creating an awareness of the challenges of handling biological data. It highlights the

---

* For more information on GenBank, refer to http://www.ncbi.nlm.nih.gov/genbank/. For more information on PubMed, refer to http://www.ncbi.nlm.nih.gov/sites/entrez?db=pubmed. For more information on EC refer to http://www.chem.qmul.ac.uk/iubmb/enzyme/. For more information on UniProt (SwissProt) refer to http://www.uniprot.org/.

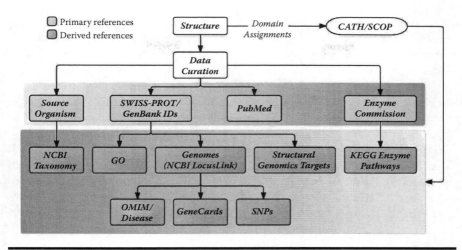

Figure 2.9 The primary and secondary references assigned to structures. The primary references are assigned during structure annotation/data curation. Secondary references are collected from external databases using the primary reference identifiers and accession numbers. This process is rerun on a weekly basis to find new structures or update information on existing structures to store in the database. (From Deshpande, N., et al., *Nucl Acids Res* 33, no. 1 (2005): D233–D237. With permission.)

Table 2.8 Information from the PDB: Sections of a Typical PDB File

Summary Reports	Features
Primary citations	A list of all PubMed citations of specific structure, along with brief abstracts
Molecular description	Information pertaining to existing classification and molecular characteristics
Derived data	Searchable features of protein from SCOP, CATH, Pfam, GO
Structure explorer	Navigation breadcrumbs, Print PDF, Toggle asymmetric and biological unit images, Ligand and ligand-structure interaction viewer, Ensemble and refinement information for NMR structures
Materials and methods	Reports customized for x-ray and NMR structures
Biology and chemistry	Detailed information including taxonomy, genome and locus, SNPs, enzyme pathways, disease, and function
Structural features	Detailed chemical bond information

KDD process and the computation techniques that are applied to data cleaning and data integration, and brings forth the concepts of data warehousing. It explains the commonly used databases in bioinformatics and the inherent design flaws and success. The chapter is also aimed at creating awareness of the degree of data integration that is used to maintain these data repositories and the need for effective integration schemes required in the future. Data warehousing is a requirement in organizations that handle vast amounts of data; however, the application of data warehousing has found limited success. This chapter enumerates the attempts to implement data warehousing for biological databases. Chapter 3 highlights the need for data transformation in high-dimensional databases and the various data transformation techniques as dimensionality reduction techniques and feature selection strategies commonly employed in data mining.

References

3rd Millennium. *Practical data integration in pharmaceutical R/D: Strategies and technologies.* White paper. 2002, pp. 1–21.

Aberer, K., and K. Hemm. A methodology for building a data warehouse in a scientific environment. In *First IFCIS International Conference on Cooperative Information Systems.* Brussels: IEEE, 1996, pp. 19–21.

Bateman, A., et al. The Pfam protein families database. *Nucl Acids Res* 30 (2002): 276–280.

Benton, D.A., I. Karsch-Mizrachi, D.J. Lipman, J. Ostell, and E.W. Sayers. GenBank. *Nucl Acids Res* 37 (2009): D26–D31.

Berman, H.M., K. Henrick, and H. Nakamura. Announcing the Worldwide Protein Data Bank. *Nature Struct Biol* 10, no. 12 (2003): 980.

Cornell, M., et al. GIMS: An integrated data storage and analysis environment for genomic and functional data. *Yeast* 20, no. 15(2003): 1291–1306.

Deelman, E., D. Gannon, M. Shields, and I. Taylor. Workflows and e-science: An overview of workflow system features and capabilities. *Future Gen Comput Syst* 25, no. 5 (2009): 528–540.

Deshpande, N., et al. The RCSB Protein Data Bank: A redesigned query system and relational database based on the mmCIF schema. *Nucl Acids Res* 33 (2005): D233–D237.

Do, H.-H., T. Kirsten, and E. Rahm. Comparative evaluation of microarray-based gene expression databases. In *10th Conference of Database Systems for Business, Technology and Web (BTW)*, 2003, pp. 482–501.

Donoho, D.L. High dimensional data analysis: The curses and blessings of dimensionality. 2000. http://www-stat.stanford.edu/~donoho/Lectures/AMS2000/Curses.pdf (accessed February 17, 2010).

Dowell, R.D., R.M. Jokerst, A. Day, S.R. Eddy, and L. Stein. The distributed annotation system. *BMC Bioinformatics* 2 (2001): 7.

Edgar, R., M. Domrachev, and A.E. Lash. Gene Expression Omnibus: NCBI gene expression and hybridization array data repository. *Nucl Acids Res* 30, no. 1 (2002): 207–210.

Etzold, T., and P. Argos. SRS: Information retrieval system for molecular biology data banks. *Methods Enzymol* 266 (1996): 114–128.

Fayyad, U., G. Piatetsky-Shapiro, and P. Smyth. From data mining to knowledge discovery in databases. *Am Assoc Artif Intell* 17 (1996): 37–54.

Flicek, P., et al. Ensembl 2008. *Nucl Acids Res* 36 (2008): D707–D714.

Fox, J., D. Glasspool, D. Grecu, S. Modgil, M. South, and V. Patkar. Argumentation-based inference and decision making—A medical perspective. *IEEE Intell Syst* 22, no. 6 (2007): 34–41.

Friedman-Hill, E.J. *JESS, the java expert system shell.* 1999. http:/herzberg.ca.sandia.gov/jess (accessed June 5, 2010).

Fuxman, A., E. Fazil, and R. J. Miller. ConQuer: efficient management of inconsistent databases. *2005 ACM SIGMOD International Conference on Management of Data.* Baltimore, Maryland, 2005, 155–166.

Gentleman, R.C., et al. Bioconductor: Open software development for computational biology and bioinformatics. *Genome Biol* 5 (2004): R80.

Haas, L.M., P.M. Schwarz, P. Kodali, E. Kotlar, J.E. Rice, and W.C. Swope. DiscoveryLink: A system for integrated access of life science data sources. *IBM Syst J* 40, no. 2 (2001): 489–511.

Hernandez, M. A., and S. J. Stolfo. The merge/purge problem for large databases. *1955 ACM SIGMOD International Conference on Management of Data.* San Jose, California, 1995, 127–138.

Hubbard, T., et al. The Ensembl genome database project. *Nucl Acids Res* 30, no. 1 (2002): 38–41.

International HapMap Consortium. A second generation human haplotype map of over 3.1 million SNPs. *Nature* 449 (2007): 851–861.

Judice, L.Y.K. *Correlation based methods for biological data cleaning.* Singapore: National University of Singapore, 2007.

Kasprzyk, A., et al. EnsMart: A generic system for fast and flexible access to biological data. *Genomic Res* 14, no. 1 (2004): 160–169.

Lee, M.L., T.W. Ling, and W.P. Low. IntelliClean: A knowledge-based intelligent data cleaner. *Proceedings of the Sixth ACM SIGKDD International Conference on Knowledge Discovery and Data mining.* Boston: ACM, 2000, 290–294.

Li, A. Facing the challenges of data integration in biosciences. *Eng Lett* 13, no. 3 (2006): EL_13_3_13.

Low, W.L., M.L. Lee, and T.W. Ling. A knowledge-based approach for duplicated elimination in data cleaning. *Inf Syst* 26 (2001): 585–606.

McLeod, K., and A. Burger. Towards the use of argumentation in bioinformatics: A gene expression case study. *Bioinformatics* 24 (2008): i304–i312.

McLeod, K., G. Ferguson, and A. Burger. Using argumentation to resolve conflict in biological databases. In *Proceedings of computation models of natural argument*, Pasadena, CA, 2009, pp. 15–23.

Muller, H., and J.-C. Freytag. *Problems, methods and challenges in comprehensive data cleaning.* Technical Report. Berlin: Humboldt-Universitat zu Berlin, Institut fur Informatik, 2003.

Ng, S.-K., and L. Wong. Accomplishments and challenges in bioinformatics. *IT Professional*, February 2004, pp. 44–50.

Potamias, G., et al. Knowledge discovery scientific workflows in clinico-genomics. In *19th IEEE International Conference on Tools with Artificial Intelligence ICTAI.* Paris, France: IEEE, 2007, pp. 91–95.

Rahm, E., and H.-H. Do. Data cleaning: Problems and current approaches. *Data Engineering*, December 2000.

Reich, M., T. Liefeld, J. Gould, J. Lerner, P. Tamayo, and J.P. Mesirov. Gene Pattern 2.0. *Nature Genet* 38, no. 5 (2006): 500–501.
Rete, C.F. A fast algorithm for the many patterns/many objects match problem. *Artif Intell* 19, no. 1 (1982): 17–37.
Schierz, A.C., L.N. Soldatova, and R.D. King. Overhauling the PDB. *Nature Biotechnol* 25 (2007): 437–442.
Smedley, D., et al. BioMart-biological queries made easy. *BMC Genomics* 10, no. 22 (2009): 1–12.
Smith, C.M., et al. The mouse gene expression database (GXD): 2007 update. *Nucl Acid Res* 35 (2007): D618–D623.
Swertz, M.A., E.O. de Brock, and S.A.F.T. van Hijum. Molecular Genetics Information System (MOLGENIS) alternatives in developing local experimental genomics databases. *Bioinformatics* 20 (2004): 2075–2083.
UniProt Consortium. The Universal Protein Resource (UniProt) in 2010. *Nucl Acids Res* 38 (2010): D142–D148.
Venkataraman, S., et al. Emage: Edinburgh mouse atlas of gene expression: 2008 update. *Nucl Acid Res* 36 (2008): D860–D865.
Verheij, B. Dialectical argumentation with argumentation schemes: An approach to legal logic. *Artif Intell Law* 11, no. 1–2 (2003): 167–195.
Webb, E.C. *Enzyme nomenclature.* San Diego: Academic Press, 1992.
Wong, L. Kleisli, a functional query system. *J Funct Program* 10 (2000): 19–56.
Wong, L. Technologies for integrated biological data. *Briefings Bioinformatics* 3 (2002): 389–404.

Chapter 3

Knowledge Discovery in Databases

In Chapter 2, we provided a synopsis of the various databases and BioMarts prominently used in the area of bioinformatics. The chapter also sheds light on the role of knowledge discovery in databases (KDD) in bioinformatics. In this chapter, our objective is to familiarize the reader with key data mining techniques that can be used to clean and preprocess the data obtained from these databases for analysis.

3.1 Introduction

In the last 20 years, genomic and proteomic databases have grown exponentially, causing existing computational systems to suffer from the constantly evolving nature of the data. In such cases, the data changes can result in legacy data not conforming to newly added information in databases. Further challenges arise when data from various sources are integrated into a common schema, as witnessed in data warehousing.

In this chapter, we introduce the process known as knowledge discovery in databases (KDD). KDD is used to develop methods, techniques, and tools that aid analysts in discovering useful information and knowledge in databases (Fayyad et al. 1996). Like data, KDD is constantly evolving as research from pattern recognition, databases, statistics, artificial intelligence, machine learning, data visualization, and high-performance computing is incorporated into the schema. In nonprofessional terms, the KDD process is interactive and iterative and provides an abstraction of low-level data (datasets) that enable better understanding (knowledge) for better

82 ■ *Data Mining for Bioinformatics*

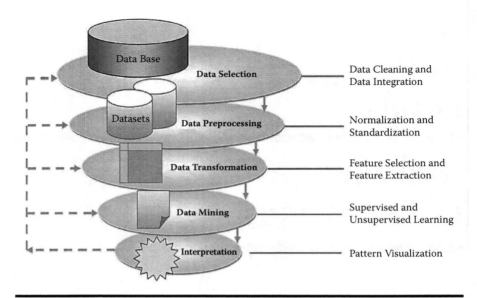

Figure 3.1 The process of KDD and the steps involved.

decision support. Thus, KDD is used to discover information from data (Han and Kamber 2006). This information includes data storage and access records, such as how the data are stored and accessed, and algorithm data, such as how algorithms can be scaled for use on massive datasets (Fayyad et al. 1996). KDD is a multistep process that is best described as shown in Figure 3.1.

As shown in Figure 3.1, KDD is a five-step process that begins with data selection, includes data processing, data transformation, and data mining, and ends with data interpretation. Also note the emphasized interactive nature of KDD. Below, we provide a general outline of KDD as a systematic process captured using the following steps:

1. As a prelude to the initial steps of the KDD process, it is imperative that the user/developer have a clear understanding of the application domain. A large amount of time is invested in identifying and laying out the goals and objective(s) of the process.

    Apart from outlining the goals and objectives of the KDD process, the user/developer should create or identify the data over which discovery is to be performed. The data can be an entire database, a targeted dataset, or a large subset of variables that are part of a larger database. The selection of these data forms the first step of the KDD process.

2. Once the goals and datasets have been identified, the second objective of the KDD process is to perform data cleaning and data preprocessing. In data cleaning, the data are subjected to operations that remove the noise that is an

integral part of large real-world datasets. These cleaning operations include creating models that account for the overall noise in the data, handling missing values of data features (attributes), and predicting changes in the data. As part of the data preprocessing, the data after cleaning are subject to normalization and standardization strategies that are vital when using from disparate sources.
3. With the completion of data preprocessing, the resultant data are then subject to data transformation operations. Typically, large databases/datasets are plagued with data (records) that have a large or small number of features (attributes). Traditional computational techniques are deemed computationally expensive when handling any data that possess a large number of features. Thus, as part of the third step of the KDD process, dimensionality reduction and transformation techniques are applied to the data to reduce the number of features in the data without altering the data quality. In situations where there are fewer features, feature extraction techniques are applied to extract features that are inherent in the data.
4. Once the data are transformed, the fourth step of the KDD process is the mining of data, or data mining. Data mining requires a model for mining. There are several mining strategies from which the user/developer can choose. These strategies include unsupervised, supervised, and semisupervised techniques. Apart from determining the data mining scheme, the user/developer is expected to create a hypothesis for mining. This step is vital, as it helps the user/developer decide which models and features (of the data) fit the overall criteria of the KDD process. In this way, the user/developer can understand the model and its predictive capabilities. Typically, data mining involves searching patterns in an abstraction of the transformed data. For example, supervised classification approaches can find similar patterns in rules or trees.
5. The fifth and final step of the KDD process is the interpretation of mined patterns. Here, statistical and visualization techniques are applied to validate the knowledge discovered from the data mining models applied. Typically, if the results are not as good as anticipated, the KDD process, which is iterative, enables the user/developer to repeat steps 1 through 4.

Much of the time, the results obtained either support or conflict with previously held beliefs and inferred notions. Thus, the user/developer is expected to document and validate the discovered knowledge before incorporating the knowledge into another system to avoid conflicts.

The KDD process can involve a significant number of iterations and can contain loops between any two steps. The basic flow of steps (although not the potential multitude of iterations and loops) is illustrated in Figure 3.1. Relevant literation that use KDD in bioinformatics has focused on step 4, data mining. However, the other steps are also important for the successful application of KDD. Having defined the basic notions and introduced the KDD process, we will now focus on the data mining component, which has received the most attention in the literature.

3.2 Analysis of Data Using Large Databases

Data quality is primarily used to characterize database data and associated schemas. Data quality is the mapping of the data to its corresponding conceptual model. It determines whether the data in a database or databases are accurate, complete, and consistent. The three methods of ensuring data quality include data cleaning, data quality monitoring, and data integration. In this chapter, we elaborate on the different data cleaning and data integration methods and steps.

3.2.1 Distance Metrics

Before we address the steps and problems associated with data cleaning and data integration, we will introduce commonly used distance measures in data mining. In this chapter, we refer to each data record as a data point, in which the attributes of a data record are coined as features. Thus, data record x consisting of n attributes can be viewed as data point x in an n-dimensional feature space. To measure the similarity between two data points, various distance metrics are employed. The commonly used distance metrics are Euclidian distance and Mahalanobis distance and are defined as follows:

Euclidian distance: This distance metric is also referred to as vector distance. To measure the distance between two data points x and y each having the same n features, Euclidian distance is given by the following equation:

$$\sqrt{\sum_{i=1}^{n}(x_i - y_i)^2}. \qquad (3.1)$$

Mahalanobis distance: Unlike Euclidean distance, Mahalanobis distance calculates the distance of a data point from a common reference point in the n-dimensional space, and is represented by the following equation:

$$\sqrt{(x-\mu)^T C^{-1}(x-\mu)}. \qquad (3.2)$$

Here, the common reference point on which the distance is measured is the centroid (μ). The Mahalanobis metric utilizes the correlation between features using a covariance matrix (C). This metric is thus more effective in capturing the distance between points based on the distribution of data. However, Mahalanobis distance

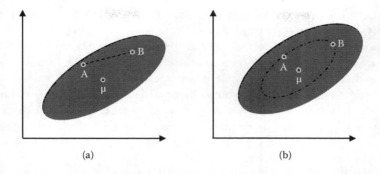

Figure 3.2 (a) The representation of Euclidean distance and (b) Mahalanobis distance between points A and B that belong to a distribution with mean μ.

requires a complete pass of the entire dataset to estimate correlation between features, before determining the distance between points. It is thus computationally more expensive for large high-dimensional datasets than Euclidean distance (Figure 3.2).

3.2.2 Data Cleaning and Data Preprocessing

Data cleaning improves the quality of data to make them fit for use (Chapman 2005). The objective of data cleaning is to reduce errors in data before the data are used in processing. This cleaning invariably helps to increase the learning component of the KDD process. The cleaning also makes the data easier to document, present, and interpret (see Chapman 2005).

Data stored in large databases are error-prone; a user/developer can expect a field error rate of 1–5%, and it is important to decrease this error rate. The uncertainties of data, especially in biological databases, lay the foundation to understanding the effects of error propagation in data. Thus, it is imperative that the steps of data cleaning should actively manage and improve overall data quality. This improvement is sometimes difficult to achieve because correcting and eliminating erroneous data is a tedious and time-consuming process that cannot be overlooked. The simple deletion of erroneous records is not the solution. Rather, the correction and documentation of corrections is suggested; this documentation ensures the tracking of changes.

Data cleaning is the outcome of a twofold process, in that it is used to identify inaccurate, incomplete, or unreasonable data, and it improves the quality of data by correcting identified errors and inconsistencies. Good data cleaning requires good existing data. Apart from replacing faulty data records, the process entails format checks, completeness checks, limit checks, outlier detection, and the assessment of data by domain experts or end users. Validation checks may include applicable standards, rules, and conventions. These processes usually result in flagging, documenting, and the subsequent checking and correction of suspect records.

Data cleaning, in the field of data warehousing, is applied when data from different sources need to be merged. Here, records that refer to the same entity but are represented differently in their formats require cleaning before being stored in the data warehouse. The merge/purge problem refers to the issues faced in the identification and elimination of such duplicate records.

Data cleaning requires data decomposition and data reassembly. The process is best described in six steps: tokening of data, standardizing, verifying, matching, householding, and documenting (Maletic and Marcus 2000). The activities of data cleaning are domain specific, and thus have many forms. Though there are several approaches for data cleaning, the generalized framework is as follows (Maletic and Marcus 2000):

1. Define and determine error types.
2. Search and identify error instances.
3. Correct the errors.
4. Document error instances and error types.
5. Modify data entry procedures to reduce future errors.

While the efforts of data integration and data warehousing are heavily dependent on the success of data cleaning, it is difficult to identify errors that involve relationships between fields. Thus, various methods identify errors in databases also referred to as outlier detection techniques and described briefly below.

3.3 Challenges in Data Cleaning

The problem of errors in data stems from but is not limited to user input errors. Therefore, the first problem encountered with data cleaning is the detection of erroneous data. User input errors could be attributed to inconsistency in input values, misspellings, missing values, improper generation of data, and data differences that are transferred from legacy databases. There are also errors attributed to the presence of duplicate records. Thus, the second problem with data cleaning is the need to detect duplicate/redundant records. Typically, the duplication of records in very large relational databases is regulated using duplication control algorithms (Williams et al. 2002). These duplication control algorithms are based on string matching and identity matching records in relational database schemas. However, as the databases have evolved and grown more complex, these duplication control algorithms have failed, as duplication becomes harder to identify. Typically, large databases are not confined to relational database schemes and include a mixture of file systems that contain legacy data along with relational models. The questions concerning whether two similar documents are duplicates is also pressing. Similarity detection, performed alone by sorting and joining records within a database, has facilitated the detection of more complex duplications in relational databases. Thus, duplication errors can occur, but they are not easy to detect.

To overcome this problem, outlier detection has been proposed. Outliers are defined as patterns (records) that do not conform to an expected behavior. Thus, outlier detection approaches must be able to define a region representing the normal behavior and declare an observation as an outlier (anomaly) if it does not conform to the normal behavior (Chandola et al. 2009). These methods are challenging to implement because of the following factors:

1. It is difficult to define a normal region because of the diversity of databases. Capturing the boundary between a normal region and an outlier is a further challenge and is not precise. An observation close to a boundary can be termed an outlier when it is actually normal and vice versa.
2. In large evolutionary databases, such as biological databases, normal behavior cannot be assumed to be constant and keeps evolving. Thus, the normal behavior might not be sufficient to represent the future.
3. An outlier needs prior definition. Defining an outlier requires domain knowledge that is not always available or straightforward.
4. In many cases, the quality of data affects what areas are determined normal and what areas are determined outliers. In large, noisy databases, the detection of outliers is a challenge.
5. Keeping these challenges in mind, effective outlier detection poses its challenges in the KDD.

However, there are three fundamental approaches of outlier detection (Hodge and Austin 2004): determining the outliers with no prior knowledge of the data, modeling both normal and abnormal data, and modeling only normal or abnormal data.

To determine outliers with no prior knowledge of the data, a learning approach that is analogous to unsupervised clustering is required. As in all clustering algorithms, this kind of outlier detection algorithm considers each record as a point in an *n*-dimensional space. It then groups the points into clusters based on their proximity and flags the remote points as outliers. Those methods that are dependent on the distance metric used and the distance of each point from a reference point (the mean or median) are categorized into this approach. In these approaches, points that are separated by large distances from the reference point are treated as outliers. The algorithms in this category require that all the data are available before processing. Thus, each data point is treated as static (Hodge and Austin 2004) and compared with every other data point in the dataset. This approach can be further divided into two methods, diagnosis and accommodation, based on the way in which the researchers choose to treat outliers.

Once the outliers are identified, the diagnosis approach iteratively prunes the outliers until no more outliers are identified, and the system model is fitted to the remaining data that represent the normal data distribution. On the other hand, the accommodation approach uses all the data points, including the outliers. It then uses a robust classification approach that induces a boundary of normal data around a majority of data points that represents normal behavior. The goodness of the accommodation method thus depends on the robustness of the classification approach used and determines the flexibility of the boundaries obtained. It is believed that more flexible boundaries lead to less computationally expensive classifiers.

In contrast to the previous approach, modeling both normality and abnormality for outlier detection is analogous to supervised classification, and thus each data point for this approach is required to possess class labels. Modeling normality and abnormality for outlier detection is best suited for online classification in which the classifier learns from a portion of the data and classifies new records as outliers. If the new record falls into the region of normality, it is treated as normal; otherwise, it is flagged as an outlier. Since the technique is a classification approach that requires classifier training, the training data should contain an equal representation of both normal data and outliers to enable generalization by the classifier. New records may be classified correctly if the classifier is limited to a known distribution, and records from unknown regions may be classified incorrectly unless the training set is generalized.

Modeling only normal or abnormal data in a few cases is better known as novelty detection or novelty recognition. The methods in this category are analogous to semisupervised detection. Here, the algorithm is trained based on samples that are believed to be normal. The algorithm uses the information from the normal samples to detect outliers. These approaches thus require training data that are preclassified as normal. The methods are suited for both static and dynamic data, as learning is based on only one class (i.e., the normal class). In these approaches, the learning is considered incremental. As new data arrive, the model is tuned to improve the fit of the normal boundary. Since this approach is semisupervised, it requires all the data for training its normal class to permit generalization. However, the need for data belonging to the abnormal class is not required.

Generally, all records in the database are treated as vectors. These vectors consist of both numeric and symbolic attributes that represent continuous, discrete (ordinal), categorical (unordered numeric), and ordered symbolic or unordered symbolic data. Vectors can be monotypes (single data types) or multitypes (mixed data types). The following list contains the categorizations of outlier detection techniques. All of these techniques are governed by suitable distance matrices that are used to measure the closeness of vectors. The two fundamental considerations when selecting an appropriate methodology for an outlier detection system are the accuracy of modeling the data distribution and defining an appropriate neighborhood of interest for an outlier.

1. **Accuracy of modeling the data distribution:** While selecting an algorithm for outlier detection, it is imperative to select an algorithm that can accurately model the distribution of the data studied. Typically, the algorithm should be able to scale up or scale down depending on the number of data points processed.
2. **Defining an appropriate neighborhood of interest for an outlier:** Selecting a neighborhood of interest is a nontrivial task. Typically, algorithms model

data distributions with the pretext of defining boundaries around the points that form a cluster, by inducing a threshold. However, these approaches are parametric; i.e., they often force a predefined distribution (model) over the points or require the number of clusters to be defined in advance. Other approaches require predefined parameters of size or density of neighborhoods for outlier thresholding. Thus, choosing the exact values of the parameters that define the neighborhood should be applicable for all density distributions likely to be encountered and can potentially improve or weaken the effectiveness of the method.

Popular approaches include statistical, neural network, machine learning, and hybrid system models. These approaches, described below, encompass distance-based, set-based, density-based, depth-based, model-based, and graph-based algorithms.

1. *Statistical models* use derived statistical variables of mean and standard deviation to detect outliers. Based on Chebyshev's theorem of inequality (Amidan et al. 2005), the upper bounds and lower bounds of the confidence interval around the mean are calculated. If a data point falls out of bounds, it is treated as an outlier.
2. *Neural network models* are generally nonparametric models that use neural networks (Haykin 1998) for training to create boundaries around data points. Data points that do not fall within the boundaries are flagged as outliers. Since they are neural network-based algorithms, they require both phases of training and testing, and thus are also considered supervised models.
3. *Machine learning models* use categorical data, unlike statistical and neural network models, which are heavily dependent on the data types of the datasets (mainly continuous real-valued or ordinal data). The methods of this category are generally tree-based algorithms used for outlier detection.
4. *Hybrid system models* are used to overcome the limitations of the above three categories. Hybrid system models are typically a combination of any two of the above three categories (statistical, neural network, or machine learning based).

3.3.1 Models of Data Cleaning

One of the first outlier detection models is the statistical models. The models in this category are applicable to 1D datasets (univariate models) as well as to datasets that have multiple dimensions (multivariate models). The foundation for these models is based on the Chebyshev theorem of inequality and is suited for datasets of real-valued data and ordinal data.

The Chebyshev's inequality theorem, or simply Chebyshev's theorem, was designed to determine the lower bound of data with k number of standard deviations from the mean of the data. Typically, datasets are assumed to possess a normal distribution (bell shaped), for which it is known that 95% of the data will fall between two standard deviations from its mean. In this assumption, 5% of the data will fall outside of the two standard deviations. Information such as the mean and standard deviation are extracted from the dataset studied.

The simplest and one of the oldest statistical outlier detection techniques is box plot analysis. Proposed by Laurikkala et al. (2000), box plot analysis provides graphical representation to pinpoint outliers using box plots. This technique can be applied to both univariate and multivariate data. Using box plots, a user/designer can plot both the upper and lower extremes of the data. The parameter of the lower quartile, median, and upper quartile are derived from analysis of the box plots. Points that fall out of the upper and lower extreme values of the box plots are flagged as outliers. The upper and lower bounds of the limits in this method are dependent on the datasets and vary with the number of records in the dataset. It is noteworthy that this method does not make any assumptions about data distribution; however, it is heavily dependent on human interpretation of the outliers.

Moreover, outlier detection models are susceptible to the curse of dimensionality, and it is therefore imperative that the outlier detection models scale up to the large number of dimensions. The curse of dimensionality is based on the observation that the computational time of algorithms scales up exponentially as the number of dimensions increases. It is believed that as the dimensionality increases, the data points are spread through a larger volume and the data distribution becomes less dense. The most effective statistical techniques focus on the selection of salient dimensions (or attributes) and, by doing so, process a larger number of data points at a time. The process of attribute selection is a precursor to outlier detection. It is believed that a subset of attributes contributes to the deviation of data, while the other attributes are believed to add to the inherent noise in the dataset. An alternate technique is to project the data onto a lower-dimensional subspace, thereby containing the density of the distribution of data points.

Statistical techniques can be further divided into the following categories.

3.3.1.1 Proximity-Based Techniques

Proximity-based techniques are simple to implement and make no prior assumptions about the data distribution model. They are suitable for both unsupervised and supervised methods of outlier detection. In these techniques, each record of the dataset is treated as an independent point in an n-dimensional space, and the distance between each point and every other point in the dataset is computed (see Figure 3.3). Points that fall within a specified threshold of a reference point are considered neighbors to the point for which the threshold was calculated. An example of such an algorithm is the k-nearest neighbor (kNN) algorithm. Though reliable, proximity-based techniques suffer from exponential computational growth, as they are based on the calculation of the distances between all data points. The computational complexity of the algorithms in this technique is directly proportional to both the dimensionality of the data m

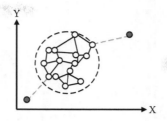

Figure 3.3 A representation of the kNN algorithm in 2D space. In this illustration, k is set at 14, which results in a cluster of 14 closely populated points and 2 outliers (in blue) that do not satisfy the distance criteria.

and the number of data points n, and their complexity is of the order $O(n^2 m)$. This computational complexity indicates that they are not feasible with high-dimensional data.

3.3.1.2 Parametric Methods

Many of the methods in the proximity-based techniques do not scale well unless modifications and optimizations are made to the standard algorithm. In this section, we introduce a new category of techniques known as parametric methods. These methods allow the rapid evaluation of models for every new instance of data and are well suited for large databases, as the complexity of the model is independent of the data size. However, the drawback of this technique is that a predefined model distribution is enforced to fit the data. Theoretically, if the data fit the model, then the results obtained are accurate. However, this condition does not always hold true for real-life datasets. An example of this technique is minimum volume ellipsoid (MVE) estimation. The objective of this algorithm is to fit the smallest possible ellipsoid around a maximum number of data points in an n-dimensional space. It is believed that the points within the ellipsoid represent a densely populated region.

An alternate approach that is similar to MVE is the convex peeling (CP) algorithm. In this approach, a convex hull is placed around all the data points so that the hull covers the maximum points (see Figure 3.4). Each data point is then assigned a weight (known as depth) that corresponds to the distance of the point from the mean of the data distribution. The points closest to the boundary (defined by a convex hull) are considered to have the lowest depth and are peeled away; i.e., points further away from the distribution are considered outliers. The process of convex hull generation and peeling is iteratively carried out until a predefined number of data points are retained within the convex hull. This method is suitable for both unsupervised clustering outlier detection techniques. Unfortunately, this method is susceptible to peeling away a large number of points that form a chunk of normal data points.

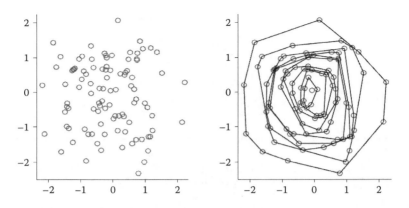

Figure 3.4 An illustration of convex hull peeling of points distributed in 2D space. Convex hulls are placed at varied depths around the data points to facilitate removal of outliers in a layered fashion from lowest depth (outermost hull) to highest depth (innermost hull).

The robustness of both MVE and CP to fit a convex hull around the data points is not dependent on the sparseness of the outlying region, and thus does not skew the boundary formed. However, the fit of the convex hull around the data points is dependent on the data point distribution. The fitting of the convex hull around the data points implies that both MVE and CP suffer from the curse of dimensionality and work best with datasets that have only a few dimensions, as more dimensions add to the sparseness of the data.

To overcome the curse of dimensionality, principal component analysis (PCA) is used for high-dimensional data. The principal components extracted using PCA have the highest variance as the corresponding eigenvalues, which have magnitudes that correspond to the variance of the points from the principal components. The extracted principal components ensure that the subspaces determined are compact, and thereby overcome the limitations of MVE and CP in their applicability, particularly for sparse distributions.

Hierarchical approaches such as decision tree and cluster trees are also included in this category. The representation of data distributions in a hierarchy provides for a multilevel abstraction of data. This hierarchy enables data points to be compared for novelty at different levels of the hierarchy—from a coarse grain (higher up in the hierarchy) to a fine grain (lower down the hierarchy). Expectation maximization (EM), in conjunction with deterministic annealing (DA), is an example of the algorithms that fall into this category. Using the maximum likelihood and information theory, the DA constructs a hierarchy using divisive clustering of data points. In this method, nodes are split into subnodes, until a top-down hierarchy is created. Outliers in this case are detected by the hierarchy, when new data points are added that do not conform to any of the existing clusters. When EM is used in conjunction with DA, the computational efficiency of the algorithm is improved by removing some initialization dependencies. DA can also avoid the local minima, which plague the EM problem. However, avoiding the local minima can produce suboptimal results.

3.3.1.3 Nonparametric Methods

Though effective, the methods in the above two categories are controlled by parameters or are data specific. In the case of kNN, the algorithm is dependent on the parameter k, and in the case of PCA, it is dependent on the number of principal components p. The algorithm is confined by assumptions in the initial iteration of processing, which is not feasible in real-world datasets and could turn out to be computationally infeasible. To overcome these generic limitations, we can use nonparametric approaches. These approaches are more practically applicable for outlier detection, especially when data are expected to grow in time and when limited computational resources are required, thereby providing more autonomy and flexibility. Algorithms in this section include multilayer perceptrons (MLPs) and adaptive resonance theory (ART) for outlier detection.

3.3.1.4 Semiparametric Methods

Semiparametric methods are used to build on the speed and complexity of parametric methods using the model flexibility of nonparametric methods. These methods are brought about by the application of local kernel methods instead of a common global distribution model. Kernel-based methods, such as Gaussian mixture models (GMMs), estimate the density distribution of the input space to identify outliers as data points that lie in regions of low density.

3.3.1.5 Neural Networks

Approaches in the second group are known as neural network approaches. These approaches are nonparametric (Reif et al. 2008) and model based, as they generalize to unseen patterns and can learn complex class boundaries. These methods, though susceptible to the curse of dimensionality, are far less likely to suffer from such problems than statistical approaches are. Since these methods are supervised, each method requires a training dataset that is spread across both normal and outlier samples to effectively fine-tune the model and determine necessary thresholds. Moreover, these approaches require the entire dataset to be traversed numerous times to allow the network to settle and model the data correctly. Just as in previous methods, the models in this category attempt to fit a surface over the data points. For effective surface generation, there must, however, be sufficient data density. By default, neural networks

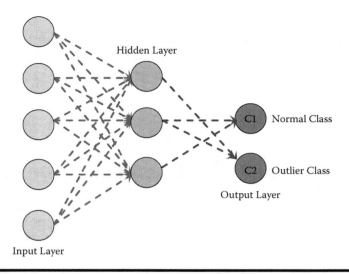

Figure 3.5 Supervised multilayer perceptron, with three layers: input, hidden, and output.

reduce the feature space by using only key features. Nonetheless, it is beneficial to use feature selection or dimensionality reduction techniques to make the algorithms more effective.

3.3.1.5.1 Supervised Neural Methods

Supervised neural networks use the classification of the data to drive the learning process. The class labels enable the neural network to adjust its weights and thresholds to correctly classify new input data (Rumelhart et al. 1986). The input data are effectively modeled by the whole network, as they are distributed across all nodes, and the output represents the classifications as shown in Figure 3.5. For example, the multilayer perceptron is a supervised neural network, which interpolates well but performs poorly for extrapolation, and thus is ineffective in classifying points that fall outside of the boundary of a class defined by the training set.

3.3.1.5.2 Unsupervised Neural Methods

Learning in supervised neural networks is driven by a predefined training set that contains equal representation of both normal and outlier data points. In situations where the training set is unavailable, unsupervised neural networks provide an alternative. In unsupervised neural networks, nodes compete with each other to represent distribution characteristics of the data points. Multilayer perceptron-based neural networks that consist of three layers with the same number of output and input neurons are trained to create a model (Williams, Baxter et al. 2002), as

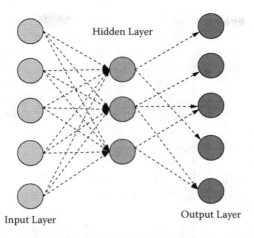

Figure 3.6 Unsupervised multilayer perceptron consisting of three layers having the same number of input and output nodes.

seen in Figure 3.6. Then, this network autonomously clusters the input data points based on the data distribution modeled, which enables the differentiation of points as normal or outlier based on how close a point resembles a modeled distribution. Assuming that related data points share common features, these features are used to topologically model the data distribution.

Self-organizing maps (SOMs) (Muruzalbal and Munoz 1997) are effective for the clustering and visualization of high-dimensional data. SOM is equivalent to a two-dimensional (2D) neural network, where each node in the network is assigned a weight vector that points to data in the input space. Thus, for a given data matrix consisting of n rows and p features, the pointers in the SOM capture the distribution of the data and are constrained by the relation between the data features. As with the k-means algorithm, self-organizing maps rearrange the points within distribution based on their proximity to the input point, as each data point is fed into the network. In this manner, the SOM consists of pointers with a density that is equal to the overall distribution of the data. Thus, in the case of outlier detection, each data point is gauged by its proximity to its immediate neighbors.

3.3.1.6 Machine Learning

Much outlier detection has only focused on continuous real-valued data attributes; there has been little focus on categorical data. Both statistical and neural approaches require cardinal or, at the least, ordinal data types to enable the calculation of distances between data points. They do not have any mechanism to handle categorical data with no implicit ordering. To this end, machine learning algorithms can handle categorical data as well. Of the machine learning approaches that are used in outlier detection, the decision tree algorithm C4.5 (John 1995; Reif et al. 2008) is the most reliable. Initially proposed in 1995, decision tree-based outlier detection is not governed by fitting a model over the data distribution, and thus is immune to the curse of dimensionality. Decision tree algorithms define simpler class boundaries and work well on noisy data. Decision tree-based outlier detection is supervised and dependent on the training set. It is scalable for handling larger datasets with high dimensionality. However, decision trees are susceptible to overfitting, as their ability to generalize is inferior to other neural network or statistical techniques that can be overcome by feature selection or pruning.

As with decision tree-based outlier detection, rule-based machine learning techniques can be exploited for outlier detection (Chandola et al. 2007). These rule-based techniques are similar to decision trees, as they test a series of conditions known as antecedents before determining the conclusion or appropriate class. The flexibility of adding new rules without disturbing existing rules proves advantageous in this technique, especially for outlier detection. Typically, the rule-based technique can be treated as a classification scheme, with both normal and abnormal instances used for training, and the scheme distinguishes between data points located in normal areas and outliers. The scheme could consist of rules generated by the rule-based classifiers that capture the behavior of the normal data points, and any instance that does not confine to the rules of the normal class are treated as outliers.

Other machine learning outlier detection strategies include those based on clustering. Clustering algorithms such as BIRCH (Zhang et al. 1996) and DBSCAN (Ester et al. 1996) that are robust in handling large datasets can be exploited for outlier detection. The BIRCH clustering algorithm uses a hierarchical tree structured index called a clustering feature tree (CFt) to cluster data points dynamically. In this method, all the data points are first scanned and inserted into the CFt. When all the data points are scanned, a global clustering scheme is employed to condense the CFt to a desired size. At this point, clusters are merged to other clusters in a hierarchical fashion, which can be effectively used to remove the outliers. The time complexity of BIRCH is of the order $O(n^2)$ and operates incrementally but is limited to handling numeric data.

A more elaborate clustering scheme, such as the density-based DBSCAN clustering algorithm, can also be employed for outlier detection. Here, outliers are estimated based on the density of data points within a predefined neighborhood. The extension of the DBSCAN clustering algorithm for outlier detection is found in the DB-outlier algorithm (Berunig et al. 2000). Its time complexity is of the order $O(n \log n)$ and is based on the R* tree structure to cluster and identify kNN.

3.3.1.7 Hybrid Systems

The most recent development in outlier detection technology is hybrid systems. These systems incorporate algorithms from at least two of the statistical, neural, and machine learning methods. Hybridization is used variously to overcome deficiencies with one particular classification algorithm to exploit the advantages of multiple approaches while overcoming their weaknesses or using a meta-classifier to reconcile the outputs from multiple classifiers.

3.4 Data Integration

Now that we have covered an overview of the different outlier detection strategies employed in data cleaning, in this section we describe the process of data integration, by which data from disparate sources are integrated to enable mining for information.

Data integration is viewed as the process that entails the merging of data from various sources that correspond to an entity of interest. In reality, it is difficult to find information pertaining to an entity in a single database; it is therefore vital to consolidate information from various databases. This need for specific information poses a challenge as different databases adopt diverse schemas and formats to store their data. Methods for overcoming these challenges are described below.

3.4.1 Data Integration and Data Linkage

The need for joining data from multiple heterogeneous databases into a single coherent data warehouse is of growing importance in the KDD process (Figure 3.7).

In reality, information pertaining to an entity of interest is found in multiple databases. Thus, we are forced to resort to integrating information from an array of databases to create a consolidated representation of an entity. The key to integrating two heterogeneous databases is to find commonalities between records in the database. Integration, in this step, is similar to performing a simple join operation; however, it is not a trivial task due to the complexity of heterogeneous databases.

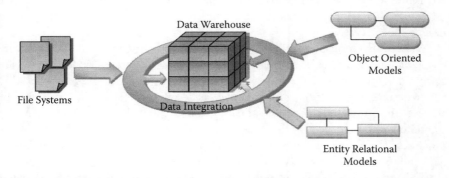

Figure 3.7 Data integration is considered to be a precursor of data warehousing. Typically data integration entails the integration of data obtained from disparate sources categorized based on the nature by which data are stored, namely: file systems, objected-oriented databases, and relational databases.

With the differences in syntax and nomenclatures adopted, finding methods of identifying similarities is still an open challenge.

Record linkage in files is used to identify duplicate identifiers in situations where the unique identifiers are unavailable. This technique works by matching the attribute fields and other fields that are not unique identifiers of entities. Record linkage is synonymous with object identification, data cleaning, approximate matching or approximate joins, fuzzy matching, and entity resolution.

There are two types of data heterogeneity, structural heterogeneity and lexical heterogeneity. Structural heterogeneity occurs when the fields of the data records in the database are structured differently in different databases. Lexical heterogeneity occurs when the data records have identically structured fields across databases but the data use different representations to refer to the same real-world object (Elmagarmid et al. 2007).

Data mining challenges have been surveyed in order to help identify lexical heterogeneity. Record linkage is also referred to as a record matching problem in statistics. The motivation for using record linkage is to identify records in the same or different databases that refer to the same real-world entity, even if the records are not identical. The same problem may have multiple names across research communities, such as the merge/purge problem, data duplication, or instance identification in the data database community.

3.4.2 Schema Integration Issues

The difficulties in integrating different database schemas stem from commonly observed problems. For example, the same attribute may have different names in different schemas, or an attribute may be derived from another attribute, different attributes might represent the same information causing redundancy, values in attributes might be different, and records may be duplicated (under different keys).

Various data integration and record linkage schemes have been proposed to handle these issues.

As an illustration of data integration, let us consider data from different relational databases. The objective here is to ensure that the data entries or records are stored in a uniform manner in a common database, resolving (at least partially) the structural heterogeneity problem by considering only relational databases. To achieve uniformity, the data are subject to parsing and data transformation and standardization. Extraction transformation loading (ETL) broadly describes these processes. The steps in this process ensure improvement in the quality of the in-flow data and make the data records comparable and more usable.

The first step of the ETL process is parsing. Parsing aids in the identification and isolation of individual data elements in the source data tables or files. It enables easier correction and standardizing and matching of data, as it allows for comparison of individual components, rather than of long complex strings of data. Multiple parsing methods are currently available, and parsing remains an active field of research.

The second step of the ETL process is data transformation, the simple conversion applied to the data to conform them to standard data types. Data transformation involves manipulating one field of the data record at a time and treating it independently without taking into account the values of the related fields. This step is generally applied to conform legacy data to specific data types pertaining to specific applications. Renaming a field is also a form of data transformation. Range checking is yet another kind of data transformation, which involves examining data in a field to ensure that they fall within the expected range. Dependency checking is a slightly more evolved form of data transformation, since it requires comparing the value in a particular field to the values in another field to ensure a minimal level of consistency in the data.

Finally, data standardization refers to the standardizing process involved in converting the data from one format to an application-specific format. Standardization is applied to data fields that are stored in different formats across different data sources, such that they are converted to a uniform representation before being subjected to the duplicate detection process. Standardization drastically reduces errors by reducing duplicate entries in the databases. Once the data have been standardized, the next step of data preprocessing is to identify which fields should be used for comparison. It is desirable to identify fields that have limited redundancy in their records.

Human errors that result in misspellings and different conventions for recording data result in multiple representations of a unique object in the database. Thus, significant research has been pursued for identifying techniques for measuring the similarity of individual fields, and techniques for measuring the similarity of entire records.

3.4.3 Field Matching Techniques

Duplicate detection is an important step in the data integration. The objective for using this step is to identify redundant fields or whole records across different databases. Mismatches caused by human typographical variations of string data are the most common source of errors in databases. Accurately completing this step invariably affects the outcome of duplicate detection techniques, as they rely on string comparison. To this end, various string matching techniques have been developed over the past decade. These techniques include character-based similarity metrics and token-based similarity metrics. Various methods within these techniques are explained below.

3.4.3.1 Character-Based Similarity Metrics

Character-based similarity metrics have been designed to handle typographical errors. Typically, the following distance metrics are used to measure the degree of similarity or dissimilarity between two strings (the objective of this section). They could also be modified and used to find similarity between different complex data structures, such as trees, graphs, etc. The following sections describe the different metrics that are prominently used in the field of bioinformatics.

Edit distance: The edit distance is one of the simplest algorithms for determining the similarity between two strings. Given strings σ_1 and σ_2, the edit distance measure is the similarity between them as the least number of single character edit operations required to transform the string σ_1 into σ_2 (Ristad and Yianilos 1998). The edit operations are confined to the following:
1. Insert a character into a string.
2. Delete a character from a string.
3. Replace one character with a different character.

In the simplest form, the edit distance is also referred to as Levenshtein distance (Ristad and Yianilos 1998), in which each edit operation is assigned a cost of zero. Typically, the computational complexity of measuring the edit distance of two strings is of the order $O(|\sigma_1|.|\sigma_2|)$, with $|\sigma_1|$ and $|\sigma_2|$ representing the length of the two strings.

Affine gap distance: An extension of the edit distance is the affine gap distance. It is believed that the edit distance metric is not effective when one of the two strings compared is truncated or shortened. The extension to edit distance includes the addition of two new operations:

1. Open gap operation
2. Extend gap operation

A solution for comparing strings of unequal length is to align the two strings. By aligning two strings, gaps are inserted into either of the two strings to ensure that the strings are of the same length, enabling easy comparison. The incorporation of gaps in strings for comparison is treated differently in different methods. The insertion of gaps is a weighted operation, in which a weight is assigned for every insertion of a gap—known as a penalty. It is not desirable to have a high gap score or alignment score. The affine gap distance metric has a variation in assigning weights to the gaps. The cost of extending the gap is usually smaller than the cost of opening a gap, which results in smaller cost penalties than the cost obtained using the edit distance metric. The time complexity of the affine gap algorithm is of the order $O(a.|\sigma_1|.|\sigma_2|)$ when the maximum length of a gap is a $\ll \min\{|\sigma_1|,|\sigma_2|\}$. In general, the algorithm runs approximately $O(a^2.|\sigma_1|.|\sigma_2|)$ steps.

Smith-Waterman distance: Smith and Waterman (1981) described an extension of edit distance and affine gap distance, in which mismatches at the beginning and the end of strings have lower costs than mismatches in the middle. This metric allows for better local alignment of the strings. The algorithm requires $O(|\sigma_1|.|\sigma_2|)$ time and space for two strings of length $|\sigma_1|$ and $|\sigma_2|$. Several improvements have thus been proposed, as in the case of the BLAST algorithm.

Jaro distance metric: The Jaro distance metric is the basic algorithm for the comparison of two strings, σ_1 and σ_2, and is based on the following steps. First, compute the string lengths $|\sigma_1|$ and $|\sigma_2|$. Second, find the "common characters" c in the two strings. By common, we refer to all the characters $\sigma_1[j]$ and $\sigma_2[j]$ for which $\sigma_1[i] = \sigma_2[j]$ and $|i - j| \leq \frac{1}{2}\min\{|\sigma_1|,|\sigma_2|\}$. Third, find the number of transpositions t. The number of transpositions is computed as follows: we compare the ith common character in σ_1 with the ith common character in σ_2. Each nonmatching character is a transposition. The Jaro comparison value is

$$Jaro(\sigma_1,\sigma_2) = \frac{1}{3}\left(\frac{c}{|\sigma_1|} + \frac{c}{|\sigma_2|} + \frac{c - \frac{t}{2}}{c}\right). \quad (3.3)$$

The Jaro algorithm requires $O(|\sigma_1|.|\sigma_2|)$ time for two strings of length $|\sigma_1|$ and $|\sigma_2|$ due to step 2, which computes the common characters in the two strings.

q-Grams: The q-grams are short character substrings of length q of the database strings. The purpose of using q-grams as a foundation for approximate string matching is that when two strings σ_1 and σ_2 are similar, they share a large number of q-grams. Given a string σ, its q-grams are obtained by "sliding" a window of length q over the characters of σ. Since q-grams at the beginning and the end of the string can have fewer than q characters from σ, the strings are conceptually extended by padding the beginning and the end of the string with $q - 1$ occurrences of a special padding character, not found in the original alphabet. With the appropriate use of hash-based indexes, the average time required for computing the q-gram overlap between two stings σ_1 and σ_2 is $O(\max\{|\sigma_1|,|\sigma_2|\})$. Letter q-grams, including trigrams, bigrams, and unigrams, have been used in a variety of applications.

3.4.3.2 Token-Based Similarity Metrics

The different character-based similarity metrics defined above aid in the detection of typographical errors. However, databases often use varied conventions that lead to the rearrangement of words. In such cases, we use the token-based metrics to measure the similarity between varied conventions.

Atomic strings: An atomic string is a sequence of alphanumeric characters delimited by punctuation characters. Two atomic strings match if they are equal, or if one is the prefix of the other. Otherwise, they do not match. The similarity of two fields is the number of their matching atomic strings divided by their average number of atomic stings.

WHIRL: An alternate to the atomic strings similarity metrics is the WHIRL algorithm (Cohen 1998). The WHIRL algorithm is based on the vector space model. A vocabulary T of atomic terms that can include words, phrases, or word stems (word prefixes) is built. A text document is represented as a document vector, and each component corresponds to terms $t \in T$ denoting the component of v that corresponds to $t \in T$ by vt.

In this algorithm, the weighting scheme used is the term frequency-inverse document frequency (TF-IDF) weighting that is normalized between 0 and 1. Once the document is represented by vector

$$\hat{v}^t = \left(\log(TF_{V,t}) + 1\right) \cdot \log(IDF_t) \tag{3.4}$$

where the term frequency $TF_{V,t}$ is the number of times the term t occurs in the document represented by v, and the inverse document frequency IDF_t is $\frac{|C|}{|C_t|}$, where C_t is the subset of documents in the collection of documents C that contains the term t.

The similarity between two documents u and v is computed using

$$sim(u,v) = \sum_{t \in T} u^t \cdot v^t \tag{3.5}$$

which is interpreted as the cosine of the angle between u and v, and which ranges between 0 and 1, as every document is of unit length.

The magnitude of the vector vt corresponds to the related importance of the term t in the document represented by v. Two documents are similar when they share many important terms. The TF-IDF weighting scheme assigns lower weights to frequently occurring terms in the collection C. However, the drawback of this method is that the vectors tend to be sparse, i.e., if a document contains only k-terms, then all but k components of the vector representation will have zero.

3.4.3.3 Data Linkage/Matching Techniques

This section addresses methods that are used to match records with multiple fields. These methods, according to Elmagarmid et al. (2007), are broadly divided into two categories: learning approaches and distance-based approaches.

1. *Learning approaches* use training data to learn how to match records from different sources. They include probabilistic approaches and supervised machine learning techniques.
2. *Distance-based approaches* match records using domain knowledge or generic distance metrics. In these approaches, special declarative languages are used to detect duplicate records.

Of the above two categories, this section focuses on the first category. For example, let us assume that tables A and B, each having n comparable fields, are expected to be matched. To this end, we define two classes M and U, where class M contains record tuples $\langle\alpha,\beta\rangle$, ($\alpha \in A, \beta \in B$) that represent the same entity (match), and class U contains the record tuples that represent different entities (nonmatch).

For matching, a pair of tuples, $\langle\alpha,\beta\rangle$, is represented as a vector $\hat{x} = [x_1,\ldots,x_n]^T$ with n components that correspond to n comparable fields of A and B. With each x_i showing the degree of agreement between the ith field of records α and β. Typically, the matches are represented by binary values 0 and 1 for the values of x_i, where $x_i = 1$ if field i agrees and $x_i = 0$ if field i disagrees.

3.4.3.3.1 Probabilistic Matching Models

The initial mathematical model based on Bayesian inference (Newcombe and Kennedy 1962) was proposed by Fellegi and Sunter (1969). In this model, two tables A and B, are matched using a vector $\hat{x} = [x_1,\ldots,x_n]^T$ as input for the creation of a decision rule. This decision rule assigns $\hat{x}$ to either class M or U. In the probabilistic approach, we assume that $\hat{x}$ is a random vector that has a density function that is different for each of the two classes M and U. If the density function for the classes M and U is known, the duplicate detection problem can be equated to the Bayesian inference problem, where observations are used to update or newly infer what is known about underlying parameters or hypotheses.

From the following equation,

$$\langle\alpha,\beta\rangle \in \begin{cases} M & if\ p(M|\hat{x}) \geq p(U|\hat{x}) \\ U & otherwise \end{cases} \quad (3.6)$$

vector $\hat{x}$ is classified to class M, if the probability of class M is greater than the nonmatch class U, and vice versa. On applying Bayes' theorem, the above equation can be expressed as

$$\langle\alpha,\beta\rangle \in \begin{cases} M & if\ l(\hat{x}) = \dfrac{p(\hat{x}|M)}{p(\hat{x}|U)} \geq \dfrac{p(U)}{p(M)} \\ U & otherwise \end{cases} \quad (3.7)$$

where $l(\hat{x})$ is the likelihood ratio with a threshold $\frac{p(U)}{p(M)}$ for the decision.

However, this approach is true only when the posterior probabilities $p(\hat{x}|M)$, $p(\hat{x}|U)$ and the prior probabilities $p(M)$ and $p(U)$ are known, which is rarely the case.

To overcome this problem, the naïve Bayes approach to compute the posterior probabilities based on a conditional independence is used. The conditional independence assumes that $p(x_i|M)$ and $p(x_j|M)$ are independent if $i \neq j$. This assumption results in the following:

$$p(\hat{x} \mid M) = \prod_{i=i}^{n} p(x_i \mid M) \qquad (3.8)$$

and

$$p(\hat{x} \mid U) = \prod_{i=i}^{n} p(x_i \mid U), \qquad (3.9)$$

where the values of $p(x_i \mid M)$ and $p(x_i \mid U)$ can be computed using a training set of known class labels.

3.4.3.3.2 Supervised and Semisupervised Learning

In supervised and semisupervised learning, probabilistic models base the classification of records using Bayesian approaches on classes M and U. Other commonly used duplication techniques are based on traditional classification techniques, where the system relies on the existence of training data in the form of record pairs, labeled as matching or not matching.

Similar to the probabilistic approach defined in the previous section, the Classification as Regression Trees (CART) classification algorithm could be extended to match records. For this function, the algorithm generates regression trees, based on which a linear discriminant algorithm generates all possible combinations of parameters to separate the data into their respective classes. Grouping records is brought about using the vector quantization approach. Similarly, support vector machines (SVMs) can be extended to match records.

3.5 Data Warehousing

A data warehouse is a subject-oriented, time-varying, nonvolatile collection of data used primarily in organization decision making.

As the information stored in a warehouse is focused on one subject related to an organization, a warehouse is termed subject oriented. When a warehouse is being built, useful pieces of this information from disparate data sources are gathered in one universally accepted format for storage. The data are integrated as per requirements and not simply transferred from source to warehouse. A data warehouse is time varying, as every piece of data has a time stamp associated with it that is derived from the source. The data from the source vary with time, as appropriate modifications and updates take place. However, the data warehouse is nonvolatile, and once it is stored in the warehouse, it will not change. This static status means that the data in the warehouse do not reflect the changes made to the data at the source. For the changes to be reflected, the data warehouse must be refreshed at regular intervals. For every refresh cycle, the updates are incorporated into the warehouse and outdated information is purged.

Though the data in the warehouse do not store updated information, their primary function is to enable high-level decision making, rather than store day-to-day information. Because of the difficulty of updating information, data warehousing is not applicable to all fields of applications where up-to-date information is required (Figure 3.8).

3.5.1 Online Analytical Processing

As previously discussed, data warehouses emphasize integration and decision support. Thus, the overall focus of these storage systems is on consolidating information so that it is available at a glance, rather than on clarifying the specific details of individual transactions. Using data warehousing is like visualizing a forest, rather than the individual trees in the forest. Thus, the performance metrics in data warehousing are be related to query throughput and response times. The technology well suited to these metrics is online analytical processing (OLAP). OLAP performs data consolidation and complex analysis of information and is apt for use in warehouse creation.

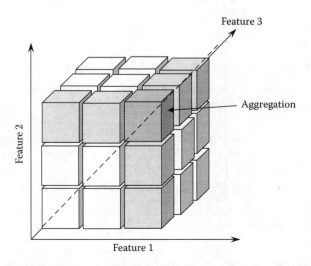

Figure 3.8 The representation of a data warehouse in three dimensions.

Table 3.1 Comparison between OLTP and OLAP

Characteristic	OLTP	OLAP
Main purpose	To support day-to-day operations; control and run fundamental business tasks	To support managerial, strategic planning, and problem solving; decision support
Queries	Short transactions; relatively simple structured query language (SQL)	Longer transactions; complex SQL with analysis
Updates	Random updates; few rows accessed	Sequential/bulk updates; many rows accessed
Processing speed and response times	Subsecond response time	Seconds to minutes response time
Database model	ER modeling; minimizes redundancy	Dimensional modeling; okay to have redundancy
Data normalization	Normalized data (5NF); minimizes duplicates	De-normalized data (3NF); duplicates are okay
Indexes	Few indexes; avoids index maintenance cost in writes	Okay to have more indexes; mostly read-only operations
Workload predictability and tuning	Precompiled queries; repeated execution of queries	Ad hoc queries; unpredictable load

3.5.2 Differences between OLAP and OLTP

Online transaction processing (OLTP) is a system that is used in clerical processing tasks that emphasize fast query processing times. These systems are characterized by a large volume of short online transactions, such as data entry, deletion, update, and retrieval in typical databases. In contrast, OLAP systems are characterized by a low volume of transactions. In OLAP, the queries are complex and involve aggregations of data from multiple data sources. These queries are performed ad hoc. Differences between these two systems are elaborated in Table 3.1.

3.5.3 OLAP Tasks

Since OLAP tasks do not require constant updates for transactions, it is assumed that the data required by OLAP systems are stored in a data warehouse, which separates the input from both the operational databases and the output. This separation renders a structured approach to storing information in a systematic timely fashion. However, to suit a more realistic model of evolving data, the OLAP systems are required to be more dynamic, and their design is continuous. The following are the tasks that make this possible.

Roll-up task: The roll-up task is the process of getting a higher level of aggregation in the integration process, i.e., reducing detail. Here the aggregation function provides an abstraction of the data by reducing the lower-level details.
Drill-down task: The drill-down task is the opposite of the roll-up task. In the drill-down task, emphasis is given to highlighting the lower-level details of the data. It can be visualized as the process of drilling down an aggregation of data, implying an increase in detail.
Slice and dice task: The slice and dice task is analogous to the select and project operations in regular databases.
Pivot task: The pivot task enables the transformation of data to enable easy interpretation. This task involves reorientation and visualization of data.

Based on the above operations, we can see that OLAP functions at a higher level. In order to implement OLAP technology, relational OLAP (ROLAP) is used as the original model. ROLAP uses the simple relational database management system (RDBMS) model, where data are stored in tuples, and bears attributes. However, in warehousing, there is also a trend toward multidimensional OLAP (MOLAP). MOLAP adds more dimensions in which to store information, so we are not restricted to 2D tuples. We now have more sophisticated data structures, for example, 3D data cubes, complex arrays, and more. The emphasis in MOLAP is on multiple facts and multiple dimensions.

3.5.4 Life Cycle of a Data Warehouse

Now that we have briefly covered the concepts of OLAP and its operations, in this section we describe the life cycle of a data warehouse to enable its functioning. A warehouse is an extensive structure that has several phases of development. A fully developed warehousing system typically has the following parts: information sources, wrappers, integrators, and warehouses, each of which is described below.

- Information sources are the building blocks of the warehouse. They are the original sources of the data, like flat files, RDBMS, and object-oriented database management system (OODBMS). The raw data that exist here must be integrated into the final warehouse.
- Wrappers are responsible for data transformations prior to data integration. Each information source has a wrapper. The functions of a wrapper during warehouse creation include data gathering from the sources, data cleaning, and format conversions. Once the data are available in the generic format required by the warehouse, they can be consolidated. After the warehouse

is functional, periodic updates are needed to refresh the information. The wrapper is also responsible for the tasks involved in obtaining these updates.
- An integrator is a filter within the warehouse. Each warehouse has one integrator module that filters, summarizes, and merges the data from the individual wrappers and then dumps the desired information into the warehouse.
- A warehouse is the actual storage of integrated information.

The above are the necessary components of a warehouse. The data warehousing lifecycle includes warehouse creation and warehouse maintenance. The modules required for these steps include requirement analysis, architecture, data modeling, layout, meta-data, extraction, transformation, and load, monitoring, and administration, user interface, view maintenance, and purging. Each of these modules is explained below.

1. Requirement analysis is similar to the initial step in any software life cycle model, where facts are collected, needs are outlined, and detailed specifications are documented, so as to serve as a guideline for development.
2. Architecture indicates whether a data warehouse is centralized or distributed and whether it requires a dedicated server, among other issues. Users/designers often select servers and tools using architectural information.
3. Data modeling helps users to logically analyze raw data. These data models help in identifying the entities in the system and their relationships with each other, their properties, common features, and the like. Data modeling includes terminology, like E-R diagrams, star and snowflake schemas, and materialized views.
4. Layout is determined after the logical analysis is performed. Users/designers identify the individual information sources and apply open database connectivity (ODBC) connections in this step.
5. Meta-data are data about data. An example of meta-data is a library where the books form the data and the catalog forms the meta-data giving information about the books, i.e., about the data. In a data warehouse, meta-data are crucial, because the focus is on analysis rather than transactions. Therefore, we need to identify a meta-data repository, find a suitable location for it (centralized, distributed, etc., depending on warehouse architecture), and proceed to build access mechanisms for it.
6. Extraction, transformation, and loading are the three steps involved in the physical data transfer. In these steps, data are extracted from the underlying information sources in a raw form, transformed to the desired format by the wrapper module, and then loaded to the warehouse by the integrator module in a consolidated manner.
7. Monitoring and administration are the first steps in warehouse maintenance. The above steps all dealt with warehouse creation. Once the warehouse is built, users/designers set up support to keep it running. Users/designers have DWAs (data warehouse administrators) that are analogous to DBAs (database

administrators). Users/designers have to develop mechanisms for fault detection and correction, recovery from breakdown, and the like to ensure functional reliability.
8. A user interface is integral to a data warehousing system. A warehousing system could have a simple character interface, but such an interface would make access cumbersome, thus defeating the purpose of a warehouse. Therefore, a considerable amount of time and effort is spent on building an interface to cater to the needs of the user.
9. View maintenance is the way in which updates are reflected in a warehouse. In regular databases, this updating happens automatically, but in a warehouse, the updates need to be physically transferred. We refresh the warehouse in batches at certain intervals. During such updates, the warehouse ceases to be functional. The system is shut down, updates are performed, and then the system is restored. On the other hand, the system could reflect the changes as they occur without causing a system shutdown. This change has to be done one transaction at a time. The former approach is called batch updates, and the latter is incremental updates. Each method has pros and cons.
10. Purging helps keep information in the warehouse up-to-date. Once the warehouse data get old, they have to be driven out to make room for new information. In addition, users need the latest up-to-date information for correct analysis and decision making. This process of removing the old data is essential for effective functioning of the warehouse.

3.6 Conclusion

In this chapter we have provided a description of the challenges in handling biological databases with respect to data cleaning, data integration, and data warehousing through the various techniques used. In the following chapters we provide an overview of the different data transformation techniques and their implications on specific research endeavors in the area of bioinformatics.

References

Amidan, B.G., T.A. Ferryman, and S.K. Cooley. Data outlier detection using the Chebyshev theorem. In *2005 IEEE Aerospace Conference*. Manhattan Beach: IEEE, 2005, pp. 1–6.
Berunig, M.M., H.-P. Kriegel, R.T. Ng, and J. Sander. LOF: Identifying density-based local outliers. In *Proceedings of International Conference on Management of Data*. Dallas TX: ACM SIGMOD, 2000, pp. 1–12.
Chandola, V., A. Banerjee, and V. Kumar. *Outlier detection—A survey.* Minneapolis: ACM Computing Surveys, 2007.
Chandola, V., A. Banerjee, and V. Kumar. Anomaly detection: A survey. *ACM Comput Surv* 41, no. 3 (2009): 1–72.

Chapman, A.D. *Principles of data quality: Conference on Management of Data* (SIGMOD '98), 1998, pp. 201–212.

Cohen, W. W. Integration of heterogeneous databases without common domains using queries based on textual similarity. *SIGMOD International Conference on Management of Data.* (SIGMOD'98), 1998, 201–212.

Elmagarmid, A.K., P.G. Ipeirotis, and V.S. Verykios. Duplicate record detection: A survey. *IEEE Trans Knowledge Data Eng* 19, no. 1 (2007): 1–16.

Ester, M., H.-P. Kriegel, J. Sander, and X. Xu. A density-based algorithm for discovering clusters in large spatial databases with noise. In *Proceedings of the Second International Conference on Knowledge Discovery and Data Mining (KDD-96).* Portland, OR AAAI Press, 1996, pp. 226–231.

Fayyad, U., G. Piatetsky-Shapiro, and P. Smyth. From data mining to knowledge discovery in databases. *Am Assoc Artif Intell* 17 (1996): 37–54.

Fellegi, I.P., and A.B. Sunter. A theory for record linkage. *J Am Stat Assoc* 64, no. 328 (1969): 1183–1210.

Han, J., and M. Kamber. *Data mining: Concepts and techniques.* San Francisco CA: Morgan Kaufmann, 2006.

Haykin, S. *Neural networks: A comprehensive foundation.* 2nd ed. Lebanon, IN: Prentice Hall, 1998.

Hodge, V.J, and J. Austin. A survey of outlier detection methodologies. *Artif Intell Rev* 22, no. 2 (2004): 85–126.

John, G.H. Robust decision trees: Removing outliers from databases. In *KDD-95 Proceedings.* Montreal: AAAI, 1995, pp. 174.

Laurikkala, J., M. Juhola, and E. Kentala. Informal identification of outliers in medical data. In *Fifth International Workshop on Intelligent Data Analysis in Medicine and Pharmacology IDAMAP-2000.* Berlin, Germany: 14th European Conference on Artificial Intelligence ECAI-2000, 2000, pp. 20–24.

Maletic, J.I., and A. Marcus. Data cleansing: Beyond integrity analysis. In *Information quality.* Boston: Massachusetts Institute of Technology, 2000, pp. 200–209.

Muruzalbal, J., and A. Munoz. On the visualization of outliers via self-organizing maps. *J Comput Graph Stat* 6, no. 4 (1997): 355–382.

Newcombe, H.B., and J.M. Kennedy. Record linkage: Making maximum use of the discriminating power of identifying information. *Commun ACM* 5, no. 11 (1962): 563–566.

Reif, M., M. Goldstein, A. Stahl, and T.M. Breuel. Anomaly detection by combining decision trees and parametric densities. In *19th International Conference on Pattern Recognition (ICPR 2008).* Tampa, FL: IEEE, 2008, pp. 1–4.

Ristad, E.S., and P.N. Yianilos. Learning string-edit distance. *IEEE Trans Pattern Anal Machine Intell PAMI* 20, no. 5 (1998): 522–532.

Rumelhart, D.E., G.E. Hinton, and R.J. Williams. *Learning internal representations by error propagation.* Cambridge, MA: MIT Press, 1986.

Smith, T.F., and M.S. Waterman. Identification of common molecular subsequences. *J Mol Biol* 147, no. 1 (1981): 195–197.

Williams, G., R. Baxter, H. He, S. Hawkins, and L. Gu. A comparative study of RNN for outlier detection in data mining. *ICDM* (2002): 709.

Williams, P.H., C.R. Marguiles, and D.W. Hilbert. Data requirements and data sources for biodiversity priority area selection. *J Biosci* 27, no. 4 (2002): 327–338.

Zhang, T., R. Ramakrishnan, and M. Livny. BIRCH: An efficient data clustering method for very large databases. *Proc ACM SIGMOD* (1996): 103–114.

FEATURE ANALYSIS II

Chapter 4

Feature Selection and Extraction Strategies in Data Mining

In Chapter 4 we focus on the different data preparation and transformation strategies in the knowledge discovery in databases (KDD) process. The chapter contains a list of widely used data normalization strategies for processing raw data and lists various data transformation techniques. We explain feature selection and feature extraction/construction strategies in lieu of their application to biological data. We contain our discussion to a selected set of algorithms that encapsulate the diversity of the various feature selection techniques of filter-based and wrapper-based approaches. We also describe the various feature construction/extraction techniques that are described in the following sections.

4.1 Introduction

The purpose of data preparation in the KDD process is to potentially improve the quality of real-world data that are potentially incomplete, noisy, and inconsistent (Zhang et al. 2003). These inconsistencies reduce the discovery of useful patterns. Missing values contribute to a large percentage of issues in databases; thus it is imperative to define methods that address the missing values. Problems associated with missing values are amplified when attributes are missing or datasets have attributes in the form of aggregates (i.e., attributes that are a combination of other attributes). As stated above, biological data are considered noisy, as they are plagued by noise and outliers (Furey et al. 2000). These errors and outliers are attributed to data inconsistencies in codes and nomenclature (refer to Chapters 2 and 3 for data cleaning strategies).

Data transformation and preparation result in a refined form of the original data that is smaller and free of noise. These methods are used with the objective of improving both the accuracy and the computational efficiency of data mining. Data transformation strategies listed in this chapter are used to ensure that all the data are free of noise and inconsistencies. These strategies thus enhance the effective comparisons between data points. Data preparation includes strategies of feature/attribute selection in which various filtering and wrapper approaches are used to select relevant features/attributes that enhance the prediction accuracy of the learning algorithm applied in the data mining step later in the KDD process.

To avoid learning biases and simultaneously overcome computational bottlenecks with respect to resources and algorithm efficiency, data-nested validation strategies play an important role in preparing data in which various iterative sampling and instance selection strategies have been applied to handle the large number of data effectively, avoid learning biases, and estimate the performance of feature selection. Thus, data transformation and preparation is viewed as a guided process focused on generating quality data, which leads to the discovery of relevant patterns.

4.2 Overfitting

In the quest of fitting a statistical or learning model to the data, we typically run in to the problem of overfitting. Overfitting occurs when the intended learning model captures the inherent noise in the data instead of the underlying relationship between attributes of the data. Overfitting can be correlated to the learning algorithm's ability to give more importance to redundant and irrelevant attributes than to the amount of data available, making it overly complex and decreasing its predictive capacity.

Thus, the data must be subjected to data preparation to overcome overfitting. Data preparation can be used to select features that exhibit a causal relation to the class labels (target function) of the data records. This process is called dimensionality reduction. In addition to increasing the predictive accuracy, there are two goals for performing dimensionality reduction: to increase the speed of the algorithm and to utilize space effectively. Typically, dimensionality reduction falls into the third step of data transformation (see Chapter 2, Figure 2.1). Thus, in this chapter, we elaborate on the various data transformation techniques and the various feature selection and feature extraction schemes.

4.3 Data Transformation

Data transformation, a key concept of data preparation, ensures that data are transformed or consolidated (prepared) into a form in which learning can be applied. Typically, data transformation includes smoothing, a process in which noise and inconsistencies are removed from data. This process typically involves discretization of data features/attributes. Data generalization is another strategy of data transformation, which is applied to data when abstraction of data is required. In such cases, the low-level raw data are generalized to higher-level concepts such that resultant knowledge after mining can provide a better understanding of data. Just like data smoothing and generalization, data normalization is important in data transformation, as it facilitates an effective comparison of data points. Typically, real-world data are recorded at different scales, and through normalization, those data are converted to a universal form for comparison. These techniques are detailed as follows.

4.3.1 Data Smoothing by Discretization

Data smoothing is a data transformation strategy that is based on data discretization. In this method, data are categorized into intervals or bins to capture characteristics that could potentially be used to handle data inconsistencies. This process of dividing the data into intervals is commonly referred to as data discretization. Data discretization employs various binning strategies to remove inherent noise present in the data. This noise in data takes many forms, specifically the form of missing and inconsistent data values. Simple alternatives can be employed for handling missing values without going through the tedious procedure of manual updates. These include substituting all missing values with a global constant. Though easy to implement, these methods do affect the learning from data, and thus we do not recommend using them. Other strategies include substituting the missing values with the feature/attribute mean for a given class. Other approaches fill missing values based on inference derived from probabilistic Bayesian approaches or induction-based decision trees such as C4.5 and CART.

In this section, we explain how binning methods are used to handle noisy data that are present as inconsistent values for a given feature/attribute. Binning methods can be categorized as unsupervised or supervised methods. Unsupervised binning includes sorting data for a specific feature/attribute and dividing them into equal-sized intervals called bins. Using these bins, one can transform or smooth data smoothing by bin means, smoothing by bin median, or smoothing by bin boundaries.

Unsupervised binning methods include equi-width and equi-depth binning, which is controlled only by a predetermined number of bins N. The equi-width binning strategy is described in the following steps:

1. Sort the values of attribute/feature f in ascending or descending order.
2. Determine the range of values of f, and divide the range into N intervals of equal size.

3. Determine the width of each bin by finding max(f) and min(f) of f, using the following relation:

$$width = \frac{(\max(f) - \min(f))}{N}$$

4. Allocate values to their corresponding bins based on the range in which they fall into.
5. Smooth by means, median, or boundaries.

Though equi-width binning is most straightforward, it is sensitive to outliers and cannot handle skewed data. The alternative unsupervised approach is equi-depth binning, which is based on frequency partitioning. In this approach, the range is determined by the number of data samples in the dataset and a predefined number of bins. For example, if a dataset consists of 30 samples and 3 bins, then each bin is populated by 10 samples per bin. This method effectively handles data scaling.

4.3.1.1 Discretization of Continuous Attributes

The discretization of continuous attributes requires slicing a domain into a finite number of intervals. The minimum description length (MDL) principle is an original approach used to minimize the quantity of information contained in both the model and the exceptions to the model.

Unlike the equi-depth and equi-width discretization approaches, Khiops discretization (Boulle 2004) is a supervised approach that discretizes attributes using the chi-square (χ^2) test.

In brief, the Khiops discretization is a bottom-up approach to discretization that searches for the best place to merge adjacent intervals by minimizing the χ^2 criterion applied locally to two adjacent intervals; i.e., they are merged if they exhibit statistical similarity. The χ^2 threshold is user defined, and χ^2 statistics are parameterized by the number of explanatory values (related to the degrees of freedom). To compare two discretizations with different interval numbers, we use the confidence level instead of the χ^2 value.

Considering the contingency table as shown in Figure 4.1, let $e_{ij} = n_{i.}.n_{.j}/N$ be the expected frequency for cell (i,j), if the explanatory and class attributes are independent. In this case, the χ^2 value is a measure of the contingency table of the difference between observed frequencies and expected frequencies and can be interpreted as a distance to the hypothesis of independence between attributes. The numerical representation is shown below:

$$\chi^2 = \sum_i \sum_j \frac{(n_{ij} - e_{ij})^2}{e_{ij}} \tag{4.1}$$

Feature Selection and Extraction Strategies in Data Mining

	A	B	C	Total
a	n_{11}	n_{12}	n_{13}	$n_{1.}$
b	n_{21}	n_{22}	n_{23}	$n_{2.}$
c	n_{31}	n_{32}	n_{33}	$n_{3.}$
d	n_{41}	n_{42}	n_{43}	$n_{4.}$
e	n_{51}	n_{52}	n_{53}	$n_{5.}$
Total	$n_{.1}$	$n_{.2}$	$n_{.3}$	N

Figure 4.1 A schematic representation of the contingency table used to compute the χ^2 value.

The Khiops algorithm minimizes the confidence level between the discretized explanatory attributes by using χ^2 statistics. The χ^2 value is not reliable for testing the hypothesis of independence if the expected frequency in any cell of the contingency table falls below a defined minimum value. The algorithm is described by the following steps:

1. Initialization:
   1.1. Sort the explanatory attribute values.
   1.2. Create an elementary interval for each value.
2. Optimization of the discretization:
   2.1. Repeat the following steps.
   2.2. Search for the best merge. Search among the merges with at least one interval that does not meet the frequency constraint if one exists; merge. Otherwise, merge interval that maximizes the χ^2 value.
   2.3. Evaluate the stopping criterion. Stop if all constraints are respected and if no further merge decreases the confidence level.
   2.4. Merge and continue if the stopping criterion is not met.

The Khiops method is based on a greedy bottom-up algorithm. It starts with initial single-value intervals and then searches for the best merge between adjacent intervals that contain two levels of merging. At the first level of merging, the Khiops method merges with at least one interval that does not meet the constraint; at the second level of merging, it merges with both intervals, fulfilling the constraint. The best merge candidate (with the highest χ^2 value) is chosen from among the first level of merges (in which case the merge is accepted unconditionally). Otherwise, if all minimum frequency constraints are respected, the merge candidate is selected from among the second level of merges (in which case the merge is accepted under the condition of improvement of the confidence level). The algorithm is reiterated until all minimum frequency constraints are respected and no further merge can decrease the confidence level. The computational complexity of this algorithm is of the order $O(N\log(N))$ with some optimization.

4.3.2 Normalization and Standardization

According to Guyon and Gunn (2006), data transformation is an integral part of model selection. Thus, data preparation in this chapter refers to the selection of the best normalization strategies and mathematical transformations of the feature space in the perspective of the learning machine used for processing the data.

Normalization and standardization strategies are applied to data to remove certain systematic biases that are inherent to the data. These biases are brought about by the dependencies between attributes and do not have to deal with the normal or Gaussian distribution of the data. In normalization, each attribute is treated independently. Normalization methods include min-max normalization, z-score normalization, and normalization by decimal scaling.

4.3.2.1 Min-Max Normalization

According to Han and Kamber (2006), min-max normalization is a linear transformation of the original data. Min-max normalization maps the value v of an attribute A in a record within a user-defined minimum and maximum (new_min_A and new_max_A) for the given attribute using the following expression:

$$v' = \frac{v - min_A}{max_A - min_A}(new\_max_A - new\_min_A) + new\_min_A \quad (4.2)$$

where min_A and max_A represent the minimum and maximum values of the attribute A across the entire dataset and v' is the normalized value of v.

Since the values of new_min_A and new_max_A are arbitrarily set by the user to 0 and 1, respectively, the min-max normalization is known as zero-one normalization.

4.3.2.2 z-Score Standardization

Instead of the user specifying the range through new_min_A and new_max_A in z-score standardization, the range for an attribute is determined by the mean and standard deviation possessed by the attribute across the dataset. The z-score standardization is brought about by the following expression:

$$v' = \frac{v - \mu_A}{\sigma_A}, \quad (4.3)$$

where μ_A and σ_A represent the mean and standard deviation of the attribute A. It is advantageous to use z-score standardization when it is difficult to determine the minimum and maximum values of a given attribute and when the dataset is plagued by outliers (Han and Kamber 2006).

4.3.2.3 Normalization by Decimal Scaling

Another prominently used normalization technique is normalization by decimal scaling. In this normalization strategy, normalization is carried out by proportionally reducing the value of each attribute in a record to a value of less than 1 using the following criteria, as presented in Equation 4.4:

$$v' = \frac{v}{10^j}, \qquad (4.4)$$

where j is the smallest integer such that $\max(|v'|) < 1$ (Han and Kamber 2006).

Supervised learning is typically presented with a set of training instances in which each instance is described by a vector of features (or attributes), values, and a class label. The task of the machine learning algorithm is to obtain the highest possible classification accuracy given a set of features. However, this objective is rarely achieved given a real-world scenario in which a large number of features describe a given instance, since the classification accuracy decreases proportionally as the number of features rises. For example, the accuracy for detecting data points in n-dimensional space decreases if there are a large number of features. Feature selection and feature extraction techniques are used to overcome such inherent big N small p problems. Feature selection selects an optimal subset of features from an existing set of features, while feature extraction constructs features from an existing set of features. In this chapter, we elaborate on the problems faced in feature selection and feature extraction and explain the techniques available for both. Before we delve into the various feature selection and feature extraction strategies, let us first describe the significance of features and their relevance to a dataset.

4.4 Features and Relevance

A feature (f), also referred to as an attribute, is a descriptor data point of instance. The relevance of a feature (f) is always measured by its ability to distinguish instances of the dataset with respect to the target class to which the instance belongs. Features can therefore be categorized into two types: those that are strongly relevant to the dataset/distribution and those that are weakly relevant to the dataset/distribution (Figure 4.2) (Kohavi and John 1997).

4.4.1 Strongly Relevant Features

A feature f is strongly relevant to dataset S if two instances A and B in S belong to different classes (or have different distributions of labels if they appear in S multiple times) and differ only in their value of f. Moreover, f is strongly relevant to target c

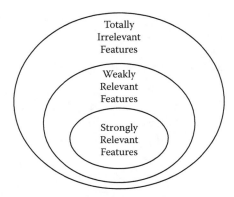

Figure 4.2 A view of feature set relevance. (From Kohavi, R., and John, G.H., *Artif Intell* 97, no. 1–2 (1997): 273–324. With permission.)

and distribution D if data points A and B have nonzero probability over D that differs only in their assignment to f and satisfy $c(A) \neq c(B)$, in which case, A and B are now required to be in S (or have nonzero probability).

4.4.2 Weakly Relevant to the Dataset/Distribution

A feature f is weakly relevant to sample X (or to target c and distribution D) if it is possible to remove a subset of the features so that f becomes strongly relevant.

4.4.3 Pearson Correlation Coefficient

Now that we know the characteristic difference between the kinds of features, it is a challenge to design algorithms to choose a set of strong features for a given dataset. Based on the definition of a strong feature above (Guyon and Elisseeff 2003), use Pearson's correlation to rank features with respect to the target outcome y. The Pearson correlation coefficient is defined as

$$\mathcal{R}(i) = \frac{cov(X_i, Y)}{\sqrt{var(X_i)var(Y)}} \qquad (4.5)$$

where cov designates the covariance and var the variance. The estimate of $R(i)$ is given by

$$R(i) = \frac{\sum_{k=1}^{m}(x_{k,i} - \bar{x}_i)(y_k - \bar{y})}{\sqrt{\sum_{k=1}^{m}(x_{k,i} - \bar{x}_i)^2 \sum_{k=1}^{m}(y_k - \bar{y})^2}} \qquad (4.6)$$

where the bar notation stands for an average over the index k.

As in linear regression, the coefficient of determination represents the fraction of the total variance around the mean value $\bar{y}$ that is explained by the linear relation between x_i and y. Therefore, using $R(i)^2$ enforces a variable ranking criterion according to how well the variable fits the linear model.

However, the correlation criteria, such as $R(i)^2$, can only detect linear dependencies between the variable and target and fail to fit a nonlinear model.

4.4.4 Information Theoretic Ranking Criteria

Many algorithms for variable selection use information theoretic criteria in the literature. Mutual information between variables and target classes is prominently expressed as

$$\mathfrak{T}(i) = \int_{x_i} \int_{y} p(x_i, y) \log \frac{p(x_i, y)}{p(x_i) p(y)} dx dy \qquad (4.7)$$

where $p(x_i)$ and $p(y)$ are the probability densities of x_i and y, and $p(x_i, y)$ is the joint density. $\mathfrak{T}(i)$ is the criterion that measures the dependency between the density of variable x_i and the density of the target y.

However, the densities $p(x_i)$, $p(y)$, and $p(x_i, y)$ are all unknown and are hard to estimate from data. To this end, it is simpler to convert the integral to a sum as below:

$$\mathfrak{T}(i) = \sum_{x_i} \sum_{y} P(X = x_i, Y = y) \log \frac{P(X = x_i, Y = y)}{P(X = x_i) P(Y = y)}. \qquad (4.8)$$

The above formulation of $\mathfrak{T}$ makes it easier to implement in a code as computing probabilities simplified to frequency counts. However, the estimation becomes harder with larger numbers of classes and variable values.

In the case of continuous variables (and possibly continuous targets), this estimation becomes even more challenging. Discretization of variables provides an immediate solution. However, using the normal distribution to estimate densities will allow us to estimate the covariance between X_i and Y, thus creating a similar criterion for the correlation coefficient.

Keeping these challenges in mind, we look into the various feature selection and feature extraction strategies available. We detail the mathematical principles involved and highlight the challenges they pose.

4.5 Overview of Feature Selection

There are four steps to feature extraction and feature selection: (1) feature construction, (2) feature subset generation, (3) evaluation criterion definition, and (4) evaluation criterion estimation.

Feature extraction uses feature construction, and feature selection schemes use the steps generating feature subsets, defining evaluation criterion, and estimating evaluation criterion. Based on these four steps, feature selection is further characterized into filter and wrapper techniques (Das 2001).

In wrapper approaches of feature selection, a feature subset selection algorithm is wrapped around the learning algorithm. The subset selection algorithm searches for an optimal subset using the learning algorithm that is independent from the final evaluator. This subset selection algorithm performs all the necessary evaluation of feature subsets. The wrapper approach is run on a dataset, which is usually partitioned into internal training and holdout sets, with sets of features removed from the data. The feature subset with the highest estimated value is chosen as the final set on which to run the classifier (Saeys et al. 2007). The resulting classifier is then evaluated on an independent test set that was not used during the search.

An important component of any feature selection technique is the projection matrix. The projection matrix is used to store weights of features that generally reflect the importance of each feature in the dataset. This matrix is multiplied by the feature vectors in order to optimize the base criterion function. Typically, the off-diagonal elements of a projection matrix are all set to zero and the diagonal elements of the projection matrix are set to {0,1} in feature selection. Given the criterion function, feature selection is equated to an exhaustive search problem. As the complexity of the search is directly proportional to the number of features in the dataset, feature selection is empirically based on forward or backward selection schemes (Pudil et al. 1994).

Alternatively, as an improvement to the feature selection schemes is feature weighting. In feature weighting, the diagonal elements of the projection matrix are not confined to just {0,1}, but rather are allowed to take real values. This modification to the projection matrix allows for the employment of more well-known optimization schemes. In this chapter, we elaborate on some of the well-known feature selection and feature extraction schemes.

4.5.1 Filter Approaches

This category of methods is closely associated with feature-ranking techniques (see Figure 4.3). Filter approaches rank features based on the correlation (degree of dependence) of individual features with respect to the target (class) label of the dataset. This process is called the relevance index.

Feature subset generation entails a category of algorithms that include, but are not limited to, a heuristic or stochastic search, exhaustive searches of features, a nested subset strategy for feature selection, forward selection/backward elimination, and single-feature ranking. The evaluation criteria for these filter methods include single-feature relevance, relevance in context, and feature subset relevance. Evaluation criteria estimation typically entails statistical tests.

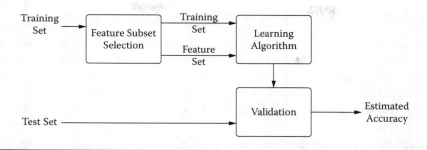

Figure 4.3 The filter approach to feature subset selection.

4.5.2 Wrapper Approaches

In wrapper methods, the performance of a learning algorithm is used to evaluate the goodness of selected feature subsets by their information content rather than by optimizing the performance of a learning algorithm directly (see Figure 4.4). Though filter methods are computationally more efficient, wrapper methods yield better results (Yijun and Dageng 2008).

Feature subset generation entails a category of algorithms that include but are not limited to a heuristic or stochastic search, an exhaustive search of features, a nested subset strategy for feature selection, forward selection/backward elimination, and single-feature ranking.

Evaluation criteria estimation involves various cross-validation and performance bounds techniques.

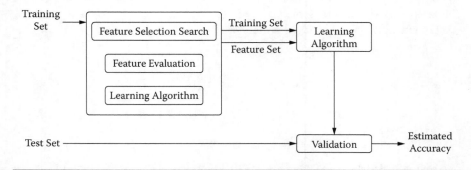

Figure 4.4 The wrapper approach to feature subset selection. The learning algorithm is used as a black box by the subset selection algorithm. (From Kohavi, R. and John, G. H., Artif Intell 97, no. 1–2 (1997): 273–324. With permission.)

4.6 Filter Approaches for Feature Selection

Filter approaches for feature selection use the predictive power of many features collectively rather than independently. This process is driven by features that are irrelevant individually but become relevant when used in combination with one another. Thus, feature selection is the problem of choosing a small subset of features that is necessary and sufficient to describe a class (or target).

4.6.1 FOCUS Algorithm

As an example of the filter approach to feature selection we describe the FOCUS algorithm. In the FOCUS algorithm (Almuallin and Dietterich 1992), the features describing a data point are a set of Boolean features and are conceptualized to work on a binary class scenario. Thus, let $\{x_1, x_2, \ldots, x_n\}$ be a set of n Boolean features and $\{C^+, C^-\}$ represent the associate classes that each data point belongs to. The ultimate goal of the algorithm is to select features based on a sufficiency test. The sufficiency test is a procedure for checking whether the selected features (Q) are sufficient to form a consistent hypothesis or are sufficient to differentiate between the two classes.

Let $\langle X_1, C^+ \rangle$ and $\langle X_2, C^- \rangle$ represent two independent samples from classes C^+ and C^-, respectively. The sufficiency test determines whether the samples have the same values for all selected features of Q. If the pair of samples has all feature matches in Q, then the selected features Q cannot discriminate all of the positive examples from all of the negative examples. On the contrary, the feature set Q is sufficient if no such matching pairs appear in the training set.

As a working example, for the two samples $\langle X_1, C^+ \rangle$ and $\langle X_2, C^- \rangle$, we define a conflict vector a of length n, $\langle a_1 a_2 \ldots a_n \rangle$, where $a_i = 1$ if X_1 and X_2 have different values for the feature x_i and $a_i = 0$ otherwise. We say that a is explained by x_i if $a_i = 1$. Using this terminology, a set Q of features is sufficient to construct a hypothesis consistent with a given training sample if every conflict generated from the sample is explained by some feature in Q.

For example, let the training sample be

$\langle 010100, C^+ \rangle$ $\langle 011000, C^- \rangle$

$\langle 110010, C^+ \rangle$ $\langle 101001, C^- \rangle$

$\langle 101111, C^+ \rangle$ $\langle 100101, C^- \rangle$

Then, the set of all conflicts generated from this sample is

$a_1 = \langle 001100 \rangle$ $a_4 = \langle 101010 \rangle$ $a_7 = \langle 110111 \rangle$

$a_2 = \langle 111101 \rangle$ $a_5 = \langle 011011 \rangle$ $a_8 = \langle 000110 \rangle$

$a_3 = \langle 110001 \rangle$ $a_6 = \langle 010111 \rangle$ $a_9 = \langle 001010 \rangle$

Double-check to ensure that subset $\{x_1.x_3, x_4\}$ is sufficient to form a consistent hypothesis (e.g., $\overline{x}_1\overline{x}_3 \vee (\overline{x}_3 \oplus x_4)$), and that all subsets of cardinality less than 3 are insufficient.

Despite the ease of using this method, there is one disadvantage to using the FOCUS algorithm. The algorithm tries all subsets of features of increasing size until a sufficient set is encountered. As seen in the above example, the FOCUS algorithm tests the $\binom{6}{0} + \binom{6}{1} + \binom{6}{2} = 22$ subsets of features of size 0, 1, 2, and some of the $\binom{6}{3} = 20$ subsets of size 3 before returning the solution. FOCUS thus does not exploit all the information given in the training sample. For example, it does not accurately exploit $a_1 = \langle 001100 \rangle$, where any associated sufficient set must contain x_3 or x_4 to elucidate the conflict. Thus, none of the sets $\{x_1\}$, $\{x_2\}$, $\{x_5\}$, $\{x_6\}$, $\{x_1, x_2\}$, $\{x_1, x_5\}$, $\{x_1, x_6\}$, $\{x_2, x_5\}$, $\{x_2, x_6\}$, $\{x_5, x_6\}$ can be solutions. Therefore, all of these sets can immediately be ruled out of the algorithm's consideration. Many other subsets can be similarly ruled out based on the other conflicts.

The FOCUS-2 algorithm is presented in Figure 4.5 (Almuallin and Dietterich 1992). This algorithm proposes the use of a first-in/first-out data structure, in which each node of the data structure represents a subspace of all feature subsets. Each node is of the form $M_{A,B}$, which denotes the space of all feature subsets that include all the features in the set A and the node of the feature in the set B. Thus, $M_{A,B}$ is formally represented as

$$M_{A,B} = \{T \mid T \supseteq A, T \cap B = \phi, T \subseteq \{x_1, x_2, \ldots, x_n\}\} \qquad (4.9)$$

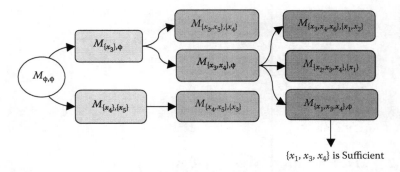

Figure 4.5 The working of the FOCUS algorithm. (Modified from Almuallin, H., and Dietterich, T.G., in *Proceedings of the Ninth Canadian Conference on Artificial Intelligence*. Vancouver, BC: Morgan Kaufmann, 1992, pp. 38–45.)

The objective of using FOCUS is to retain only the portions of the space of feature subsets that could contain a solution. Initially, the data structure contains only the element $M_{\phi,\phi}$, which represents the power set. In each iteration of the algorithm, the space represented by the head of the queue is partitioned into disjoint subspaces, and those subspaces that cannot contain solutions are pruned from the search.

In respect to the conflict $a_1 = \langle 001100 \rangle$ and the power set of features $M_{\phi,\phi}$, we know that any sufficient feature subset must contain either x_3 or x_4. This structured approach helps further refine $M_{\phi,\phi}$ into the two subspaces: $M_{\{x_3\},\phi}$, those feature subspaces that contain x_3, and $M_{\{x_4\},\{x_3\}}$, all feature subspaces that contain x_4 and not x_3. Thus, conflicts with fewer 1s in them provide more constraint for the search than conflicts with more 1s. Therefore, if the head node of the queue is $M_{A,B}$, then the algorithm searches for a conflict a such that (1) a is not explained by any of the features in A, and (2) the number of 1s corresponding to features that are not in B is minimized.

The algorithm of FOCUS, given $M_{\phi,\phi}$, is described by the following steps: Given the conflict $a_1 = \langle 001100 \rangle$, $M_{\phi,\phi}$ is replaced by $M_{\{x_3\},\phi}$ and $M_{\{x_4\},\{x_3\}}$. Next, for $M_{\{x_3\},\phi}$, the conflict $a_8 = \langle 000110 \rangle$ is selected, and $M_{\{x_3,x_4\},\phi}$ and $M_{\{x_3,x_5\},\{x_4\}}$ are added to the queue. $M_{\{x_4\},\{x_3\}}$ is then processed with $a_9 = \langle 001010 \rangle$ and $M_{\{x_4,x_5\},\{x_3\}}$ is inserted.

1. If all the examples in the sample have the same class, then return ϕ.
2. Let G be the set of all conflicts generated from the sample.
3. Queue = $\{M_{\phi,\phi}\}$.
4. Repeat.
    a. Pop the first element in queue. Call it $M_{A,B}$.
    b. Let $OUT = A$.
    c. Let a be the conflict in G not explained by any features in A, such that $|Z_a - B|$ is minimized, where Z_a is the set of features explaining a.
    d. For each $x \in Z_a - B$,
        i. If sufficient $(A \cup \{x\})$, return $(A \cup \{x\})$.
        ii. Insert $M_{A \cup \{x\}, OUT}$ at the tail of queue.
        iii. $OUT = OUT \cup \{x\}$.

Finally, when $M_{\{x_3,x_4\},\phi}$ is processed with $a_3 = \langle 110001 \rangle$, the algorithm terminates before adding $M_{\{x_1, x_3, x_4\},\phi}$ to the queue since $\{x_1, x_3, x_4\}$ is a solution.

4.6.2 RELIEF Method—Weight-Based Approach

Other feature selection methods assign weights to features that have a high degree of relevance. One the most prominent such methods, the RELIEF algorithm, takes into consideration inherent relations between features. In this section, we describe the RELIEF algorithm.

Given a training dataset, the RELIEF algorithm (Kira and Rendell 1992; Yijun and Dageng 2008) iteratively estimates feature weights according to the weight's ability to discriminate between neighboring patterns. The approach used in RELIEF is based on instance-based learning (Kira and Rendell 1992). For the given training set T, consisting of samples of length m, the algorithm aims to detect relevant features that correlate to the target class (binary class). The training set T is initially split into positive and negative samples. The iterative RELIEF algorithm then uses a weight vector W of length m equal to the number of features in a sample. This weight vector W is initialized to zero before the first iteration of the algorithm. The following steps are performed iteratively for each attribute m. At first, a random instance X is chosen. For comparison, two of the closest samples are chosen, one from the class of positives (T^+) and the other from the class of negatives (T^-). Using Euclidean distance, the RELIEF algorithm selects either T^+ or T^- as its near hit (NH) or near miss (NM). Once the NH and NM have been determined, the weight vector W is updated to reflect the weight of each attribute. The weight vector is averaged and then used to identify the relevance of each attribute based on the values of W. The algorithm selects those features that have a weight above threshold τ. The following are the steps of the RELIEF algorithm, for the given dataset T, with m attributes and threshold τ.

1. Separate T into $T^+ = \{\text{positive instances}\}$ and $T^- = \{\text{negative instances}\}$.
2. Initialize the weight vector $W = \langle 0, 0, ..., 0 \rangle$.
3. For $i = 1$ to m,
    a. Pick at random an instance $X \in T$.
    b. Pick at random one of the *positive instances* closest to X, $t^+ \in T^+$.
    c. Pick at random one of the *positive instances* closest to X, $t^- \in T^-$.
    d. If X is a positive instance,
        i. Then *near hit* = T^+; *near miss* = T^-,
        ii. otherwise *near hit* = T^-; *near miss* = T^+.
    e. Call *Update - Weight*(W, X, *Near hit*, *Near miss*)
4. Compute *Relevance* = $(1/m).W$.
5. For $i = 1$ to p,
    a. If ($relevance_i \geq \tau$)
        i. Then $feature_i$ is a relevant feature.
        ii. Otherwise $feature_i$ is an irrelevant feature.
6. *Update - Weight*(W, X, *Near hit*, *Near miss*)
    a. For $i = 1$ to p,
        i. $W_i = W_i - \textit{diff}(x_i, \textit{Near hit}_i)^2 + \textit{diff}(x_i, \textit{Near miss}_i)^2$

Based on the above algorithm, there are two important components of the RELIEF algorithm, the relevance (averaged weight vector) and the threshold τ (Kira and Rendell 1992). Relevance is the averaged value of the weight vector W having the values $W_i - \textit{diff}(x_i, \textit{Near hit}_i)^2 + \textit{diff}(x_i, \textit{Near miss}_i)^2$ for each feature $feature_i$ over m sample triplets. Each element of a relevance vector corresponds to a feature that shows its relevance with respect to its corresponding target class. The relevance threshold τ is used to determine whether the feature should be selected. The RELIEF algorithm is valid only when (1) the degree of relevance is large for relevant features and comparatively small for irrelevant features, and (2) the relevance threshold τ retains relevant features and discards irrelevant features.

4.7 Feature Subset Selection Using Forward Selection

Many high-throughput bioinformatics applications are required for computational techniques in order to handle high-dimensional datasets. In such situations, methods like FOCUS and RELIEF are not computationally effective. Nested feature subset selection approaches have shown computational prowess in handling these high-dimensional datasets that give FOCUS and RELIEF. There are two kinds of nested approaches: (1) forward selection approaches and (2) backward elimination approaches. It is often argued that forward selection is computationally more efficient than backward elimination for generating nested subsets of variables. However, the defenders of backward elimination argue that weaker subsets are found by forward selection because the importance of variables is not assessed in the context of variables that have not been included yet (Guyon and Elisseeff 2003). In this section, we focus on the forward feature subset selection approach for selecting the most discriminatory features.

Forward selection refers to a search that begins with an empty set of features and thus has a maximum error. At each step, the feature that decreases the error the most is added one at a time until any feature addition does not significantly decrease the error. On the contrary, backward elimination proceeds initially with all the features and iteratively eliminates features that are least useful. Both techniques are robust toward overfitting and provide a nested subset of features.

4.7.1 Gram-Schmidt Forward Feature Selection

This feature selection method was intended to be applied directly to models that have linear parameters that are independent of the learning machine method employed. It is based on the Gram-Schmidt orthogonalization (Chen et al. 1989; Stoppiglia et al. 2003) procedure for ranking variables.

Consider a dataset that consists of N data points and their associated classes, represented as a vector consisting of Q features. We represent a data point in the dataset X as a vector $x_i = \{x_i^1, x_i^2, \ldots, x_i^Q\}$, with the associated class label y_i. Similarly, the vector $x^i = \{x_1^i, x_2^i, \ldots, x_N^i\}^T$ represents feature i across the dataset and is considered the input to this algorithm. Thus, dataset X is represented as a matrix of dimensions (N, Q).

The Gram-Schmidt procedure is an iterative process; in the first iteration, we search for the feature vector that best explains the concept, i.e., the feature vector that has the smallest angle with the process output vector in the N-dimensional space of observations. To this end, the following quantities are computed as

$$\cos^2\left(x^k, y_i\right) = \frac{\left(x^k \cdot y_i\right)^2}{\|x^k\|^2 \|y_i\|^2}, k = 1 \text{ to } Q \qquad (4.10)$$

and the vector x^k of largest magnitude is selected. Once the largest vector is selected, the remaining vectors (that represent other features) are projected onto a null subspace of the selected feature. In that subspace, the projected input vector that best explains the projected output is selected, and the $Q - 2$ remaining feature vectors are projected onto the null space of the first two ranked vectors. The procedure terminates when all Q input vectors are ranked or when a stopping criterion is met.

To determine an effective stopping criterion, the algorithm proceeds with the computation of a cumulative distribution function of the squared cosine of the angle between a given vector and a random vector. This cumulative distribution function is used to determine the rank of the feature.

The first step in this method is to compute the probability distribution function of the squared cosine of the angle φ between a fixed vector and a vector that has components that are normally distributed, in a space of dimension v. The step can be expressed as

$$f_v(x) = \frac{\Gamma\left(\frac{v}{2}\right)}{\Gamma\left(\frac{1}{2}\right)\Gamma\left(\frac{v-1}{2}\right)} \frac{(1-x)^{\frac{v-3}{2}}}{\sqrt{x}} \qquad (4.11)$$

where $\Gamma(.)$ is the gamma function, with $x = \cos^2\varphi, v \geq 2$ and $0 \leq x \leq 1$. $f_v(x)$ is a beta function with $a = 1/2$ and $b = (v - 1)/2$.

The cumulative distribution function $F_v(\cos^2\varphi)$ is obtained using the above relation (Equation 4.11). From this function, the probability that the angle between a random vector and a fixed vector is smaller than a given angle φ is easily derived as

$$P_v(\cos^2\varphi) = 1 - F_v(\cos^2\varphi), \qquad (4.12)$$

for $v \geq 2$ (Figure 4.6).

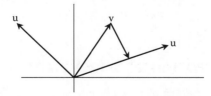

Figure 4.6 Gram-Schmidt process.

Finally, the cumulative distribution function of the rank of a random vector can be derived as follows. At iteration n, n candidate features have been ranked, and a new feature is chosen among the $Q - n$ remaining features. Using φ_n, we denote the angle (in a space of dimension $v = N - n$) between the selected projected feature and the projected output, and by Π_n the probability that the angle between a realization of the random feature and the projected output is smaller than φ_n : $\Pi_n = P_{N-n}(cos^2\varphi_n)$. We denote by G_{n-1} the probability that a realization of the random feature is less relevant than one of the $n - 1$ previous features, which is equal to $1 - G_{n-1}$. Therefore, the probability that the probe will be more relevant than the $n - 1$ previous features but less relevant than the nth feature is equal to

$$P_{N-n}(cos^2\varphi)(1-G_{n-1}). \tag{4.13}$$

Hence, the probability that a realization of the random feature is more significant than one of the n features selected after iteration n is given by

$$G_n = G_{n-1} + P_{N-n}(cos^2\varphi)(1-G_{n-1}), \tag{4.14}$$

with $G_0 = 0$.

Taking the cumulative distribution function into consideration, at each step of the Gram-Schmidt orthogonalization, four steps must be performed:

1. After orthogonalization, pick the projected candidate feature (not selected during previous steps) that has the smallest angle with the projected output.
2. Compute the value of the cumulative distribution function as described previously.
3. If the value is smaller than the rank, retain the feature and perform the next step of the Gram-Schmidt orthogonalization.
4. If that value is larger than the rank, discard the feature under consideration and terminate the procedure.

The choice of rank is problem dependent; i.e., if data are sparse, the model should be as parsimonious as possible. Hence, a low value of the rank should be chosen to make sure that only relevant inputs are present (but some features with low relevance might be missed); conversely, if data are abundant, a higher rank may be acceptable (but some irrelevant features might be kept).

4.8 Other Nested Subset Selection Methods

In feature selection, the number of subsets considered is usually very large, and a different method must be used not to overpenalize large subsets (Guyon 2009). The optimum number of features N is assessed using a cross-validation method, which includes a separate feature ranking in each fold. Then, a final ranking is performed using the entire training set and the first set of N features is selected. This method is less biased than using the ranking produced with the entire training set and selecting the best subset directly using cross-validation. The eight steps of this method are listed below:

1. Choose an algorithm $\mathcal{A}$ to create nested feature subsets.
2. Choose a learning machine $\mathcal{M}$ to evaluate the feature subsets.
3. Split the m available training samples into K training and validation subset pairs $\{D_t^j, D_V^j\}$ of dimension t and v, $t + v = m$, $j = 1{:}K$.
4. For $j = 1{:}K$,
    a. Using $\mathcal{A}$ and only the D_t^j examples, create nested subsets of the n available features:
    $$S_1^j \subset S_2^j \subset \cdots S_i^j \subset \cdots S_n^j$$
    b. For $i = 1{:}n$, train $\mathcal{M}$ on subset S_i^j using D_t^j, and test it using D_v^j. Call $r_{val}[i, j]$ the resulting estimation of performance.
5. Compute the CV scores of the nested feature subsets: $R_{CV}[i] = \sum_{j=1}^{K} r_{val}[i, j]$
6. Select the best number of features: $N = \operatorname{argmin}_i(R_{CV}[i])$
7. Using all m training examples, create nested subsets of the n available features:
$$S_1 \subset S_2 \subset \cdots S_i \subset \cdots S_n$$
8. Select S_n.

Now that we have covered the gamut of feature selection strategies, the following sections focus on the different feature extraction strategies in data mining.

4.9 Feature Construction and Extraction

It is a common practice to represent large datasets in the form of matrices, in which rows represent individual features/attributes and columns represent the data points (Berry et al. 1995). Matrix factorization has played a key role in many dimensionality reduction methods and is thus the focus of this section. To explain the relationship between dimensionality reduction and matrix factorization, let us consider a data matrix A, with d data points represented by t features, resulting in a $t \times d$ matrix. Each column of the data matrix A is thus a vector of t dimensions. The rank r_A of matrix A in linear algebra is the maximal number of linearly independent columns of A. The rank r_A of matrix A is considered to be the basis set if the rank can represent every vector in the vector space of A. Thus, the rank r_A of the data matrix A, which is equal to the size of the basis of the linear space it spans, is equal or near to *min(t,d)*. The aim of a dimensionality reduction technique is to find A', which is a good approximation of A and has a rank of k, where k is significantly smaller than r_A. For this reason, the A' matrix is often referred to as the *k*-rank approximation of A.

4.9.1 Matrix Factorization

Matrix factorization (Oh 2006), or decomposition, of matrix A is the process of breaking A into a product of two matrices U and V such that $A \approx U.V^T \approx A'$, with dimensions of the matrix $U = t \times k$ and the matrix $V = d \times k$, respectively. The columns of the U matrix are the basis vectors of the extracted lower-dimensional space, and the rows of V correspond to the coefficients that allow the approximate reconstruction back to the original data. In Equation 4.15, we show the commonly used LU decomposition of the data matrix A.

$$\underbrace{\begin{pmatrix} * & * & * & * \\ * & * & * & * \\ * & * & * & * \\ * & * & * & * \\ * & * & * & * \end{pmatrix}}_{A} \approx \underbrace{\begin{pmatrix} * & * \\ * & * \\ * & * \\ * & * \\ * & * \end{pmatrix}}_{U} \underbrace{\begin{pmatrix} * & * & * & * \\ * & * & * & * \end{pmatrix}}_{V^T} \quad (4.15)$$

4.9.1.1 LU Decomposition

Data matrix A and its associated class labels are represented by vector b. The LU decomposition method is employed to decompose the matrix without depending on elaborate computation of the inverse of A.

Thus, considering the data matrix and its associated class/target information, we can represent the data matrix as a linear form $A = b$. Since A needs to be factorized, we assume it is invertible and thus has a unique factorization. The LU decomposition works on the philosophy of splitting the data matrix A into upper and lower triangle matrices, as represented in Equation 4.16.

$$\underbrace{\begin{pmatrix} * & * & * & * \\ * & * & * & * \\ * & * & * & * \\ * & * & * & * \end{pmatrix}}_{A} = \underbrace{\begin{pmatrix} 1 & 0 & 0 & 0 \\ * & 1 & 0 & 0 \\ * & * & 1 & 0 \\ * & * & * & 1 \end{pmatrix}}_{L} \underbrace{\begin{pmatrix} 0 & * & * & * \\ 0 & 0 & * & * \\ 0 & 0 & 0 & * \\ 0 & 0 & 0 & 0 \end{pmatrix}}_{U} \quad (4.16)$$

where L is a unit of the lower triangle matrix (in which all diagonals are one) and U is the upper triangular matrix. LU decomposition employs the principle of Gaussian elimination to derive both L and U. Thus, we can substitute A by its equivalent L and U as $LU = b$. Various other methods that focus on decomposing a matrix are out of the scope of this book, but the motivation of matrix decomposition is to make computation with large matrices easier to handle. Once the matrices are decomposed, the next objective is to extract a set of vectors that capture a basis that is lower in number than the original set of vectors, and yet retains maximum information equivalent to the original matrix.

4.9.1.2 QR Factorization to Extract Orthogonal Features

Based on LU decomposition of a matrix, we introduce the QR factorization of a matrix, which is used to find the orthogonal basis vector set for a given matrix (subspace) A described by n features. QR factorization is based on the Gram-Schmidt process, which transforms a given matrix A to its orthogonal set of column vectors Q and the set of corresponding coefficient R.

This process is explained by assuming matrix A be an $n \times m$ ($n > m$) matrix with m linearly independent columns (which is the basis set for the subspace A). In this process, A can be expressed as

$$\underbrace{\begin{pmatrix} * & * & * & * \\ * & * & * & * \\ * & * & * & * \\ * & * & * & * \\ * & * & * & * \end{pmatrix}}_{A} = \underbrace{\begin{pmatrix} * & * & * & * \\ * & * & * & * \\ * & * & * & * \\ * & * & * & * \\ * & * & * & * \end{pmatrix}}_{Q} \underbrace{\begin{pmatrix} * & * & * & * \\ 0 & * & * & * \\ 0 & 0 & * & * \\ 0 & 0 & 0 & * \\ 0 & 0 & 0 & 0 \end{pmatrix}}_{R} \quad (4.17)$$

where Q represents the n orthonormal columns of dimensions $m \times n$ and R is the corresponding upper triangular matrix, containing the coefficients.

Using the Gram-Schmidt process, the given matrix A, its orthogonal matrix Q, and its corresponding n columns of orthonormal basis are obtained. Similarly, the coefficient matrix $R = Q^T A$ is obtained. Other factorization techniques are based on the concept of eigenvalues and vectors. The remainder of this section elaborates on them.

4.9.1.3 Eigenvalues and Eigenvectors of a Matrix

Some properties of eigenvalues and eigenvectors are important in feature extraction, as they provide certain properties of a matrix A and determine whether a given matrix can be factored based on a certain choice of properties.

1. A matrix with zero eigenvalues cannot be inverted.
2. Invertible matrices have all $\lambda \neq 0$, whereas singular (noninvertible) matrices include zero among their eigenvalues.
3. Eigenvectors that have distinct eigenvalues are linearly independent.
4. A full-rank matrix has a nonzero determinant, and thus has nonzero eigenvalues.
5. A triangular matrix has eigenvalues on its main diagonal.
6. For any integer n, λ^n is an eigenvalue of A^n with corresponding eigenvector x (negative integer n works when A is invertible).

We consider the above properties when we explore the use of eigenvalues and eigenvectors for the factorizations of a given matrix.

4.9.2 Other Properties of a Matrix

While employing feature extraction on a matrix of dimension $m \times n$ when $n \gg m$, it is important to reduce the matrix to its square form (i.e., map the matrix to its equivalent $n \times n$ matrix). The following section emphasizes the need for a square matrix and the properties that a square matrix entails.

4.9.3 A Square Matrix and Matrix Diagonalization

The relationship between a diagonalized matrix, eigenvalues, and eigenvectors of a square matrix A of dimension $n \times n$ is as follows:

$$A = E^{-1}DE \qquad (4.18)$$

where D is an $n \times n$ matrix that denotes a diagonal matrix, E represents a matrix of eigenvectors of matrix A, and E^{-1} represents the inverse of E. The diagonalization is feasible under the following three equivalent conditions:

1. n distinct eigenvectors are linearly independent.
2. The union of the basis of the eigenspace of A contains n eigenvectors.
3. The algebraic multiplicity of each eigenvalue equals its geometric multiplicity (algebraic multiplicity >= geometric multiplicity).

Above, the diagonal elements of the diagonal matrix D are the eigenvalues of A, and the rows of the matrix E represent the corresponding distinct eigenvectors. The diagonalized form of A can be used to speed up the computation of $A^k = E^{-1}D^k E$, respectively. The remaining problem is to obtain the (eigenvalue, eigenvector) pairs.

Note that an $n \times n$ full-rank matrix A does not necessarily have n linearly independent eigenvectors.

4.9.3.1 Symmetric Real Matrix: Spectral Theorem

One of the great achievements of linear algebra is the proof that a real $n \times n$ symmetric A has n distinct orthogonal eigenvectors (not necessarily distinct eigenvalues) if it satisfies: (1) a real symmetric matrix has real eigenvalues, and (2) in the case of symmetric matrices, the eigenvectors that correspond to distinct eigenvalues are orthogonal.

In such symmetric real matrices, we may encounter eigenvalues with multiple associated eigenvectors. In such cases, we can transform the eigenvalues into an orthogonal basis of the corresponding eigenspace using the Gram-Schmidt process (where the real matrix is transformed to its corresponding eigenvectors). Additionally, it has been proven that a real symmetric matrix has a complete set of eigenvectors, which implies that a real symmetric matrix always has a complete orthogonal basis.

Hence, the following decomposition is always possible for a symmetric real matrix, known as the spectral theorem:

$$A = Q^T D Q. \tag{4.19}$$

Above, the diagonal matrix D in Equation 4.19 has eigenvalues on its diagonal, and matrix Q has eigenvectors as its rows. The spectral decomposition (Equation 4.19) is a special case of the diagonalization (Equation 4.18) in which the most strict orthogonality is enforced in symmetric matrices. Also, the inverse matrix in Equation 4.18 is replaced with a transpose matrix in Equation 4.19 because the inverse of an orthogonal matrix is its transpose. The spectral decomposition form in Equation 4.19 is often expressed as follows as well (which is also called the projection form of the spectral theorem):

$$A = \sum_{i=1}^{n} \lambda_i e_i e_i^T. \tag{4.20}$$

Now that we know how a symmetric matrix is decomposed to its corresponding eigenvalues and eigenvectors, the remainder of the sections describe key feature extraction strategies that use the extracted eigenvalues and eigenvectors.

4.9.3.2 Singular Vector Decomposition (SVD)

Factorization methods such as QR and matrix diagonalization, as discussed in previous sections, are applicable to only limited classes of matrices with linearly dependent columns and real symmetric squares. Singular value decomposition (SVD) (Laudauer et al. 1998) breaks an $m \times n$ matrix A into its components, as shown below, and can be applied to all kinds of matrices.

$$\underbrace{\begin{pmatrix} * & * & * \\ * & * & * \\ * & * & * \\ * & * & * \\ * & * & * \\ * & * & * \end{pmatrix}}_{A} = \underbrace{\begin{pmatrix} * & * & * & * & * \\ * & * & * & * & * \\ * & * & * & * & * \\ * & * & * & * & * \\ * & * & * & * & * \\ * & * & * & * & * \end{pmatrix}}_{U} \underbrace{\begin{pmatrix} \circ & & \\ & \circ & \\ & & \circ \\ \hline & & \end{pmatrix}}_{D} \underbrace{\begin{pmatrix} * & * & * \\ * & * & * \\ * & * & * \end{pmatrix}}_{V^T}$$

$$= [u_1 u_2 \ldots u_n \ldots u_{m-1} u_m]_{m \times m} \begin{bmatrix} d_1 & 0 & \cdots & 0 & 0 \\ 0 & d_2 & \cdots & 0 & 0 \\ \vdots & \vdots & \ddots & \vdots & \vdots \\ 0 & 0 & \cdots & d_{n-1} & 0 \\ 0 & 0 & \cdots & 0 & d_n \\ 0 & 0 & \cdots & 0 & 0 \\ 0 & 0 & \cdots & 0 & 0 \end{bmatrix}_{m \times n} \begin{bmatrix} v_1^T \\ v_2^T \\ \cdots \\ v_{n-1}^T \\ v_n^T \end{bmatrix}_{n \times n} \quad (4.21)$$

here the rows of V^T are the eigenvectors of a product (symmetric) matrix $A^T A$. The elements of diagonal matrix D in the middle are the square roots of the corresponding eigenvalues of $A^T A$. Finally, the columns of the first factor matrix U are defined as follows:

$$u_i \triangleq \frac{1}{d_i} A v_i. \quad (4.22)$$

It is also important to note that the factor matrices U and V are both orthogonal.

4.9.4 Principal Component Analysis (PCA)

Principal component analysis (PCA) (Maitra and Yan 2008) is a linear dimensionality reduction technique. Linear techniques result in each of the $k \leq p$ components of the new variable being a linear combination of the original variables:

$$s_i = w_{i,1} x_1 + \cdots + w_{i,p} x_p, \quad \text{for} \quad i = 1, \ldots, k,$$

or

$$s = Wx$$

where $W_{k \times p}$ is the linear transformation weight matrix, expressing the same relationship as

$$x = As. \quad (4.23)$$

With $A_{p \times k}$, we note that the new variables s are also called hidden, or the latent variables. In terms of an $n \times p$ observation matrix X, we have

$$S_{i,j} = w_{i,1}X_{1,j} + \cdots + w_{i,p}X_{p,j}, \quad for \quad i = 1,\ldots,k, \quad and \quad j = 1,\ldots,n, \quad (4.24)$$

where j indicates the jth realization, or equivalently, $S_{k \times n} = W_{k \times p} X_{p \times n}$, $X_{p \times n} = A_{p \times n} S_{p \times n}$. Such linear techniques are simpler and easier to implement than more recent methods that consider nonlinear transforms.

A traditional multivariate statistical method (Anderson 1984), PCA is commonly used to reduce the number of predictive variables and finds linear combinations of variables, thereby summarizing the data without losing too much information. This method of dimensionality reduction is also known as parsimonious summarization of the data.

Considering a data matrix $X_{n \times p}$, with n observations as rows represented by p predictive variables, $X_1, X_2, \ldots X_p$ represent a random observation from this data matrix. The objective here is to select the subset of the above variables (columns) that holds most information for matrix X.

Let σ_{ij} denote the covariance between two observations X_i and X_j of data matrix X. The covariance between all observations of X and the resultant covariance matrix is denoted as Σ. The σ_{ij}s may be estimated by observations of standard deviation s_{ij} calculated from the data. If standard deviations are used in the matrix, then the matrix is denoted by S. The resultant Σ or S is a $p \times p$ square and symmetric matrix.

A linear combination of a set of vectors $\{X_1, X_2, \ldots X_p\}$ is the sum of the product of the vectors with scalar constants $\propto_i$ given through the following expression: $\Sigma \alpha_i X_i, i = 1 \; to \; p$. The absolute sum of the scalars in a linear combination is set to be equal to 1, i.e., $\Sigma |\alpha i| = 1$, which normalizes or standardizes the linear combination. In cases in which $\Sigma |\alpha i| = 0 \rightarrow \alpha_i = 0$, the set of vectors $\{X_1, X_2, \ldots X_p\}$ is thus said to be linearly independent. In such cases the set of vectors can be written as a linear combination of any other vectors in the set. Statistically, correlation is a measure of linear dependence among variables, and the presence of highly correlated variables indicates a linear dependence among the variables. The rank of a matrix, as discussed previously, denotes the maximum number of linearly independent rows or columns of a matrix. As our data matrix will contain many correlated variables that we seek to reduce, the rank of data matrix $X_{n \times p}$ is less than or equal to p.

4.9.4.1 Jordan Decomposition of a Matrix

Now that we have the covariance matrix $\Sigma_{p \times p}$, a square symmetric matrix representing the covariance between n input vectors, we decompose this matrix using

Jordan decomposition, a well-known spectral decomposition technique formalized as follows:

$$\sum_{p\times p} = \Gamma D \Gamma^T$$

$$= \sum \lambda_i \gamma'_{(i)} \gamma_{(i)} \qquad (4.25)$$

where $D_{p\times p}$ is a diagonal matrix and $\Gamma_{p\times p}$ is an orthonormal matrix, i.e., $\Gamma\Gamma' = I$. The diagonal elements of D are denoted by λ_i ($i = 1$ to p) and the columns of Γ are denoted by $\gamma_{(i)}$ ($i = 1$ to p). In matrix algebra, $\lambda'_i s$ represents the eigenvalues of X, and $\gamma_{(i)} s$ represents the corresponding eigenvectors.

It should be noted that if X is not a full-rank matrix, i.e., $rank(X) = r < p$, then there are only r nonzero eigenvalues in the above decomposition, with the rest of the eigenvalues being equal to zero.

4.9.4.2 Principal Components

The objective of using principal component analysis (PCA) is to obtain a suitable linear combination of the data matrix X. This objective is met using the Jordan decomposition of the covariance matrix Σ of X (or the correlation matrix S of X). Thus, a random vector in the data matrix X is represented as $x_{i\times p} = (x_1, x_2, \ldots, x_p)$ having mean $\mu_{i\times p}$ and covariance matrix Σ.

A principal component in the PCA is a transformation of the form

$$x_{i\times p} \to y_{i\times p} = (x - \mu)_{i\times p} \Gamma_{p\times p}, \qquad (4.26)$$

where Γ is obtained from the Jordan decomposition of Σ, i.e., $\Gamma^T \Sigma \Gamma = D = diag(\lambda_1, \lambda_2, \ldots, \lambda_p)$, with $\lambda'_i s$ being the eigenvalues of the decomposition.

Each element of $y_{i\times p}$ is a linear combination of the elements of $x_{i\times p}$. Also, each element of y is independent of the other elements of y.

Thus, we obtain p independent principal components corresponding to the p eigenvalues of the Jordan decomposition of Σ. Generally, we use the first few of these principal components.

4.9.5 Partial Least-Squares-Based Dimension Reduction (PLS)

Now that we have discussed PCA in detail, it is worth noting that PCA follows an unsupervised approach to determining the linear correspondence between variables. However, at times it is desirable to determine the dependence between variables by taking into consideration the target variable. Partial least squares (PLS) is one such dimensionality reduction technique that was initially proposed as a matrix decomposition technique and then was adopted as a multivariate regression algorithm. However, more recently PLS has also been found to be an effective dimension reduction technique.

The underlying assumption of PLS is that the observed data are generated by a system or process that is driven by a small number of latent (not directly observed or measured) features. Therefore, PLS aims at finding uncorrelated linear transformations (latent components) of the original predictor features, which have high covariance with the response features. Based on these latent components, PLS predicts response features y, the task of regression, and reconstructs the original matrix X, the task of data modeling, all at the same time.

Assume X is an $n \times p$ matrix and its corresponding class label Y is an $n \times 1$ matrix. The PLS technique successively extracts factors from both X and Y such that covariance between the extracted factors is maximized.

According to (Maitra and Yan 2008), PLS attempts to find a linear decomposition of X and Y such that $X = TP^T + E$ and $Y = UQ^T + F$, where

$$T_{n \times r} = X\_scores \qquad U_{n \times r} = Y\_scores$$

$$P_{p \times r} = X\_loadings \qquad Q_{1 \times r} = Y\_loadings$$

$$E_{n \times p} = X\_residuals \qquad F_{n \times 1} = Y\_residuals.$$

The decomposition is terminated when the covariance between the X_scores and Y_scores is maximized or until X is reduced to a null matrix. Generally, the PLS algorithm is an iterative algorithm used to extract the X_scores and Y_scores, where the number of extracted factors (r) depends on the rank of X and Y, respectively.

4.9.6 Factor Analysis (FA)

Like PCA, factor analysis (FA) (Fodor 2002) is also a linear method. FA assumes that the measured variables depend on some unknown, and often unmeasurable, set of common factors. The motivation for using FA is to uncover hidden relations, and thus it can be used to reduce the dimension of datasets following the factor model.

According to the k-factor model, a p-dimensional random vector $x_{p \times 1}$ with covariance matrix Σ satisfies the k-factor model if

$$x = \Lambda f + u \tag{4.27}$$

where $\Lambda_{p \times k}$ is a matrix of constants, $f_{k \times 1}$ represents random common factors, and $u_{p \times 1}$ represents specific factors, respectively. Moreover, according to the k-factor model, the factors are all uncorrelated and the common factors are normalized such that variance is equal to 1:

$$E(f) = 0, Var(f) = I$$
$$E(u) = 0, Cov(u_i, u_j) = 0 \text{ for } i \neq j \quad (4.28)$$
$$Cov(f, u) = 0.$$

Given these conditions, the covariance matrix Σ can be decomposed as

$$\Sigma = \Lambda\Lambda^T + \psi, \quad (4.29)$$

and the diagonal covariance matrix of u can be written as $Cov(u) = \psi = diag(\psi_{11}, \ldots, \psi_{pp})$, where x_i can be written as

$$x_i = \sum_{j=1}^{k} \lambda_{ij} f_j + u_i, i = 1, \ldots, p \quad (4.30)$$

Furthermore, the variance may be decomposed as

$$\sigma_{ii} = \sum_{j=1}^{k} \lambda_{ij}^2 + \psi_{ii} \quad (4.31)$$

where the first part, $h_i^2 = \sum_{j=1}^{k} \lambda_{ij}^2$, is called the communality and represents the variance of x_i common to all variables, while the second part, ψ_{ii}, is called the specific or unique variance, and it is the contribution in the variability of x_i due to u_i, not shared by the other variables.

The term λ_{ij}^2 measures the magnitude of the dependence of x_i on the common factor f_j. If several variables x_i have high loadings λ_{ij} on a given factor f_j, the implication is that those variables measure the same unobservable quantity, and are therefore redundant.

4.9.7 Independent Component Analysis (ICA)

Similar to PCA, the ICA is a higher-order method that finds linear projections of data. The components found by ICA are not necessarily orthogonal to each other; i.e., they are as nearly statistically independent as possible. Statistical independence has a much stronger correlation (Fodor 2002) that depends on higher-order statistics.

To explain the difference between correlation and independence, we define a lack of correlation among random variables, as $x = \{x_1, \ldots, x_p\}$ are uncorrelated. If for $\forall_i \neq j$, $1 \leq i, j \leq p$, we have

$$Cov(x_i, x_j) = E\{(x_i - \mu_i)(x_j - \mu_j)\} = E(x_i x_j) - E(x_i)E(x_j) = 0. \quad (4.32)$$

On the contrary, independence requires that the multivariate probability density function factorizes, and can be written as

$$f(x_1,\ldots,x_p) = f_1(x_1)\ldots f_p(x_p). \quad (4.33)$$

Typically, independence among variables always implies no correlation, but not vice versa; only if the distribution $f(x_1,\ldots,x_p)$ is multivariate normal are the two equivalent. For Gaussian distributions, the PCs are independent components.

The objective of the ICA model for the p-dimensional random vector x is to estimate the components of the k-dimensional vector s and the full-rank matrix $A_{p \times k}$:

$$x = As \quad (4.34)$$

$$(x_1,\ldots,x_p)^T = A_{p \times k}(s_1,\ldots,s_k)^T$$

such that the components of s are as independent as possible, based on the definition of independence above.

Noisy ICA, an extension to the typical ICA, contains an additive random noise component u as below, where its estimation is still an open research challenge, as explained in the following equation:

$$(x_1,\ldots,x_p)^T = A_{p \times k}(s_1,\ldots,s_k)^T + (u_1,\ldots,u_p)^T. \quad (4.35)$$

From the above discussion, it is clear that the objective of ICA is not dimensionality reduction. Thus, to reduce the number of dimensions using ICA, one has to resort to using PCA to find $k < p$, and then use ICA to estimate the independence of the selected features. It should also be noted that there is no specific ordering of independent components as in the case of PCA. To order the components once they are estimated, one can use the norm of the columns or some non-Gaussian measure.

We reiterate that PCA is aimed at finding uncorrelated variables, while ICA is aimed at finding independent variables. ICA finds its applications in various fields, including, but not limited to, exploratory data analysis, blind source separation, natural image processing, and feature extraction. We further emphasize that in the context of feature extraction, the columns of matrix A represent features in the data, and the components s_i give the coefficient of the ith feature in the data.

4.9.8 Multidimensional Scaling (MDS)

Unlike PCA and ICA used to obtain the linear projections of data, the main objective of MDS is to represent the dissimilarities between pairs of objects as distances between points in a low-dimensional space (Groenen and van de Velden 2004). Given the data matrix $X_{n \times p}$, consisting of n instances, each instance is defined by p distinct features (dimensions). We defined the dissimilarity between a pair of instances i and j of X as δ_{ij}. The dissimilarity between instances of X is measured using the Euclidean distance and is defined as

$$d_{ij}(X) = \left(\sum_{s=1}^{p} (x_{is} - x_{js})^2 \right)^{1/2} \tag{4.36}$$

In Equation 4.36, d_{ij} is the shortest line joining instances i and j.

As mentioned above, the objective of MDS is to find a matrix $\hat{X}$ of the lower dimension, as compared to X, such that $d_{ij}(\hat{X})$ matches δ_{ij} as closely as possible. Various methods can achieve this objective. Users should, however, refrain from using the definition of raw stress $\sigma^2(X)$ (J. B. Kruskal 1964a, 1964b), as shown below:

$$\sigma^2(X) = \sum_{i=2}^{n} \sum_{j=1}^{i-1} w_{ij} (\delta_{ij} - d_{ij}(\hat{X}))^2. \tag{4.37}$$

The method above is better known as the least-squares MDS model. As the dissimilarities between instances are symmetric, the summation only involves the pairs i and j, where $i > j$. w_{ij} is a user-defined weight that must be nonnegative.

The objective of MSD is to minimize the stress function $\sigma^2(X)$, which is rather complex to solve using closed systems. To this end, MDS algorithms employ various iterative techniques to find a matrix $\hat{X}$ for which $\sigma^2(X)$ is minimum.

As Euclidean distance is not susceptive to change in rotation, translation, and reflection, orations on the matrix d_{ij} may be applied freely without altering the raw stress $\sigma^2(X)$. Thus, many of the MDS algorithms exploit this property so that the dimensions coordinate to zero, and the solution is oriented on the principal axis. That is, the axes are rotated in such a way that the variance of X is maximal along the first dimension, the second dimension is uncorrelated to the first and has maximal variance as well, and so on.

4.10 Conclusion

In conclusion, Chapter 4 provides the description of various data preparation and data transformation techniques. Aptly titled "Feature Selection and Extraction Strategies in Data Mining," the chapter provides the application of these techniques to bioinformatics data. The reader should familiarize himself or herself with the workings of these techniques, as they lay the foundation for data mining and knowledge discovery techniques described in future chapters.

References

Agarwal, A., J.M. Phillips, and S. Venkatasubramanian. Universal multi-dimensional scaling. In *KDD'10*. Washington, DC: ACM, 2010, pp. 1–10.

Almuallin, H., and T.G. Dietterich. Efficient algorithms for identifying relevant features. In *Proceedings of the Ninth Canadian Conference on Artificial Intelligence*. Morgan Kaufmann, 1992, pp. 38–45.

Anderson, T.W. *An introduction to multivariate statistical analysis*. New York: John Wiley & Sons, 1984.

Berry, M.W., S.T. Dumais, and G.W. O'Brien. Using linear algebra for intelligent information retrieval. *SIAM Rev* 37 (1995): 573–595.

Boulle, M. Khiops: A statistical discretization method of continuous attributes. *Machine Learn* 55, no. 1 (2004): 53–69.

Chen, S., S.A. Billings, and W. Luo. Orthogonal least squares methods and their applications to non-linear system identification. *Int J Control* 50, no. 5 (1989): 1873–1896.

Das, S. Filters, wrappers and a boosting-based hybrid for feature selection. In *Proceedings of 18th International Conference of Machine Learning*, 2001, pp. 74–81.

Fodor, I.K. A survey of dimension reduction techniques. Technical paper. Livermore: Lawrence Lovermore National Laboratory, 2002.

Furey, T.S., N. Cristianini, N. Duffy, D.W. Bednarski, M. Schummer, and D. Haussler. Support vector machine classification and validation of cancer tissue samples using microarray expression data. *Bioinformatics* 16, no. 10 (2000): 906–914.

Groenen, P.J.F., and M. van de Velden. *Multidimensional scaling*. Econometric Institute Report EI 2004-15, 2004.

Guyon, I. A practical guide to model selection. In *Machine learning summer school*, J. Marie, 1–37. Springer, to appear.

Guyon, I., and A. Elisseeff. An introduction to variable and feature selection. *J Machine Learn Res* 3 (2003): 1157–1182.

Guyon, I., S. Gunn, M. Nikravesh, and L.A. Zadeh. *Feature Extraction: Foundations and Applications (Studies in Fuzziness and Soft Computing)*. Secaucus, NJ: Springer-Verlag New York, Inc., 2006.

Han, J., and M. Kamber. *Data mining: Concepts and techniques*. 2nd ed., Morgan Kaufmann Series in Data Management Systems. San francisco, CA: Morgan Kaufmann, 2006.

Kira, K., and L.A. Rendell. The feature selection problem: Traditional methods and a new algorithm. *Proc AAAI-92*, San Jose, (1992), 129–134.

Kohavi, R., and G.H. John. Wrappers for feature subset selection. *Artif Intell* 97, no. 1–2 (1997): 273–324.

Kruskal, J.B. Multidimensional scaling by optimizing goodness of fit to a nonmetric hypothesis. *Psychometrika* 29 (1964a): 1–27.

Kruskal, J.B. Nonmetric multidimensional scaling: A numerical method. *Psychometrika* 29 (1964b): 115–129.

Landauer, T.K., P.W. Foltz, and D. Laham. Introduction to latent semantic analysis. *Discourse Proc* 25 (1998): 259–284.

Maitra, S., and J. Yan. Principle component analysis and partial least squares: Two dimension reduction techniques for regression. 2008 Discussion Paper Program. Arlington, VA: Casualty Actuarial Society, 2008, pp. 79–90.

Oh, S.M. Note on matrix factorization. Georgia Institute of Technology, January 29, 2006.

Pudil, P., J. Novovicova, and J. Kittler. Floating search methods in feature selection. *Lett Pattern Recog* 15, no. 11 (1994): 1119–1125.

Saeys, Y., I. Inza, and R. Larranaga. A review of feature selection techniques in bioinformatics. *Bioinformatics* 23, no. 19 (2007): 2507–2517.

Stoppiglia, H., G. Dreyfus, R. Duboism, and Y. Oussar. Ranking a random feature for variable and feature selection. *J Machine Learn Res* 3 (2003): 1399–1414.

Yijun, S., and W. Dageng. A RELIEF based feature extraction algorithm. In *Proceedings of the 2008 SIAM International Conference on Data Mining*. Atlanta: SIAM, 2008, pp. 188–195.

Zhang, S., C. Zhang, and Q. Yang. Data preparation for data mining. *Appl Artif Intell* 17, no. 5–6 (2003) 375–381.

Chapter 5

Feature Interpretation for Biological Learning

Feature selection techniques have become an integral part of many bioinformatics applications and have thus added to the collection of existing well-known techniques discussed in Chapter 4. This chapter provides an overview of the application of the various feature selection and feature extraction techniques commonly used in bioinformatics. The key areas touched upon describe the issues and challenges faced during the analysis of high-dimensional data, whether gene expression data, protein sequence, or structural data.

5.1 Introduction

The objectives of using feature selection and extraction are manifold. The most important of these objectives are:

1. To avoid overfitting and improve the model performance, i.e., prediction performance in the case of supervised classification and better cluster detection in the case of unsupervised clustering
2. To provide faster and more effective computational models
3. To gain a deeper insight into the underlying process that generated the data

Apart from the above-mentioned benefits of using feature selection techniques in bioinformatics, feature selection and extraction techniques can be used to find an optimal set of features that performs best with the chosen learning technique.

As described in Chapter 4, feature selection strategies are subdivided into filter-based and wrapper-based approaches. The focus of this chapter is to enable the reader to understand how different data preprocessing techniques are applied to address the challenges of bioinformatics data (Kuonen 2003).

5.2 Normalization Techniques for Gene Expression Analysis

High-throughput real-time quantitative reverse transcriptase polymerase chain reaction (qPCR) is widely considered to be the gold standard for the analysis of micro-RNA (miRNA) expression (Ach et al. 2008). qPCR is useful for acquiring and profiling (50 to a few thousand) expression patterns on a microarray. Because of the large number of genes available, qPCR is considered to be highly susceptive to noise. In this section we therefore focus on the use of appropriate normalization techniques for qPCR expression data.

Nearly all normalization techniques are based on the assumption that one or more control genes are constitutively expressed at near-constant levels under all experimental conditions. The most widely used control genes are those selected from among an assumed set of housekeeping genes. Housekeeping genes are those that are constantly expressed through different samples so as to maintain basic cellular function. The expression levels of remaining target genes in a sample are adjusted with respect to the selected control genes. In most qPCR experiments, a single housekeeping gene is chosen and added to the collection of experimental target genes to be assayed for each sample. The control gene is then compared between samples and a sample-specific scaling factor is calculated to equalize their expression (Robinson and Oshlack 2010). This sample-specific scaling factor is applied to all genes in the sample. However, this approach has numerous limitations. The primary limitation is that many of the experimental conditions may alter the expression of the control genes. Moreover, evidence postulates that housekeeping genes may not always be expressed constantly across all samples. Therefore, more sophisticated normalization techniques are needed. These techniques use multiple housekeeping genes where each of their expressions is combined to represent a virtual housekeeping gene. It is believed that this approach is more robust than a single-control gene approach. However, it is important to note that this virtual housekeeping gene is also confounded by the same assumption that its expression does not vary across samples.

5.2.1 Normalization and Standardization Techniques

Microarray technology provides researchers with the ability to measure the expressions of thousands of genes for a given sample. Biologically relevant expression patterns between these genes are identified by comparing the expression levels of genes between samples of different states on a one-on-one (gene-by-gene) basis.

The selection of a significant set of differentially expressed genes between samples obtained from different states is sensitive to errors brought about by measurement of intensity values of genes. It is therefore important to eliminate questionable or low-quality measurements of intensity values of genes through appropriate transformations on the data.

To facilitate this process, it is imperative to understand how the microarray is generated. Typically, RNA is first isolated from different tissues, developmental stages, disease states, or samples that have been subjected to appropriate treatments. The RNA is then labeled and hybridized to arrays. After hybridization, the arrays are measured. These measurements enable the conversion from raw data to processed data through the implementation of three steps.

First, the arrays are scanned to create grayscale images (Chen et al. 1997). Once the images are generated, then image analysis is performed to identify the arrayed spots and to measure their corresponding relative fluorescence intensities. The quantification of florescence intensities is brought about using quantification matrices based on image analysis. Several commercial and freely available software packages generate high-quality, reproducible measures of hybridization intensities (see Figure 5.1).

Second, these images are subjected to various image preprocessing techniques, such as mean, median, or average difference operations, to provide the required background correction, thereby reducing possible equipment errors. Finally, the obtained quantified data from the images are consolidated into a matrix of expression values that are then subjected to normalization.

Several normalization strategies are used to normalize microarray data. These normalization strategies are categorized into global normalization strategies and intensity-dependent normalization strategies (Park et al. 2003).

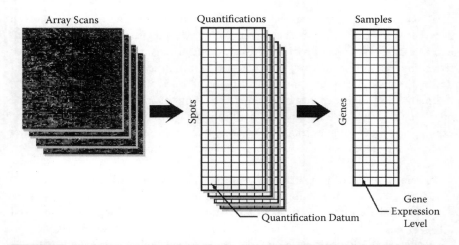

Figure 5.1   Microarray gene expression data processing. (From Brazma, A., et al., *Nature Genet* 29 (2001): 365–371.)

5.2.1.1 Expression Ratios

Every gene within a sample in a microarray is represented by a ratio (T) of the intensities of the colors R (red) and G (green). Therefore, the ratio T_i of the *ith* gene in a sample is represented as follows:

$$T_i = \frac{R_i}{G_i}. \tag{5.1}$$

Though the ratio T provides a measure of expression change of a gene with respect to its R and G intensities, it is ineffective in capturing if the gene is upregulated or downregulated. For instance, if a gene is upregulated (R) by a factor of 2, its resultant expression ratio will have a value of 2, whereas if the gene is downregulated (G) by a factor of 2, the resultant expression ratio will have the value of −0.5.

To overcome this drawback, the most widely used alternative transformation of the ratio of the logarithm base 2 is used. The logarithm base 2 has the advantage of producing a continuous spectrum of values and treating up- and downregulated genes.

We know that logarithms treat numbers and their reciprocals symmetrically: $log_2(1)=0$, $log_2(2)=1$, $log_2\left(\frac{1}{2}\right)=-1$. The logarithms of the expression ratios are also treated symmetrically, and those that are expressed at a constant level have a $log_2(T)$ equal to zero.

5.2.1.2 Intensity-Based Normalization

Analysis involving gene expression data is sensitive to the changes in fluorescent dyes between samples of the analysis. Moreover, measurements from different hybridizations may occupy different scales, and to ensure meaningful and effective comparisons of thousands of genes, it is common to adjust the expression values of the genes between samples using normalization (Kreil and Russel 2005). Normalization strategies that focus on normalizing the intensity values of the genes on a single slide are referred to as within-slide normalization. However, before explaining within-slide normalization of gene expression, we denote the common assumptions made about the samples and genes of microarray data.

1. **The number of genes in each sample is the same:** All the samples in the study have the same number of genes.
2. **There are equal quantities of RNA for the two samples being compared:** Given millions of individual genes in each sample, we assume that, on average, equal quantities of RNA (the mass of each molecule) use approximately the same quantities of RNA for each gene.

3. **Arrayed elements represent a uniform random sampling of genes across samples:** Nearly all normalization strategies are based on the assumption that one or more genes are expressed at near-constant levels under all experimental conditions, and the expression levels of all genes in a sample are adjusted to satisfy that assumption.

5.2.1.3 Total Intensity Normalization

Considering the above assumptions, we ensure that approximately an equal number of genes from each sample are hybridized. Therefore, the total hybridization intensities summed over all elements in the array should be the same for each sample.

For total intensity normalization (Quackenbush 2002), we compute a normalization factor by summing the measured intensities (both R and G) as follows:

$$N_{total} = \frac{\sum_{i=1}^{N_{array}} R_i}{\sum_{i=1}^{N_{array}} G_i}, \tag{5.2}$$

where G_i and R_i are the measured intensities for the *ith* array element and N_{array} is the number of genes represented in the microarray. Once the total intensity N_{total} is computed, we use this value to normalize the expression ratio of a gene as follows:

$$T_i = \frac{R_i}{G_i} = \frac{1}{N_{total}} \frac{R_i}{G_i}. \tag{5.3}$$

This total intensity normalization in effect adjusts T_i such that the mean is equal to 1, rendering the mean $log_2(ratio)$ equal to 0. Thus, the various normalization strategies in microarray data analysis, including scaling the individual intensities, are aimed at rendering the mean or median intensities uniform within a single array or across all arrays. These strategies include linear regression analysis, log centering, ranking invariant methods, and Chen's ratio statistics (Quackenbush 2002).

5.2.1.3.1 Global Normalization (LOWESS)

Microarray data are plagued by inconsistencies in the way data are recorded. These inconsistencies are manifestations of noise in the form of inconsistencies in intensity values of the spots on the microarray chip. Though there are several methods of intensity normalization, these methods do not take into consideration systemic biases that are inherent in the data (Yang et al. 2002). These systemic biases include the $log_2(ratio)$ values that represent low-intensity spots on the microarray that appear as a minor deviation from zero. Researchers use the $log_{10}(R*G)$ by

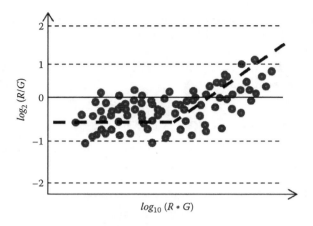

Figure 5.2 The distribution of genes using the R-I plot.

$log_2(R/G)$ plot, better known as the ratio-intensity (R-I) plot, to visualize the intensity-dependent effects of the genes in a microarray.

As shown in the Figure 5.2, the R-I plot is a plot of the $log_2(ratio)$ on the y-axis to the $log_{10}(intensity)$ (i.e., the product of intensities R and G) on the x-axis for each gene in the microarray. This R-I plot can reveal intensity-specific artifacts in the $log_2(ratio)$ measurements.

Locally weighted linear regression (LOWESS) analysis has been used as a normalization approach to remove intensity-dependent effects in the $log_2(ratio)$ values (Hijum et al. 2008). Using the R-I plot, LOWESS detects systemic deviations for each point in the R-I plot. Using a local weighted linear regression function of $log_{10}(intensity)$, the correction of intensity values is carried out by subtracting the calculated best-fit average $log_2(ratio)$ from the experimentally observed ratio for each data point.

By performing this function, the LOWESS deemphasizes the contribution of genes that are far (on the R-I plot) from densely populated data clusters. We illustrate the process as follows:

Let $x_i = log_{10}(R_i \times G_i)$ and $y_i = log_2(R_i/G_i)$. We then use LOWESS to create a function $y(x_k)$ that estimates the dependence of the $log_2(ratio)$ on the $log_{10}(intensity)$. Using this functional estimate, each $log_2(ratio)$ of every point in the R-I plot is subject to the following correction:

$$log_2(T_i') = log_2(T_i) - y(x_i) = log_2(T_i) - log_2(2^{y(x_i)}) \qquad (5.4)$$

or equivalently,

$$log_2(T_i) = log_2\left(T_i \times \frac{1}{2^{y(x_i)}}\right) = log_2\left(\frac{R_i}{G_i} \times \frac{1}{2^{y(x_i)}}\right). \qquad (5.5)$$

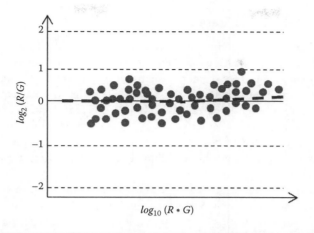

Figure 5.3 The normalized representation of the spots, where the new mean is normalized to zero.

The application of the LOWESS correction normalizes the intensity of values of each gene closer to the new mean set at $log_2(R/G)$ values at zero, as reflected in Figure 5.3.

Similar to the R-I plot, we have the M-A plot, which is used to identify spot artifacts and detect intensity-based patterns. The M-A plot, first proposed by Dudoit et al. (2002), is a plot of intensity values of independent spots with the x-axis representing M, the log ratio, on the overall log intensity, and the y-axis representing A, where $M = log_2(R/G)$ and $A = log_2\sqrt{(R \times G)}$, as shown below in Figure 5.4.

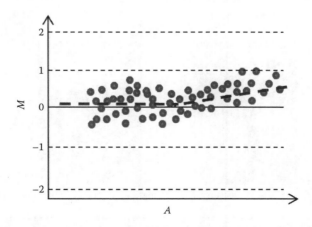

Figure 5.4 The distribution of genes using an M-A plot.

5.2.1.3.2 Local Normalization

The above-described normalization strategy can be applied either globally or locally. By local normalization, we refer to the application of a normalization strategy to a subset of array elements deposited by a single spotting pen. Local normalization proves advantageous, as it can aid in the simultaneous correction of spatial variations in a microarray chip, for example, local differences in hybridization conditions across the microarray. As in the case of global normalization, the satisfaction of all assumptions must be satisfied. For example, a sufficiently large number of elements should be included in each pen group for the approach to be validated.

Local normalization strategies take into consideration subsets of array elements; normalization is then performed on independent subsets. During this stage, we encounter variations in the $log_2(ratio)$ measurements across the different subsets.

As all normalization strategies are aimed at establishing a uniform global mean across all elements of the array, it is imperative that local normalization strategies take the variance between the subsets of array elements. Numerous computational approaches address this challenge of regulating variance between subsets of array elements (Workman et al. 2002; Papana and Ishwaran 2006).

Variance regularization is accomplished by adjusting the $log_2(ratio)$ measures of each subset such that the global variance is the same throughout the array (Quackenbush 2002).

Let us consider a single microarray that is divided into distinct subgrids (subsets). Figure 5.5 provides a schematic representation. Let each subgrid be normalized independent of each other (i.e., local normalization is performed). Our objective is therefore to determine a factor that can be used to scale the measurements within each subgrid. A commonly used scaling factor is the geometric mean of the independent variances of all subgrids.

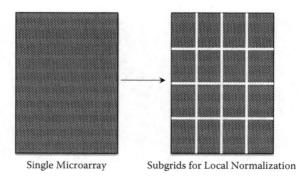

Single Microarray   Subgrids for Local Normalization

Figure 5.5 The division of a single microarray into subgrids, to carry out local normalization. Here, normalization is carried out on each grid independently.

If we assume that each subgrid has M elements, because we have already adjusted the mean of the $log_2(ratio)$ values in each subgrid to be zero, each subgrid variance in the nth subgrid is

$$\sigma_n^2 = \sum_{j=1}^{M} \left[log_2(T_j) \right]^2 \qquad (5.6)$$

where the summation runs over all the elements in that subgrid.

If the number of subgrids in the array is N_{grids}, then the appropriate scaling factor for the elements of the kth subgrid on the array is

$$a_k = \frac{\sigma_k^2}{\left[\prod_{n=1}^{N_{grids}} \sigma_n^2 \right]^{1/N_{grids}}}. \qquad (5.7)$$

We then scale all of the elements within the kth subgrid by dividing by the same value a_k computed for that subgrid,

$$log_2(T_i) = \frac{log_2(T_i)}{a_k}. \qquad (5.8)$$

This step is equivalent to taking the a_k^{th} root of the individual intensities in the kth subgrid,

$$G_i' = [G_i]^{1/a_k} \quad and \quad R_i' = [R_i]^{1/a_k}. \qquad (5.9)$$

It should be noted that other variance regularization factors have been suggested, and a similar process can be used to regularize variances between normalized arrays.

5.2.1.4 Intensity-Based Filtering of Array Elements

Due to the large number of array elements on a microarray, it is often required that array elements be removed if their measured intensity is indistinguishable from background noise (Jenssen et al. 2002). This method is commonly referred to as filtering.

On close examination of the representative R-I plots, it is observed that as the variability in the measure $log_2(ratio)$ values increases, the corresponding intensity

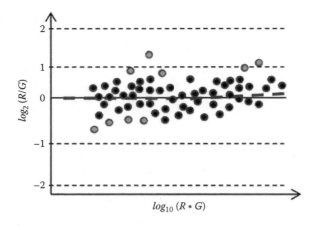

Figure 5.6 R-I plot that represents the elements that could be filtered out as outliers.

values decrease (Chen et al. 2005) (Figure 5.6). This variability in the measured $log_2(ratio)$ is attributed to the relative error in measurement increases when the intensities are low (i.e., when the intensity of a spot on the microarray matches the intensity of the background).

It is therefore a common practice to use only array elements with intensities that are statistically significantly different from the background. Thus, as a simple filtering strategy, we first measure both the average local background near each array element and its corresponding standard deviation. As a rule of thumb, it is believed that the elements with respectably good intensity values fall in the 95.5% confidence range and have intensities of more than two standard deviations above the background (refer to Figure 5.7 for more information). By following this rule, we ensure that we increase the reliability of the measurements. This measurement is represented using the following relation:

$$G_i^{spot} > 2 \times \sigma\left(G_i^{background}\right) \quad and \quad R_i^{spot} > 2 \times \sigma\left(R_i^{background}\right). \quad (5.10)$$

Similar approaches to filter out elements of low-intensity values include absolute lower thresholding for acceptable array elements (also referred to as floors) and percentage-based cutoffs in which a fixed fraction of elements is discarded.

This strategy can also be applied to filter out elements that have a very high saturation of fluorescence intensity. Typically, when elements have reached their highest intensity, the comparisons are no longer meaningful. In such situations, it is viable to filter out those elements using the similar approach by setting a maximum acceptable value (also referred to as a ceiling).

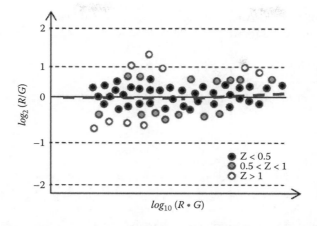

Figure 5.7 The variation of intensity and the differentiation brought about using z-score normalization.

5.2.2 Identification of Differentially Expressed Genes

One of the primary applications of microarray technology is to analyze genes from different samples and identify differentially expressed genes between samples. Considering the sheer amounts of data, data mining analysis has played a pivotal role in the endeavor of identifying differentially expressed genes over the past decades. Several clustering approaches (as described in Chapters 6 and 7) have achieved considerable success (Qin et al. 2008; Zhu et al. 2008). The objective of using these clustering techniques has been motivated by the hypothesis of reducing the number of genes to those that are variably expressed across samples.

Based on the above object of identifying differentially expressed genes, researchers traditionally rely on filtering genes using a statistical derived fixed-fold-change cutoff on expression values. In general the default number of folds is set to two, as this is where genes that satisfy this fold change are believed to be the most significant.

Similarly, the global filtering approach computes the mean and standard deviation of the distribution $log_2(ratio)$ of all microarray values. Using the computed mean and standard deviation, the global fold-change difference and confidence are computed and used to filter out genes that are not differentially expressed. This global filter is equivalent to using a z-score for the dataset. However, this approach may be inaccurate in capturing the inherent spatial differences in microarray data. Specifically, in low intensities, where the data vary more, the technique runs the risk of wrongly identifying genes as differentially expressed and vice versa.

Localized approaches take into consideration the local structure of the dataset to identify differential expressed genes. They use a sliding window and calculate the mean and standard deviation of data points within a window to define an intensity-dependent z-score threshold and identify differential expression. In this step, z measures the number of standard deviations a particular data point is from the mean.

The following relation is used to calculate a localized standard deviation $\sigma^{local}_{log_2(T_i)}$ of the $log_2(ratio)$ of a region in the R-I plot.

Thus, the normalized value of a particular array element i is

$$Z_i^{local} = \frac{log_2(T_i)}{\sigma^{local}_{log_2(T_i)}}. \tag{5.11}$$

It is believed that all differentially expressed genes fall in the 95% confidence level and would be within a value of $|Z_i^{local}| > 1.96$. This approach enables the discretization of the elements of a microarray for the identification of differentially expressed genes that are naturally more variable.

For a more refined discretization process, the quantile normalization algorithm (Mar et al. 2009) can be used. This quantile normalization approach makes the distribution of elements of each sample the same across many arbitrary samples. Each quantile of intensities is projected to lie along a unit diagonal in the M-A plot.

The following procedure generates the quantile normalization:

1. Let $X(i,k)$ be the gene expression intensity of the ith gene and the kth sample.
2. Each sample set of intensities $X(.,k)$ is first sorted by a permutation π_k according to intensity values. This permutation is then sorted, and the resultant sorted sample is represented as $X'(.,k)$.
3. The intensity value $X'(i,k)$ is then substituted by the mean across all samples $mean(X'(i,.))$.
4. The inverse permutation $inv(\pi_k)$ is then applied to each sample set to produce the normalized set of gene expression intensities.

5.2.3 Selection Bias of Gene Expression Data

In gene expression analysis, we face the problem of constructing an accurate prediction rule R using a dataset consisting of a relatively small number of microarray samples, with each sample containing the expression data of many (possibly thousands of) genes (Ambroise and McLachlan 2002). For data miners, this large sample reinstates the challenges that the small n large P problem poses on classification and prediction.

Traditional statistical approaches for prediction, such as standard discriminant analysis, are used to determine an optimal prediction rule R. However, these statistical approaches work well when the number of training observations n is much larger than the number of feature variables p (i.e., large n small p). In the context of microarray data, the number of tissue samples n is far lower than the number of genes p. This small n large p situation presents a number of problems.

First, it may not be possible to form the prediction rule R *by using all* p *available genes.* In the case of Fisher's linear discriminant function, the pooled within-class sample covariance matrix would be singular when $n \ll p$.

Second, the discriminatory power of the rule R *would be negligible.* Let us consider a situation in which we use all the genes to create a prediction model using a support vector machine (SVM). As previously discussed, not all the genes have the discriminatory potential to aid in classification. In fact, using all the genes allows the noise associated with genes of little or no discriminatory power to inhibit and degrade the performance of the rule R in its application.

This problem increases the generalization error of R when a sufficiently large number of genes are used. Therefore, researchers rely on feature selection to reduce the number of genes to be used in constructing the rule R.

Several approaches have been proposed to feature subset selection (Díaz-Uriarte and Alvarez de Andrés 2006). These approaches use either wrapper or filter techniques of feature selection to search for an optimal or near-optimal subset of features that can be used to generate the most discriminatory rule R.

As discussed in Chapter 4, feature subset selection can be classified into two categories based on the use of a learning algorithm used to construct the prediction rule. If a subset of features is chosen independently of a learning algorithm, the method is said to follow a filter approach, and if the feature subset selection depends on a learning algorithm, the method is said to follow a wrapper approach.

Regardless of how the performance of the rule is assessed during the feature selection process, it is common to assess the performance of the rule R for a selected subset of genes by its leave-one-out cross-validation (CV) error. However, if R is calculated within the feature selection process, then there will be a selection bias in it when it is used as an estimate of the prediction error. Cross-validation should be undertaken subsequently to the feature selection process to correct for this selection bias. Alternatively, the bootstrap can be used.

5.3 Data Preprocessing of Mass Spectrometry Data

In this section we focus on the data transformation strategies used in the data-rich field of mass spectrometry (MS). MS is a prominent technique that biologists use for studying the role of various proteins in a biological sample (Veltri 2008). MS consists of generating a signal (spectrum) of values that represent the presence of a protein measured by the mass-to-charge ratio (*m/z*) and abundance (intensity) in the sample.

Myriad tools exist to analyze a sample and generate its corresponding MS. The analysis of samples using MS is, however, challenging, as each spectrum potentially occupies a gigabyte of memory. Apart from the sheer volume of data generated from these experiments, the analysis of MS signals is affected by errors introduced during sample curation that take the form of noise in the MS. The different manifestations of noise are attributed to peak broadening, instrument distortion and saturation, miscalibration, and contaminants in the samples. Considering these two challenges, it is not surprising to see the importance of data preprocessing and data transformation techniques before the analysis of the MS from different samples.

In the following sections, we would highlight the predominantly used data preprocessing schemes applied to MS data (datasets). These include various binning, alignment, and baseline subtraction techniques that are used to improve the quality of raw MS data prior to their analysis.

5.3.1 Data Transformation Techniques

The effect of data analysis on MS data across multiple samples entails the use of data preprocessing. The objective of using data preprocessing on MS data is to (1) reduce the spectral noise that manifests itself in a single sample and (2) reduce the number of dimensions across multiple samples. Therefore the data preprocessing strategies used in MS data preprocessing focus on correcting the intensity and *m/z* values in order to reduce noise, reduce the amount of data, and enable effective comparison of spectra across different samples. Noise in MS takes the form of variations along the *m/z* axis across different fractions of the spectra. This very nature of the noise requires specialized normalization strategies that can be applied on MS data. There are different noise reduction and normalization strategies that are used on MS data. The following sections describe the data preprocessing steps applied to MS data.

5.3.1.1 Baseline Subtraction (Smoothing)

Baseline subtraction or smoothing is the first step of data preprocessing applied to MS data. The objective of applying baseline subtraction is to remove systematic artifacts that are caused by clusters of ionized matrix molecules that hit the detector at the early portions of the experiment. Baseline subtraction entails the use of an iterative algorithm to remove the baseline slope and offset from a spectrum by iteratively calculating the best-fit straight line through a set of estimated baseline points. The baseline points are determined by fitting the line through the spectrum and then discarding all data points with intensities above a threshold from the fitted line. The number of points above and below the line is then counted. If there are fewer points above the line than below, they are considered peaks and discarded. Then, a new line is fit through the remaining data points. This process is repeated until the number of points above the line are less than or equal to those below the line. This final line is subtracted from the spectrum to get the baseline-corrected spectrum.

5.3.1.2 Normalization

The next step of data preprocessing of MS data is normalization. The objective of performing normalization is to correct systematic differences in the total amount of protein desorbed and ionized from the sample plate. Furthermore, normalization is done to make the data independent of experimental variations. Thus, normalization facilitates the comparison of different samples since the absolute peak values of different fractions of the spectrum may be incomparable. Spectrum normalization identifies and removes sources of systematic variation between spectra due to, for instance, varying numbers of samples or variation within instrument detector sensitivity. There exist different normalization techniques as suggested by Bachmayer (2007), of which the following intensity-based normalization techniques are prominently used.

Direct normalization: This normalization technique is similar to the min-max normalization technique suggested in Chapter 4. It is formulated as follows: the direct normalization primarily rescales the intensity values (I) of every expression value in a sample based on its corresponding minimum (I_{min}) and maximum (I_{max}) values.

$$I_{jnorm} = \frac{I_j - I_{min}}{I_{max} - I_{min}} \quad (5.12)$$

Inverse normalization: Similar to the direct normalization, the inverse normalization considers the inverse of the rescaled intensity value by subtracting the direct normalization from 1. This is represented as follows:

$$I_{jnorm} = 1 - \frac{I_j - I_{min}}{I_{max} - I_{min}}. \quad (5.13)$$

Canonical normalization: This normalization strategy is the simplest form of normalization, where the intensity values are rescaled by the sum of the intensity values in a sample. Therefore canonical normalization ensures that the rescaled values are relative to all the intensity values in that sample.

$$I_{jnorm} = \frac{I_j}{\sum I_j} \quad (5.14)$$

Other normalization strategies commonly used include the logarithmic normalization, to transform the values if the distribution of intensity values in a sample is skewed.

5.3.1.3 Binning

The next data preprocessing step of MS data is binning. The objective of performing binning on MS data is to bring about a reduction in the volume or dimensions inherent in the data. Dimensionality reduction is performed by grouping measured data into bins. During this process adjacent values are grouped, and a representative member is elected for each group. The binning algorithm takes a subset of N peaks from a spectra, represented by the couples $[(I_1, m/z_1), (I_2, m/z_2), \ldots, (I_N, m/z_N)]$, and substitutes all of them with a unique peak $(I, m/z)$. The unique peak has an intensity I, which is an aggregate function of the N original intensities (e.g., their sum), and the mass m/z is usually chosen among the original mass values (e.g., the median value or the value corresponding to the maximum intensity). Such a basic operation is conducted by scanning all spectrums using a sliding window.

5.3.1.4 Peak Detection

With the completion of normalization and binning, researchers rely on the detection of peaks to compare samples. Peak detection is therefore one of the most important steps in MS analysis (Barla et al. 2008). The methods used for peak detection focus on identifying those peaks that are clearly detectable in a sample (spectra). It is believed that these clearly detectable peaks correspond to those peptides/proteins in the sample that have the most discriminatory potential to distinguish between samples. In order to achieve the effective identification of these peaks, the method should account for the variation in the m/z location and heights of the same peak across different samples; i.e., it should be able to be stable enough to manage systemic variations brought about by the instrumentation used.

However, the detection of peaks is not trivial, as the result is greatly affected by severe spectrum variations (Zhang et al. 2009). Most of the techniques of peak detection rely on peak alignment to identify and quantify the discriminatory power of all peaks across samples.

5.3.1.5 Peak Alignment

Peak alignment focuses on aligning corresponding peaks across samples. Without alignment, the same peak (e.g., the same peptide) can have different values of m/z across samples. To allow an easy and effective comparison of different spectra, peak alignment methods find a common set of peak locations in a set of spectra, in such a way that all spectra have common m/z values for the same biological entities.

Several methods have been proposed for peak alignment, as in the case by Tibshirani et al. (2004), who effectively used complete linkage hierarchical clustering to align peaks across samples. They effectively used a distance function along the log m/z axis to retain location information of the peaks across the samples. The idea is that tight clusters should represent the same biological peak that has been horizontally shifted in different spectra. The centroid (mean position) of each cluster will therefore represent the consensus position for that peak across all spectra.

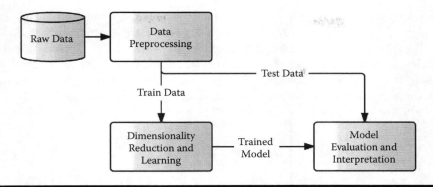

Figure 5.8 A schematic representation of the knowledge discovery (KD) process in analyzing MS data.

5.3.2 Application of Dimensionality Reduction Techniques for MS Data Analysis

In the previous section we provide an overview of the various data preprocessing strategies used on MS data for effective peak alignment. In this section we provide an overview of the use of the knowledge discovery (KD) process for the analysis MS data, as seen in Figure 5.8 (Hilario and Kalousis 2008).

Dimensionality reduction strategies, both feature extraction and feature selection, have played a key role in the analysis of MS data. Driven by the objective of identifying the most discriminatory peaks across multiple samples (spectra), there are several techniques presented in this area of bioinformatics. For instance, the feature extraction technique principal component analysis (PCA) is the most commonly used method on MS data. PCA aims to find the best linear transformation that captures the variance in the data (as described in Chapter 4) (Bair et al. 2006). Other popular feature extraction schemes include the Fourier and wavelet transformations that depict a signal as a linear combination of prespecified basis functions like the Debauches wavelet functions (Qu et al. 2003).

As shown in Figure 5.8, another important component of the KD process to analyzing MS data is the supervised classification techniques. The preprocessed data are split into train and test sets. The train set is then subject to both dimensionality reduction and learning. Typically the dimensionality reduction is kept independent of the class labels of the train set. One of the drawbacks of having the dimensionality reduction independent of class labels is that they fail to exploit the information provided by class labels. As a result, the transformations these techniques generate may not reflect the underlying class structure, as the maximum variance directions do not guarantee maximum discrimination.

On the contrary, techniques that take into consideration the class labels to reduce the dimensions are known as supervised feature extraction schemes. The most popularly used supervised feature extraction scheme is Fisher's linear discriminant analysis (LDA) (Lilien et al. 2003). Though viewed as a classification method, LDA projects the initial data onto a $k - 1$ linear subspace, where k represents the number of classes. Like PCA, LDA effects a linear transformation of the form $Z = XW$, where the projection dimensions, i.e., the linear discriminants, simultaneously maximize between-class distance and minimize within-class variance.

This projection results in the solution of a generalized eigenvalues problem:

$$S_B W_{.i} = \lambda_i S_W W_{.i} \tag{5.15}$$

where S_B is the between-class scatter matrix, S_W the within-class scatter matrix, and the ith column, $W_{.i}$ of W represents the generalized eigenvector that corresponds to the ith largest eigenvalue λ_i. Note that scatter matrices are essentially unscaled covariance matrices.

The resultant solution of LDA requires the inversion of the within-class scatter matrix S_W; however, when $p > n - k$, as is typical with mass spectral data, the scatter matrix is not invertible. One way to solve this problem is to reduce the feature set size to less than $n - k$ prior to LDA, using feature selection or other feature extraction techniques such as PCA.

Similar to LDA, an alternate supervised feature extraction technique is the partial least squares (PLS) (Boulesteix and Strimmer 2006). PLS is a regression method that incorporates feature extraction, but it is equally applicable to classification problems. Contrary to LDA, PLS is not bound by any $p < n$ constraint and is therefore better adapted to high-dimensional small samples. Furthermore, it can handle highly correlated features.

Like PCA, PLS finds linear combinations of the input features that maximize variance. However, unlike PCA, PLS finds the linear combinations of the input features while simultaneously maximizing correlations with the class labels. Thus, PLS usually performs better than PCA for prediction problems. Furthermore, PLS is considerably more efficient than PCA with its computational cost $O(np)$, i.e., linear in the number of cases n and the number of original predictors p, whereas that of PCA is on the order of $min(np^2 + p^3, pn^2 + n^3)$, i.e., cubic in n or p, whichever is smaller.

5.3.3 Feature Selection Techniques

Just as feature extraction techniques have played a prominent role in MS data analysis, there is a gamut of feature selection techniques that have achieved considerable success. Typically, feature selection techniques for MS data analysis are categorized as univariate or multivariate, based on whether they evaluate individual features or feature subsets. Both univariate and multivariate methods can be used as filters prior to learning or can be embedded in the learning algorithm.

5.3.3.1 Univariate Methods

These methods assume that all the features are mutually independent of each other. In univariate methods each feature is scored or ranked based on its individual relevance, i.e., in isolation from all other features. The final feature subset is determined by a user-defined threshold (cutoff) on the computed scores or ranks. In MS data each feature (representing a peptide or protein) is selected when it is shown to be differentially expressed at a statistically significant level in the classes of interest (e.g., diseased versus controlled). Standard statistical tests (such as the χ^2 test) have been widely used to gauge the significance levels. These statistical tests rely on an iterative procedure to evaluate each feature independently as follows: first partition the sample according to classes (e.g., healthy versus diseased), then compute a test statistic for an independent feature, and check for significant differences in the value of the test statistic. These standard statistical tests are categorized into parametric tests and nonparametric tests. Parametric tests assume a specific probability distribution of the data, and on the contrary, nonparametric tests do not depend on the probability distribution of the data and have been used in a filter and an embedded setup.

Parametric statistical tests have been prominently used as filters in bioinformatics applications due to the flexibility they offer. Examples of parametric tests that have been used in proteomic analyses are the t-test, F-ratio, χ^2-test, Kolmogorov-Smirnov test, and Wilcoxon rank test. Another parametric statistic is based on the measure of mutual information that is derived on the concepts of information theory. This measure quantifies the reduction in class entropy brought about by the inclusion of a specific feature. Thus, mutual information provides an effective feature ranking criterion used in MS features ranking.

Univariate methods have been used in conjunction with supervised learning schemes for effective identification of discriminatory features sets. These simple learning schemes are aimed at exploiting known class labels information along with univariate methods. Centroid shrinkage is one such feature selection method that is embedded in the nearest centroid classification algorithm (Tibshirani et al. 2004). In this learning scheme, the training samples are used to compute the class centroids; a test sample is assigned to the class with the closest centroid. Class centroid computation is strictly univariate: the ith component of the centroid of a given class k is $\bar{x}_{ik} = \sum_{j=1}^{n_k} x_{ij} / n_k$, where x_{ij} is the value of the ith variable when $j \in k$ and n_k is the number of cases in class k. Similarly, the ith component of the overall centroid is $\bar{x}_i = \sum_{j=1}^{n} x_{ij} / n$, where n is the total number of cases. To reduce the number of features, the distance between the class centroids and the overall centroid is shrunk by an amount determined by Δ, a user-tuned parameter; the class centroids move more rapidly to the overall mean when the shrinkage parameter is higher. Centroid shrinkage can reduce the distance between the class mean and the overall mean to zero for noisy or nondiscriminatory features, which are eliminated (Tibshirani et al. 2002).

The main advantage of using univariate methods is that they are computationally efficient. This computational efficiency is based on the fact that they are driven by the computing of p scores. However, these methods have a number of drawbacks, such as they cannot detect correlated or redundant features or interacting features (i.e., features that are irrelevant by themselves but highly discriminatory when combined with others).

5.3.3.2 Multivariate Methods

Multivariate methods assess the predictive power of feature subsets rather than individual features. Multivariate methods take feature dependencies into account in the feature subset selection process. The major difficulty encountered when using this method is that the number of possible subsets increases exponentially with the growth in the number of features.

Multivariate methods of feature selection are based on exhaustive search strategies that test the effectiveness of different combinations of features (i.e., the strategies of generating and evaluating all $2^p - 1$ possible subsets of p features). This is a daunting task for all but trivial datasets and is considered to be the limitation of multivariate methods. Forward selection and backward elimination in conjunction is a prominent heuristic search strategy that has been proposed to overcome the limitation of multivariate methods.

Forward selection starts with an empty feature subset S and selects the feature that maximizes a predefined scoring function. Thereafter, it searches the remaining features and selects that feature X that, when added to set S, maximizes the score of the resulting subset. The process continues until a predefined criterion is met, e.g., until the score of S ceases to improve. Once this criterion is met, backward elimination proceeds in the reverse direction; it starts with the full variable set and at each step removes the variable with the elimination that yields the highest score for the remaining subset. Both forward selection and backward elimination are greedy search strategies that are not guaranteed to achieve optimal results.

As a partial remedy to this challenge of greedy search, researchers use floating strategies that allow forward and backward selection to eliminate or add previously selected or eliminated features. Alternatively, stochastic search methods use randomization to overcome a second pitfall of greedy methods, being trapped in local optima. Among these stochastic search methods, biologically inspired techniques, which mimic mechanisms underlying the behavior or evolution of living populations, have proved to be effective strategies for finding discriminatory feature subsets.

A number of variable subset selection strategies have been used as filters prior to supervised learning. The RELIEF algorithm, as described in Chapter 4, computes the relevance of each predictive variable using a method based on k-nearest neighbors. In a binary classification problem, this method repeatedly picks a case at random and identifies the case's nearest neighbor from the positive class and its nearest neighbor from the negative class. It then adjusts feature weights by rewarding features that discriminate neighbors from different classes while penalizing those with different values for neighbors of the same class. Although feature weights are updated separately, the RELIEF algorithm is a multivariate method that computes the distance of underlying nearest-neighbor identification and takes into account all features. The RELIEF algorithm can be used as a feature selection filter for any learning algorithm.

5.3.3.2.1 Multivariate Embeddings

Rather than having a feature selection strategy prior to the supervised learning, several supervised learning algorithms have multivariate filters embedded as part of their model building process. Decision trees (DTs) like CART and C4.5 are classical examples of learning algorithms that have embedded heuristic feature selection as part of their model building process. A DT is constructed by a sequential forward search of the features in the dataset that is used to find the most discriminatory feature subset. At each leaf node of the partially built tree, the algorithm selects the feature that maximally reduces the class impurity (or entropy) of the examples associated with that node. Chapter 8 provides a description of the construction of a DT.

Information gain (IG) is a measure used by the DT C4.5 that is defined as $I(X;C) = H(C) - H(C|X)$, where C is the class variable, X is a feature, and $H(.)$ is their corresponding entropy. In other words, IG is the decrease in class entropy brought about by the feature X.

Though DTs are constructed by gauging the entropy of independent features, DTs are multivariate rather than univariate. DTs measure the *cumulative* reduction in entropy brought about by the feature subset consisting of all features along the path from the root to the current node. Though DTs are sufficient for the identification of feature subsets, it is common practice to precede DT learning by a feature selection method, such as the t-test (or any of the univariate methods).

Another embedded multivariate technique consists of building ensembles or communities of univariate classifiers, which are then combined to yield a single prediction (refer to Chapter 8 for details). A widely used ensemble learning method is boosting, which builds a sequence of classifiers from adaptively generated data. This method builds a classifier at each iteration, and the classifier's accuracy on the training data is estimated.

5.4 Data Preprocessing for Genomic Sequence Data

An important aspect to bioinformatics is the analysis of sequence data, in the form of genomic sequence data or proteomic sequence data. The objective of sequence data is the identification of motifs (short sequence signals) that are embedded in the sequence composition. It is believed that some or all of a set of promoters from coexpressed or orthologous genes may contain binding sites (signals) for the same transcription factor. Similarly, a set of proteins that interact with a single host protein may do so via similar domains (the signal). Both types of sequence signals can often be represented as motifs that are ungapped, approximate subsequence patterns. This section aims to describe the several techniques that have been proposed to identify statistically significant motifs for a given set of sequences.

Motif discovery algorithms look for a set of similar short sequences in a set of much longer sequences. This problem is easier when the motif instances are long and very similar to each other. It gets much harder when the motif instances are short or when the input sequences are very long (Bailey et al. 2006).

5.4.1 Feature Selection for Sequence Analysis

With the exponential growth of genome sequence data, there is a need for computationally effective and accurate tools to automatically identify genes from the sequences. This objective poses a challenging problem, as only a fraction of the genome sequence (miniscule in number) actually contains coded information. This makes several statistical techniques unreliable and inaccurate in identifying informational parts in the sequence (Saeys et al. 2006). Therefore, while analyzing sequences in bioinformatics, feature selection has played a key role in recent times. According to Saeys et al. (2007), there are two types of sequence analysis: content analysis and signal analysis.

> **Content analysis:** The prediction of subsequences that code for proteins (coding potential prediction) has been a focus of interest since the early days of bioinformatics. Because many features can be extracted from a sequence and most dependencies occur between adjacent positions, many variations of Markov models have been developed (Eddy 2004). Addressing the high number of possible features and the often limited amount of samples led to the introduction of the interpolated Markov model (IMM) through the implementation of the GLIMMER system (Salzberg et al. 1998). Using a two step process, the GLIMMER system is used to find coding regions in microbial genome sequences. First, it uses the IMM to interpolate between different orders of the Markov model to deal with a small number of samples. And then a filter method (χ^2 filter) is used to select only relevant features.
>
> **Signal analysis:** Many sequence analysis methodologies involve the recognition of short, more or less conserved signals in the sequence, representing mainly binding sites for various proteins or protein complexes. A common approach to finding regulatory motifs is to relate motifs to gene expression levels using a regression approach (Guyon et al. 2002). Feature selection is then used to search for the motifs that maximize the fit of the regression model (Keles et al. 2002).

Sequence features: Commonly used sequence features represent only the nucleotide or amino acid at each position in a sequence. However, there are many other features, such as higher-order combinations of the nucleotides or amino acids (e.g., k-mer patterns), that can be derived, and their number is growing exponentially with the pattern length k. Such higher-order features are described as follows (Saeys et al. 2006):

1. **Frame-dependent k-mers:** For each of the three possible reading frames, k-mer frequencies ($1 \leq k \leq 3$) can be extracted. These frame-dependent features would result in 252 [$= 3 \times (4 + 16 + 64)$] features.
2. **In-frame k-mers:** Assuming the sequence is in reading frame 1 (the start of the sequence coincides with the start codon), in-frame k-mer frequencies ($4 \leq k \leq 6$) can be extracted. These in-frame k-mer features can result in a set of 5,376 possible features.
3. **Frameless k-mers:** For each possible k-mer ($1 \leq k \leq 3$), the global frequencies of occurrence are calculated (i.e., without taking into account the reading frame). These frameless k-mers will result in possible 84 features.
4. **Fourier transform features:** This is the most common application of Fourier analysis on DNA sequences. The features, derived from the Fourier transform, include (1) the magnitude of the peak at frequency 1/3 in the Fourier spectrum, (2) the global magnitude at frequency 1/3, which is the sum of all four magnitudes of the sequence, and (3) the signal-to noise ratio of the peak frequency 1/3. This results in a possible six features.
5. **ORF feature:** Given a sequence and an assumed reading frame, this feature denotes whether there is an in-frame stop codon present.
6. **Run features:** For each of the nontrivial subsets of {A, T, G, C}, a new sequence is constructed by replacing each base present in the subset with 1 and replacing each base not in the subset with 0. Using this transform of the sequence, the number of runs of 1s of length 1, 2, 3, 4, 5, and greater than 5 are then counted. This results in a set of 84 features.

With the numerous features listed above, it is observed that many of these are irrelevant or redundant. This therefore requires feature selection techniques, which will be applied to focus on the subset of relevant features.

5.5 Ontologies in Bioinformatics

In data mining, it is common to use conceptual hierarchies to provide a certain degree of data abstraction, to describe complicated concepts that require a certain degree of heterogeneity in the data types used to describe them. To this end, conceptual hierarchies have been extensively used in various applications of data mining, and take the form of ontologies in bioinformatics. These ontologies are then used to integrate data from heterogeneous sources.

According to Bodenreider and Stevens (2006), ontologies are techniques or technologies used to represent and share knowledge about an entity by modeling the features of the entity and the relationships between those features. These relationships describe the properties of those features in the entity being modeled. Thus, the ontology represents a conceptualization of reality, or simply reality. The labels used for the features and their properties in an ontological model can provide a common language for a community to talk about their entity. By agreeing on a particular ontological representation, a common vocabulary can be used to describe and analyze data.

Newer technologies such as microarrays and the ever-changing volumes of data have necessitated the ability of computational services or algorithms to automate gene calling, the identification of individual genes across a genome. Gene calling entails using algorithms to identify biologically functional regions or exons of sequences that explicitly code for proteins commonly referred to as coding regions. These algorithms are based on machine learning, which predicts unique signatures of the genetic spectrum. Once the genes have been located, the tasks of assimilating the biological functions of the resulting proteins have to be determined. The prediction of protein functions is then determined by the sequence alignment of sequences from their homologs. This process is plagued by errors, as it is time-consuming and dependent on the accuracy of previously discovered sequences. New alternative approaches of functional genomics aim at identifying functionally significant gene sequences through the use of knowledge gained from annotated functionality, pathways, and protein-protein interactions. These approaches, however, necessitate the resolution of semantics, i.e., the differences in meaning and naming conventions between heterogeneous sources.

Semantic incompatibility: Though data from conflicting database schemas can be pooled together using simple queries, the semantics between heterogeneous biological data is not as transparent. For example, a gene is defined differently in different databases. According to the Human Genome Database (HGD) a gene is defined as a DNA fragment that is analogous to a protein, whereas in GenBank and the Genome Sequence Database (GSDB), a gene is considered to be a region of biological interest that carries a genetic trait and has an associated name. Thus, these databases are built using two theories of what a gene is. As a result, the retrieval of data from semantically different databases on the basis of the keyword *gene* would typically propagate an error. Moreover, problems also arise when two variables in disparate databases are semantically equivalent; it must be noted that their relations to other knowledge objects in the data repository may not be equivalent. These conflicts are commonly referred to as schematic incompatibility.

Context: To facilitate both semantic and schematic differences inherent in biological data, the context from which biological data originates is given more importance at the database level. These differences are elucidated by the fact that functional prediction hinges on the need to find similar cellular components that participate in similar biological processes rather than the sequence homologs. To connect sequence and cellular components, it is imperative to depend on additional data sources that support information about the diverse components entailed in a biological process. Thus, to explore the vast number of databases, it is required that the biological context of sequences be adequately encoded and machine readable.

5.5.1 The Role of Ontologies in Bioinformatics

To resolve the issue of semantics in bioinformatics databases, ontologies have provided better biological interoperability. By definition an ontology is a machine-readable model of the objects (allowed into a formal universe) and the associations (or relationships) between these objects, upon which some automated reasoning (or task) can be performed (Schuurman and Leszczynski 2008). Typically scientific ontologies contain three levels of formalization. The first level is conceptual description of elements, which is then translated into the second level, a formal model of the data elements in the ontology (e.g., proteins) and the possible relationships between the data elements. The final level is the development of code that can be run by computers that use the outputs of the second level. Ontologies like biological taxonomy are a hierarchy of concepts, with the general concepts placed at the top of the hierarchy, and the specificity of the concepts increases as we traverse down the hierarchy. Ontologies are populated by domain knowledge in the form of semantics that allows all entities declared into the ontology to be defined and for their interrelationships to be given strict parameters, enabling realistic biological models (Baker et al., 1999).

Therefore semantics enables the distinction of concepts declared into the model. To satisfy the strict criteria of formal ontology building, the semantics used to instantiate an ontology should be based on formally defined logics. These formally defined logics can be based of logical algebra such as description logics (DLs). These DLs should accommodate predetermined rules for (1) when two concepts are the same, (2) when the two concepts are one of a kind, and (3) how two concepts differ from each other. These rules must furthermore be expressed in some machine-readable syntax, such as a knowledge representation language like Web Ontology Language (OWL). Such rules govern the expression and processing of relations between concepts in the hierarchy. It should be noted that these relational expressions between concepts in a hierarchy form the basis for all modeling tasks for using any ontology (see Figure 5.9).

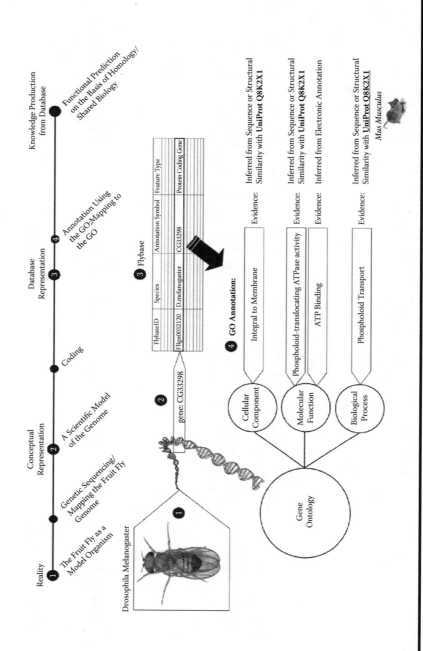

Figure 5.9 The formalization of the process of moving from a concept (of a gene) to its encoded reification and ontological representation. Note how the entity (fruit fly) becomes increasingly represented in digital database format as it is formalized, or abstracted from its real-world form. (From Schuurman, N., and Leszczynski, A., *Bioinformatics Biol Insights* 2 (2008): 187–200.)

5.5.1.1 Description Logics

Ontologies differ from data integration by their ability to define relationships between concepts. Typically, relationships are developed using a natural language that is an expression of context. In other words, relationships between concepts are captured so that they convey some semantics. Similarly, content semantics are expressed by identifying how concepts relate to each other in the hierarchical knowledge space.

The very hierarchical nature of an ontology brings about a parent-child ordering of semantic granularity of the relation between any two concepts. The hierarchical structure establishes a hyponymic (is-a) relationship between terms by their relative position to each other in the hierarchy on the basis of subsumption (where a concept is a subclass or member of the other concept) and specialization (where a concept is the superclass of or contains another concept). The semantic edges of the tree are the relationships referred to as properties that reflect the meaning of data elements by providing the context of their usage.

Ontological expressions are stated in the form of propositional triplets. The triplets consist of concepts (real-world entities that populate the model), their properties (or relationships between entities), and instances (particular occurrences of a concept) in a hierarchical model. A triplet is considered to be a definitive statement about the world. Thus, if an ontology is represented using a description logic (DL), the axioms of the logic can be used to impose restrictions on the concepts in which domains logically participate in relationships with each other. These logics thus form a content specification. Using the description logic makes implementing an ontology simpler, where each propositional triple describes a knowledge base (Schuurman and Leszczynski 2008). The following sections contain descriptions of the most prominently used ontology in bioinformatics: the Gene Ontology (GO), a derivative of the Open Biomedical Ontologies (OBO).

5.5.1.2 Gene Ontology (GO)

The Gene Ontology has been one of the most successful ontologies in the area of bioinformatics. The success of this ontology can be enumerated as follows (Bada et al. 2004):

1. **Community involvement:** The development of the GO is an open process; response is welcomed from the community that it seeks to serve. The GO is built by and for biologists, and groups join the GO because it suits their needs. Such activity is less likely than in a dictated, unresponsive organization.
2. **Clear goals:** The GO promotes consistent annotation for gene products for the three major functional attributes. While GO has been used for many other purposes, this narrow, clear goal enables focus to be maintained.

3. **Limited scope:** An ontology for the whole of biology would be useful. However, it would be impractical to develop such an ontology. A limited but useful scope was able to demonstrate utility.
4. **Simple structure:** The GO's use of a simple directed acyclic graph (DAG) is sufficient to capture the relationships between concepts derived from biology.
5. **Continuous evolution:** Our understanding of biology changes and expands. Part of the community engagement is to respond to and put in place the apparatus to cope with change.
6. **Active curation:** In addition to community input, the continuous evolution and necessary maintenance require curators to implement changes.
7. **Early use:** The evolutionary nature of genomics enabled early use and evolution of the GO. A relatively small number of gene products and consistent annotation enabled its use.

The GO is a global ontology that is a central knowledge proxy to which other ontologies or knowledge representations may be aligned. The alignment of derived knowledge representation is brought about by ontology mapping. Ontology mapping is the process of defining associations between ontologies. This method involves the formal declaration of relational links between entities, much like those involved in relating concepts in a hierarchical ontological structure. Ontologies can be either aligned, whereby the formalisms remain separate entities but are related, or merged, wherein a singular ontology is generated from the cross-products of two input ontologies. Mapping is thus unidirectional and always from the constituent database to the GO. Figure 5.10 illustrates the role that GO plays in the development of global biological ontology and the mechanics involved.

5.5.1.3 Open Biomedical Ontologies (OBO)

The success of the GO in meeting its objectives, its wide use by other databases for attributing gene product functionality, and finally the use of the GO outside its intended purpose have led to many other groups developing ontologies for database annotation. In order to provide some coordination to these efforts, the OBO consortium was established.

OBO is guided by a set of principles that are used to give coherence to wider ontological efforts across the community.

1. **Openness:** All the OBO ontologies are freely available to the community as long as the ontologies are properly attributed.
2. **Common representation:** Both the OBO format and the Web Ontology Language (OWL) provide common access via open tools. Although not mentioned as part of the criteria, this common access offers common semantics for knowledge representation.

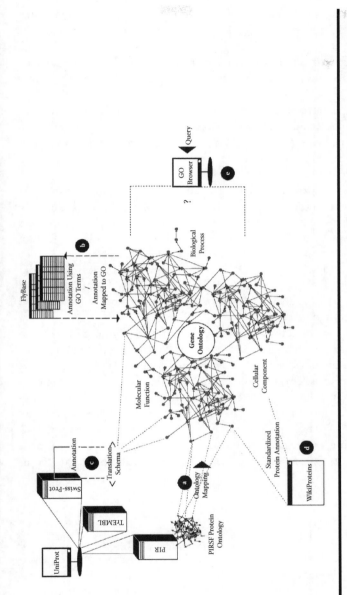

Figure 5.10 The Gene Ontology as a global ontology for bioinformatics. (a) Smaller-scale bioinformatics ontologies almost invariably map to the GO. (b) Several large databases, such as the FlyBase, contribute annotation to the GO using their semantics such that there is a direct mapping between genes/gene products at the database level and their participation in the ontology. (c) Where annotation is unique to the database, a translation program can transform annotation into a tractable GO representation. (d) The GO provides a standardized vocabulary for the description of genes and gene products not only across databases but also in emerging bioinformatics infrastructures, such as WikiProteins. (e) The consistency of semantics reduces ambiguity in the query of bioinformatics resources, and allows genes and gene products to be retrieved on the basis of common biology rather than lexical coincidence. (From Schuurman, N., and Leszczynski, A., *Bioinformatics Biol Insights* 2 (2008): 187–200.)

3. **Independence:** Lack of replication across separate ontologies encourages combinatorial reuse of ontologies and the interlink of ontologies via relationships.
4. **Identifiers:** Each term should have a semantic-free identifier, the first part of which identifies the originating ontology. These identifiers promote easy management.
5. **Natural language definitions:** Terms are often ambiguous, even in the context of their ontology, and definitions help ensure appropriate interpretation. Arguments over terms are often bitter and long, while arguments over definitions are shorter and more useful.

Through these simple criteria, the ontology community is attempting to avoid repeating the errors their ontologies have been developed to resolve, primarily the massive syntactic and semantic heterogeneity extant in bioinformatics resources. Many resources fall under the OBO umbrella, and most of these resources are shown in Figure 5.11, in which OBO have been arranged along a spectrum of genotype and phenotype.

The most significant OBO are the GO (Gene Ontology Consortium 2008) and the Sequence Ontology (Eilbeck et al. 2005). The former is used to annotate the principal attributes of gene products. The latter provides a vocabulary to describe the features of biological sequences.

Moving along the spectrum toward phenotype (refer to Figure 5.11), we see increasing numbers of species ontologies on the same subject: development and anatomy. While the descriptions of sequence features and major attributes of gene products might be core to molecular biology, these descriptions need to be placed in a context.

Other OBO ontologies include some that describe experiments that generate biological data. Foremost among these ontologies is the Microarray Gene Expression Data (MGED) ontology (Whetzel et al. 2006). This ontology provides a vocabulary for describing a biological sample used in an experiment, the treatment the sample receives in the experiment, and the microarray chip technology used in the experiment.

5.6 Conclusion

In this chapter we have described the interpretation of features obtained from bioinformatics data in context to the various data transformation and data preprocessing strategies. We have emphasized the role of various normalization techniques with their application to high-throughput gene expression data. Furthermore, this chapter describes the role of data preprocessing strategies and data transformation strategies with respect to mass spectrometry data analysis and genomic sequence data. This chapter concludes by describing the importance of ontologies and concept hierarchies that are necessary in interpreting the role of features in a computational perspective.

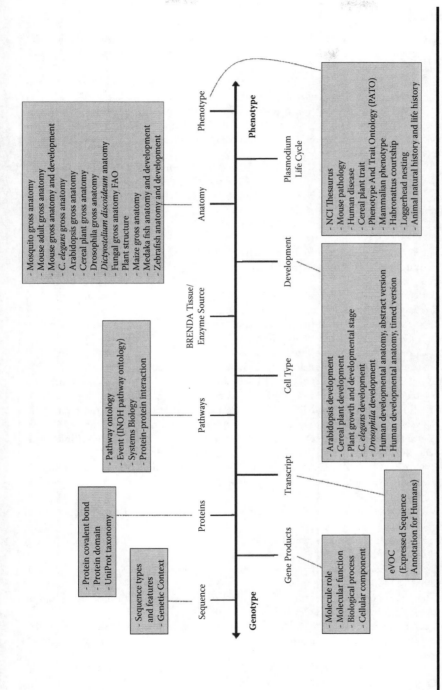

Figure 5.11 The different resources that constitute the OBO, arranged in the order from genotype to phenotype.

References

Ach, A.R., H. Wang, and B. Curry. Measuring microRNAs: Comparisons of microarray and quantitative PCR measurements, and different total RNA prep methods. *BMC Biotechnol* 8 (2008): 69.

Ambroise, C., and G.J. McLachlan. Selection bias in gene extraction on the basis of microarray gene-expression data. *Proc Natl Acad Sci USA* 99, np. 10 (2002): 1–5.

Bachmayer, P. Preprocessing of mass spectrometry data in the field of proteomics. Master Degree Program Report, University of Helsinki, Finland, 2007.

Bada, M., et al. A short study on the success of the Gene Ontology. *J Web Semantics* 1, no. 2 (2004): 235–240.

Bailey, T.L., N. Williams, C. Misleh, and W.W. Li. MEME: Discovering and analyzing DNA and protein sequence motifs. *Nucl Acids Res* 34 (2006): W369–W373.

Bair, E., T. Hastie, D. Paul, and R. Tibshirani. Prediction by supervised principal components. *J Am Stat Assoc* 101 (2006): 119–137.

Baker, P.G., C.A. Goble, B. Sean, N.W. Paton, R. Stevens, and A. Brass. An ontology for bioinformatics applications. *Bioinformatics* 15 (1999): 510–520.

Barla, A., G. Jurman, S. Riccadonna, S. Merler, M. Chierici, and C. Furlanello. Machine learning methods for predictive proteomics. *Briefings Bioinformatics* 9, no. 2 (2008): 119–128.

Bodenreider, O., and R. Stevens. Bio-ontologies: Current trends and future directions. *Briefings Bioinformatics* 7, no. 3 (2006): 256–274.

Boulesteix, A.L., and K. Strimmer. Partial least squares: A versatile tool for the analysis of high dimensional genomic data. *Briefings Bioinformatics* 8 (2006): 32–44.

Brazma, A., et al. Minimum information about a microarray experiment (MIAME)—Toward standards for microarray data. *Nature Genet* 29 (2001): 365–371.

Chen, W., F. Erdogan, H.H. Lenzner, S. Ropers, and R. Ullman. CGHPRO-A comprehensive data analysis tool for array CGH. *BMC Bioinformatics* 6, no. 85 (2005): 1–7.

Chen, Y., E.R. Dougherty, and M.L. Bittner. Ratio-based decisions and the quantitative analysis of cDNA microarray images. *J Biomed Optics* 2, no. 4 (1997): 364–374.

Díaz-Uriarte, R., and S. Alvarez de Andrés. Gene selection and classification of microarray data using random forest. *BMC Bioinformatics* 7, no. 3 (2006): 1–13.

Dudoit, S., Y.H. Yang, M.J. Callow, and T.P. Speed. Statistical methods for identifying differentially expressed genes in replicated cDNA microarray experiments. *Stat Sinica* 12 (2002): 111–139.

Eddy, S.R. What is a hidden Markov model? *Nature Biotechnol* 22, no. 10 (October 2004): 1315–1316.

Eilbeck, K., et al. The Sequence Ontology: A tool for the unification of genome annotation. *Genome Biol* 6 (2005): R44.

Gene Ontology Consortium. The Gene Ontology (GO) project in 2008. *Nucl Acid Res* 36 (2008): D440–D444.

Guyon, I., S.B. Weston, and V. Vapnik. Gene selection for cancer classification using support vector machines. *Machine Learn* 46 (2002): 389–422.

Hijum, S., et al. Supervised LOWESS normalization of comparative genome hybridization data-application to lactococcal strain comparison. *BMC Bioinformatics* 9, no. 93 (2008): 1–10.

Hilario, M., and A. Kalousis. Approaches to dimensionality reduction in proteomic biomarker studies. *Briefings Bioinformatics* 9, no. 2 (2008): 102–118.

Jenssen, T.-K., M. Langaas, W.P. Kuo, B. Simth-Sorensen, O. Myklebost, and E. Hovig. Analysis of repeatability in spotted cDNA microarrays. *Nucl Acids Res* 30, no. 14 (2002): 3235–3244.

Keles, S., M. Van der Laan, and M.B. Eisen. Identification of regulatory elements using a feature selection method. *Bioinformatics* 18, no. 9 (2002): 1167–1175.

Kreil, D.P., and R.R. Russell. There is no silver bullet—A guide to low-level data transforms and normalization methods for microarray data. *Briefings Bioinformatics* 6, no. 1 (2005): 86–97.

Kuonen, D. Challenges in bioinformatics for statistical data miners. *Bull Swiss Stat Soc* 40 (2003): 10–17.

Lilien, R.H., H. Farid, and B.R. Donald. Probabilistic disease classification of expression-dependent proteomic data from mass spectrometry of human serum. *J Comput Biol* 10, no. 6 (2003): 925–946.

Mar, J.C., et al. Data-driven normalization strategies for high-throughput quantitative RT-PCR. *BMC Bioinformatics* 10 (2009): 1–10.

Papana, A., and H. Ishwaran. CART variance stabilization and regularization for high-throughput genomic data. *Bioinformatics* 22, no. 18 (2006): 2254–2261.

Park, T., S.-G. Yi, S.-H. Kang, S.Y. Lee, Y.-S. Lee, and R. Simon. Evaluation of normalization methods for microarray data. *BMC Bioinformatics* 4, no. 33 (2003): 1–13.

Qin, H., T. Feng, S.A. Harding, C.-J Tsai, and S. Zhang. An efficient method to identify differentially expressed genes in microarray experiments. *Bioinformatics* 24, no. 14 (2008): 1583–1589.

Qu, W., et al. Data reduction using a discrete wavelet transform in discriminant analysis of very high dimensionality data. *Biometrics* 59, no. 1 (2003): 143–151.

Quackenbush, J. Microarray data normalization and transformation. *Nature Genet Suppl* 32 (2002): 496–501.

Robinson, M.D., and A. Oshlack. A scaling normalization method for differential expression analysis of RNA-seq data. *Genome Biol* 11, no. R25 (2010): 1–9.

Saeys, Y., I. Inza, and P. Larranaga. A review of feature selection techniques in bioinformatics. *Bioinformatics* 23, no. 19 (2007): 2507–2517.

Saeys, Y., P. Rouze, and Y. Van de Peer. In search of the small ones: Improved prediction of short exons in vertebrates, plants, fungi and protists. *Bioinformatics* 23, no. 4 (2006): 414–420.

Salzberg, S.L., A.L. Delcher, S. Kasif, and O. White. Microbial gene identification using interpolated Markov models. *Nucl Acids Res* 26, no. 2 (1998): 544–548.

Schuurman, N., and A. Leszczynski. Ontologies for bioinformatics. *Bioinformatics Biol Insights* 2 (2008): 187–200.

Simpson, J.T., K. Wong, S.D. Jackman, J.E. Schein, S.J.M. Birol, and I. Jones. ABySS: A parallel assembler for short read sequence data. *Genome Res* 19 (2009): 1117–1123.

Tibshirani, R., T. Hastie, B. Narasimhan, and G. Chu. Diagnosis of multiple cancer types by shrunken centroids of gene expression. *Proc Natl Acad Sci USA* 99 (2002): 6567–6572.

Tibshirani, T., et al. Sample classification from protein mass spectrometry, by "peak probability contrast." *Bioinformatics* 20, no. 17 (2004): 3034–3044.

Veltri, P. Algorithms and tools for analysis and management of mass spectrometry data. *Briefings Bioinformatics* 9, no. 2 (2008): 144–155.

Whetzel, P.L., et al. The MGED Ontology: A resource for semantics-based description of microarray experiments. *Bioinformatics* 22, no. 7 (2006): 866–873.

Workman, C., et al. A new non-linear normalization method for reducing microarray variability in DNA microarray experiments. *Genome Biol* 3, no. 9 (2002): 1–16.

Yang, Y.H., et al. Normalization for cDNA microarray data: A robust composite method addressing single and multiple slide systemic variation. *Nucl Acids Res* 30, no. 4 (2002): e15.

Zhang, P., H. Li, H. Wang, W. Stephen, and X. Zhou. Peak tree: A new tool for multiscale hierarchical representation and peak detection of mass spectrometry data. *IEEE/ACM Trans Comput Biol Bioinformatics* 99 (2009): 1.

Zhu, D., A.O. Hero, H. Cheng, R. Khanna, and A. Swaroop. Network constrained clustering for gene microarray data. *Bioinformatics* 21, no. 21 (2008): 4014–4020.

UNSUPERVISED LEARNING

Chapter 6

Clustering Techniques in Bioinformatics

We covered the different data preprocessing and transformation techniques in data mining in Chapter 4. We also described the different application areas and the significant role these techniques play in the field of bioinformatics. In this chapter we list and describe different unsupervised learning techniques in data mining, better known as clustering techniques.

6.1 Introduction

Clustering is used to divide or partition objects (or data) into groups based on their similarity or dissimilarity to one another, called clusters. It is an unsupervised learning method, as class labels or class information is not present in the beginning. Therefore clusters obtained in the output can be called classes. The quality of clustering will depend on many factors:

1. Similarity measure used by the method and its implementation
2. Ability to discover some or all of the hidden patterns

Objects in the same cluster should be similar to one another, and thus have high intraclass similarity. However, objects between other clusters should be dissimilar to each other, and thus have low interclass similarity. A good clustering method will result in a high intraclass similarity, and a low interclass similarity will produce quality clusters. A dataset may have different kinds of data points, which

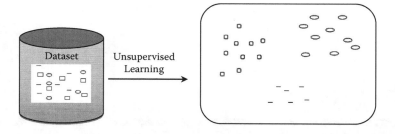

Figure 6.1 Using unsupervised learning data points, datasets are sorted into three clusters based on similarity of shape.

may belong to unknown clusters. By using unsupervised learning such as a clustering algorithm one can find potential clusters in the dataset (Cooper and Newman 2010). In Figure 6.1 data points have been sorted into three clusters where similar data points are grouped together, and dissimilar data points are separated.

Clustering consists of four steps: relevant feature selection, algorithm design, cluster validation, and visualization and evaluation (Wunsch and Xu 2005). These steps are shown in Figure 6.2.

6.2 Clustering in Bioinformatics

Clustering and cluster analysis are important techniques for bioinformatics experimentation and discovery. Clustering is widely used in microarray analysis to negate the limitations of class discovery. As mentioned above, classes are often unknown when experiments begin. For example, if a researcher is trying to determine whether a disease in a particular tissue or in a particular condition can affect a gene expression, he or she may not know whether gene expression differs between two groups (Lippert 2010). In addition, he or she may not know whether a class contains interesting subclasses until clustering is performed

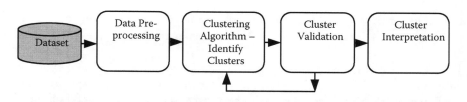

Figure 6.2 Clustering procedure. (Modified from Wunsch, D., and R. Xu, *IEEE Trans Neural Networks* 16, no. 3 (2005): 645–678.)

(Lippert 2010). For example, a subtype of a disease or a hierarchy of subclasses within a disease may not be known at implementation (Lippert 2010). For this chapter, we will present clustering methods used in gene expression. Gene expression is defined as the synthesizing of a functional gene product found in either the RNA or a protein. Gene sequence is important because genes are the fundamentals of biological inheritance in living organisms to build and maintain organism cells; genes hold the necessary information. Time series gene expression data is used to analyze underlying temporal response patterns to simplify work with nonuniform samples.

6.3 Clustering Techniques

For bioinformatics research, students and researchers can choose from a variety of clustering techniques, including distance-based clustering, hierarchical clustering, self-organizing maps, fuzzy clustering, graph clustering, kernel clustering, and model clustering. Because of the importance of these techniques, we describe each below.

6.3.1 Distance-Based Clustering and Measures

Distance-based clustering is used to find similarity or dissimilarity in terms of distance between data points of the same cluster or data points of other clusters. Distance can be found by using distance measures such as Mahalanobis distance, Minkowski distance, and Pearson correlation. Selection of these measures will depend on the characteristics or properties of attributes such as binary, continuousness, nominality, and ordinality. For example, if an attribute is numeric, one can use Mahalanobis or Mikowiski distance. Below, we outline the differences between these two methods.

6.3.1.1 Mahalanobis Distance

Mahalanobis distance is based on finding correlation between variables to measure distance, which helps classify future data belonging to a specific class. Mahalanobis distance works in the following ways:

1. Mahalanobis distance computes the covariance matrix of each class from the training data.
2. It sorts future or test data into their respective classes based on minimal Mahalanobis distance. This sorting is performed by computing the

Mahalanobis distance for each class. Mahalanobis distance can be mathematically expressed as $z = (z_1, z_2, ..., z_N)^T$ from a group of values with a mean of $\mu = (\mu_1, \mu_2, \mu_3, ..., \mu_4)^T$, and covariance matrix S is

$$D_M(z) = \sqrt{(z-\mu)^T S^{-1}(z-\mu)}, \tag{6.1}$$

where D_M is Mahalanobis distance.

6.3.1.2 Minkowiski Distance

The Minkowiski distance between two points or tuples, for example, $X_1 = (x_{11}, x_{12}, ..., x_{1n})$ and $X_2 = (x_{21}, x_{22}, ..., x_{2n})$, is

$$d(X_1, X_2) = \sqrt{\sum_{i=1}^{n} (x_{1i} - x_{2i})^t}, \tag{6.2}$$

$d(X_1, X_2)$ is the distance between two points or tuples X_1 and X_2. In Equation 6.2, when $t = 1$, distance is called a city block distance. When $t = 2$, distance is called Euclidean distance.

The Euclidean distance is used to find distance between two points that is a line segment connecting them. For example, the distance between two points A(6,3) and B(3,2) is calculated by using the Euclidean distance formula found in Equation 6.2, which is shown in Figure 6.3.

$$d(A, B) = \sqrt{(6-3)^2 + (3-2)^2} = \sqrt{10} \tag{6.3}$$

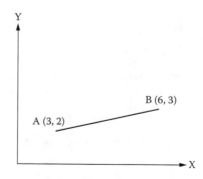

Figure 6.3 Points A and B lie in a two-dimensional plane.

6.3.1.3 Pearson Correlation

Pearson correlation is another method of finding similarity by measuring the correlation that ranges from +1 to 1 between two variables. Thus, it can be found between two variables, for example, x and y, in following way:

- For value 1: Whenever y increases, x also increases.
- For value −1: Whenever y decreases, x increases.
- For value 0: No correlation or relationship.

The Pearson correlation coefficient is symmetric and can be represented as

$$corr(x, y) = corr(y, x). \tag{6.4}$$

The Pearson correlation of genes x and y of n samples, where $\bar{x}$ is the mean of x and $\bar{y}$ is the mean of y, is

$$d_{xy} = (i - r_{xy})/2, \quad \text{where} \quad r_{xy} = \frac{\sum_{i=1}^{n}(x_i - \bar{x})(y_i - \bar{y})}{\sqrt{\sum_{i=1}^{n}(x_i - \bar{x})^2}\sqrt{\sum_{i=1}^{n}(y_i - \bar{y})^2}}, \tag{6.5}$$

where r_{xy} is the correlation between two samples x and y, and d_{xy} is the distance between two samples x and y.

The correlation coefficient $(x_i - \bar{x})(y_i - \bar{y})$ is positive if x_i and y_i are greater or less than their respective means or if they are located on the same side of their respective means. Similarly, the correlation coefficient is negative if either x_i or y_i is less than its respected mean or is located on an opposite side of its respective mean.

6.3.1.4 Binary Features

Binary features are those features that have binary values of either 0 or 1. Given two attributes, X and Y, having binary values of either 0 or 1, the total number of combinations for attributes X and Y is specified as shown in Table 6.1. In the table:

- a indicates that attributes X and Y have a value of 1.
- b indicates that attribute X is 0 and attribute Y is 1.
- c indicates that attribute X is 1 and attribute Y is 0.
- d indicates that both attributes X and Y have a value of 0.

A distance measure D for symmetric binary variables can be determined using

$$D(A, B) = \frac{b+c}{a+b+c+d}. \tag{6.6}$$

Table 6.1 Sample Binary Matrix

		Object X		
		1	0	Sum
Object Y	1	a	b	a + b
	0	c	d	c + d
	Sum	a + c	b + d	P

Likewise, a distance measure D for asymmetric binary variables can be determined using the Jaccard coefficients, which measure similarity between sample sets and are as follows.

The Jaccard similarity coefficient, $D(A, B)$, is given as

$$D(A,B) = \frac{a}{a+b+c}, \quad \text{and} \tag{6.7}$$

The Jaccard distance, $D(A, B)$, is given as

$$D(A,B) = \frac{b+c}{a+b+c}. \tag{6.8}$$

6.3.1.5 Nominal Features

Unlike binary features, nominal features can have more than two states. Thus, either nominal features must be transformed into binary, or matching criterion must be utilized to minimize the number of states. The features can be classified using the two methods below, binary transformation and simple matching.

As the name suggests, binary transformation modifies nominal features so that they can be read as binary code. For each of the M nominal states, this method will create a new binary variable that uses a 1 to indicate the occurrence of a category and a 0 to indicate the absence of a category or a nonoccurrence (Shyu 2005). Thus, for a nominal feature with C states, a set of C indicator variables can be generated as shown below.

$$\begin{aligned} S_1 &= \begin{cases} 1 \text{ if the category is 1} \\ 0 \text{ Otherwise} \end{cases} \\ S_2 &= \begin{cases} 1 \text{ if the category is 2} \\ 0 \text{ Otherwise} \end{cases} \\ S_c &= \begin{cases} 1 \text{ if the category is } c \\ 0 \text{ Otherwise} \end{cases} \end{aligned} \tag{6.9}$$

Simple matching distance methods can simplify nominal features by combining feature groups. These methods are used when two objects, for example, i and j, carry equal information. For example, marital status (married and unmarried) has a symmetry attribute because the number of married respondents and the number of unmarried respondents provide equal information. Similarly, the same information is given whether heads or tails land face up when tossing a coin.

Equation 6.10 represents simple matching as

$$d(i,j) = \frac{p-m}{p}, \tag{6.10}$$

where m is the number of matches and p is the number of variables.

6.3.1.6 Mixed Variables

A database may contain different types of features or mixed features. These features can be symmetric binary, asymmetric binary, nominal, ordinal, interval, and ratio. When objects consist of mixed variables, we can combine the variables and transform them into an interval such as (0, 1). We can then use measures such as Euclidean distance, or we can transform the variables into binary for similarity functions.

6.3.2 Distance Measure Properties

In order for a distance or similarity function to be a distance measure, it should follow all four properties, as indicated below:

1. Symmetry:

$$D(x_i, x_j) = D(x_j, x_i) \tag{6.11}$$

2. Positivity:

$$D(x_i, x_j) \geq 0 \text{ for all } x_i \text{ and } x_j \tag{6.12}$$

3. Triangle inequality:

$$D(x_i, x_j) \leq D(x_j, x_k) + D(x_k, x_i), \tag{6.13}$$

4. Reflexivity:

$$D(x_i, x_j) = 0, \text{ if } x_i = x_j \text{ holds; it is also called a metric} \tag{6.14}$$

where $D(x_n, x_m)$ is the distance between two points x_n and x_m such that $n = i, j, k, \ldots,$ and $m = i, j, k, \ldots.$

The distance measures explained in the above sections all satisfy these four conditions. These distance measures will be helpful in clustering algorithms that are described below.

6.3.3 k-Means Algorithm

k-Means clustering partitions n instances into k clusters by assigning each data point to the partition with the nearest centroid. The k-means algorithm can be performed as follows:

1. Initialize the value of k or number of partitions. This step is shown in Figure 6.4, where the value of k initialized is 3. These points which are shown in bold (□ ○ △) represent initial group centroids.
2. Assign each data point partition with the nearest centroid, as shown in Figure 6.5.
3. Calculate the positions of the k centroids again to measure the movement of objects, as shown in Figure 6.6.
4. Repeat steps 2 and 3 until movement of the centroid does not change. The final clustering results are shown in Figure 6.7.

Although the k-means algorithm is a popular and useful method, it has limitations, which are overcome by using the k-modes algorithm explained in Section 6.3.4.

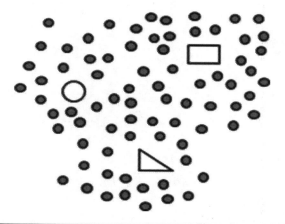

Figure 6.4   Initial group centroids.

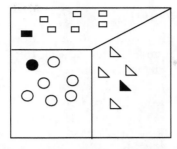

Figure 6.5 Objects in partitions.

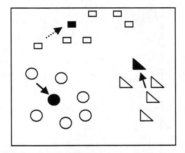

Figure 6.6 Recalculate centroids.

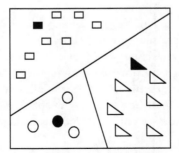

Figure 6.7 Clusters.

6.3.4 k-Modes Algorithm

Since the *k*-means algorithm works only for numerical data, its variant, the *k*-modes algorithm, can be more useful (Chiang et al. 2006). This method can be extended by calculating the median instead of the mean (Chiang et al. 2006). Using the median, *k*-means can produce accurate results for categorical data, as well as numerical data. The *k*-modes algorithm is described below.

Assume X and Y are two categorical objects with m attributes, i.e., $X = (x_1, x_2, ..., x_m)$ and $Y = (y_1, y_2, ..., y_n)$. Define the distance between X and Y as

$$d(X,Y) = \sum_{j=1}^{m} dt(x_j, y_j), \tag{6.15}$$

where dt is a function that depends on x_j and y_j, and can be represented as

$$dt(x_j, y_j) = \begin{cases} 0, & \text{if } x_j = y_j \\ 1, & \text{if } x_j = y_j \end{cases}. \tag{6.16}$$

In the above examples, $d(X, Y)$ is the distance between two objects X and Y. This algorithm is performed as follows:

Step 1: Randomly select *k*-modes for *k* clusters.
Step 2: Allocate an object to a cluster with the nearest mode according to Equation 6.16 so that

$$d(X,Y) = \sum_{j=1}^{m} dt(x_j, y_j). \tag{6.17}$$

Step 3: Update the modes of the cluster.
Step 4: Repeat steps 2 and 3 until all points in the clusters are stable.

For binary attributes, the *k*-modes algorithm, which uses the binary form for distance computation, can represent the same or different conditions corresponding to the distance values of 0 or 1. However, it is difficult to change categorical attributes into numeric form. Therefore, genetic distance measure, which can measure the distance or similarity between categorical attributes using similarity measures and is described in next section, is a preferable measure for making calculations. In Table 6.2, A, B, and C are three objects that have binary attributes, and their values are either 0 or 1.

6.3.5 Genetic Distance Measure (GDM)

The *k*-medoids algorithm can be used in the place of binary distance in a traditional *k*-modes algorithm to measure continuous distance in the genetic algorithm

Table 6.2 Distance for Conventional *k*-Modes Algorithm

	A	B	C
A	0	1	1
B	1	0	1
C	1	1	0

(Chiang et al. 2006). Given two categorical datasets with m attributes X and Y, such as $X = (x_1, x_2, \ldots, x_n)$ and $Y = (y_1, y_2, \ldots, y_n)$, we define the continuous distance between X and Y as

$$d(X,Y) = \sum_{j=1}^{m} T_j(x_j, y_j), \tag{6.18}$$

where $T_j(x_j, y_j)$ is the continuous distance table for the jth attribute under the training of genetic algorithm.

Given T objects, $X_i, i = 1, 2, \ldots T$, which belong to the partitioned set S_j, and k-modes $Q_1, Q_2, \ldots, Q_k$, which represent the corresponding clusters, the fitness function F_t is

$$F_t = \sum_{j=1}^{T} \frac{\sum_{i=1, j \neq S_j} d(X_j, Q_j)}{d(X_j, Q_{S_j})}. \tag{6.19}$$

Maximizing fitness function F_t is similar to minimizing the distance between the object and its corresponding mode in its cluster and maximizing the distance between the object and the modes in the other clusters (Chiang et al. 2006).

6.4 Applications of Distance-Based Clustering in Bioinformatics

- A new distance metric in gene expressions for coexpressed genes
- Gene expression clustering using the mutual information distance measure
- Gene expression data clustering using a local shape-based clustering

Each of these applications is described below.

6.4.1 New Distance Metric in Gene Expressions for Coexpressed Genes

Cluster analysis has been used to determine gene functions, but many clustering algorithms ignore the functionality of genes that are already known. Thus, the objective for using a distance metric in gene expression for coexpressed genes is to incorporate known gene functions and to find whether common gene functions can be shared between genes or not (Huang and Pan 2006). If common gene function can be shared, then they shrink a gene expression-based distance toward 0. The new distance metric d_{ij}^* is based on the expression-based distance metric d_{ij} and gene functional annotations, as shown in Equation 6.20:

$$d_{ij}^* = \begin{cases} rd_{ij} & \text{if there is an } f \text{ such that } 1 \leq f \leq F \text{ and } I, J \in G_f \\ \text{else } d_{ij} \end{cases}, \quad (6.20)$$

where $0 \leq r \leq 1$ is a shrinkage parameter, and the function $r = 1$ causes clustering techniques to ignore gene functions (Huang and Pan 2006). A two-step method is used to perform this equation. First, genes with known functions are clustered using a distance-based clustering method, e.g., k-medoids. Second, genes with unknown functions are clustered using an expression-based distance metric and can be assigned to clusters that are either obtained in the first step or assigned to the new cluster. An algorithm is described below to implement these two steps:

Step 1: Apply the k-medoid algorithm, using the shrinkage distance matrix D^* to the genes $\{n_0 + 1, ..., n\}$, with known functions $\{G_1, ..., G_F\}$, having k_0 clusters.

Step 2: Apply the k-medoid algorithm, which uses expression-based distance matrix D, to genes $\{1, ..., n_0\}$ with unknown function in G_0 to create k_1 new clusters while retaining information about the k_0 clusters obtained in step 1.

Step 2 can be further divided into substeps, as outlined below:

Step 2.1: Select k_1 genes from $\{1, ..., n_0\}$ as medoids at random.
Step 2.2: Update centroids and calculate cluster membership for each gene in the k_1 new clusters.
Step 2.3: Update the k_1 medoids.
Step 2.4: Repeat the above two mentioned steps until convergence.

G_0 genes are assigned to $k_0 + k_1$ clusters in step 2.1, whereas in the k_0 clustered genes in $G_1 ... G_F$, remains are obtained using the shrinkage distance matrix.

For this algorithm, the original expression-based distance matrix D is used instead of a shrinkage distance matrix D^* for step 2 due to incomplete biological knowledge.

The above method allows multiple known functions for genes because it uses only shrinkage distance matrix D^*, and shrinkage distance d_{ij}^* is well defined when one gene belongs to two or more functional groups. For example, if genes i and j belong to two or more groups, then d_{ij}^* does not change.

In conclusion, a result of $k_1 > 0$ indicates that the expressed gene has an unknown functionality and may not be assigned to clusters because of undiscovered gene functions or lack of evidence from expression profiles.

6.4.2 Gene Expression Clustering Using Mutual Information Distance Measure

A mutual information (MI) measure can be taken from different dataset sizes to provide important information for finding positive, negative, and nonlinear correlations between data (Priness et al. 2007). To accurately classify these correlations, the expression patterns should be in the form of discrete random variables. Given two random variables X, Y, such that X has a range $x_i \in A_x$ and probability distributions functions $P(X = x_i) \equiv P_i$, whereas Y has a range $y_j \in A_j$ and probability distributions functions $P(Y = y_j) \equiv P_j$, the MI between two random variables X and Y is given by

$$I(X;Y) = \sum_i \sum_j p_{ij} \log \frac{p_{ij}}{p_i p_j}. \tag{6.21}$$

If the MI is zero, then X and Y are not dependent on each other. In such a case, there is no relationship between X and Y, but it is difficult to achieve such a condition using the Pearson correlation or the Euclidean distance (Herzel 1995).

Let us assume that there are N samples in a dataset that have been correctly clustered into two groups, and l samples are misclassified. In order to find the error in clusters, samples will move from their respective clusters or true clusters. When the error increases, different solutions also increase. Hence, different possible solutions are randomly selected that have more than a single error. Then, different clustering solutions are gathered based on the number of errors in the cluster. The average homogeneity and separation scores are computed for each cluster to define the robustness of similarity measures and validate these information nodes based on the assumptions that homogeneity and separation scores are dependent on the number of errors in a solution, and that statistical methods differentiate between high-quality and low-quality clustering solutions based on statistical error. These criteria are important because homogeneity and separation scores are dependent on the number of errors in a solution. For example, if the number of errors in a solution is fewer than a threshold, then homogeneity and separation scores in the solution are good. Likewise, if the number of errors in a solution is above a threshold, then the homogeneity and separation scores are bad. Homogeneity, then, helps differentiate high-quality and low-quality clustering solutions based on their scores. Statistical methods, such as standard deviation and mean values, are more rigorous, and differentiate high-quality and low-quality clustering solutions based on statistical error. This smaller probability helps separate high-quality clustering solutions from low-quality clustering solutions based on statistical scores.

6.4.3 Gene Expression Data Clustering Using a Local Shape-Based Clustering

Clustering with local shape-based similarity (CLARITY) is used to analyze microarray time course experiments. Balasubramaniyan et al. (2005) developed CLARITY, based on Spearman rank correlation, which uses a local shape-based similarity measure and is robust toward noise. It finds similarities between gene expression profiles and includes the probability of time shifts into these relationships. CLARITY was developed using the following method. Let two gene expression profiles, X and Y, be represented by sequences $(x_1,\ldots,x_n)$ and $(y_1,\ldots,y_n)$, respectively. X and Y are similar if their respective subsequences, $X[i, j]$ and $Y[k, l]$, are also similar, where $X[i, j] = \text{def}\,(x_i, x_{i+1},\ldots, x_j)$ for $1 \leq i \leq j \leq n$.

6.4.3.1 Exact Similarity Computation

Exact similarity is defined as when all possible alignments are used to compute local, time-shifted relationships between two profiles. Thus, the similarity $SIM(X,Y)$ X and Y of length is computed as

$$SIM(X,Y) = def \max_{k_{\min} \leq k \leq n} SIM_k(X,Y), \qquad (6.22)$$

where $SIM_k(X,Y)$ measures the similarity of the best alignment of length k given by

$$\max_{1 \leq i, j \leq n-k+1} S(X[i, i+k-1], y[j, j+k-1]) \qquad (6.23)$$

6.4.3.2 Approximate Similarity Computation

The exact computation of $SIM(X,Y)$ is expensive for longer gene expression profiles. Therefore, approximate similarity computation, which is based on the basic local alignment search tool (BLAST) method, is used. The BLAST method is used to find optimal sequence alignments. Then, these optimal alignments are extended in both directions. These steps are outlined below.

Hit: Compute $SIM_k(X, Y)$ for $k = k_{min}$. This similarity degree is obtained for the best match $X[a_x, b_x], Y[a_y, b_y]$, i.e.,

$$SIM^k(X,Y) = S(X[a^x, b^x], Y[a^y, b^y]). \qquad (6.24)$$

Next, determine whether the best match is unique. If it is not unique, then go to the next step, "extend."

Extend: Derive the similarity degrees $S(X[a_x - u, b_x + v], Y[a_y - u, b_y + v])$ for all $0 \leq u \leq \min\{d, a_x - 1, a_y - 1\}$, $0 \leq v \leq \min\{d, n - b_x, n - b_y\}$, and find the best match. If more than one match meets these criteria, then choose the randomly optimal longer match.

Iterate: Update the optimal local alignment by replacing $a_x \leftarrow a_x - u^*$, $b_x \leftarrow b_x + v^*$, $a_y \leftarrow a_y - u^*$, $b_y \leftarrow b_y + v^*$, and and repeat the second step. Repeat this process until the optimal alignment does not change.

6.5 Implementation of *k*-Means in WEKA

Numerous datasets are available for applying *k*-means using the GenePattern tool (Reich et al. 2006). Of these datasets, we have selected the acute lymphoblastic leukemia (ALL)/acute myeloid leukemia (AML) dataset to run our experiments. The ALL/AML data are available through the GenePattern tool (Reich et al. 2006). The dataset consists of 71,29 gene expression profiles of two acute cases of leukemia: (1) acute lymphoblastic leukemia (ALL, 47 samples, ALL-B, 38 samples, and ALL-T, 9 samples) and (2) acute myeloblastic leukemia (AML, 25 samples, AML-BM, 21 samples, and AML-PB, 4 samples). To implement *k*-means using the opensource data mining software WEKA (Hall et al. 2009), we use four input parameters: distance measure, number of clusters, seed points, and terminating point.

In WEKA, we have chosen Euclidean distance to compute distances between instances and clusters. Two clusters are selected by default, but we can change the number based on individual requirements.

k-Means is limited in its sensitivity to how clusters are initially assigned and in its inability to determine a termination point. The seed value is used to assign instances to clusters. The terminating point may occur if there is no change in the position of the centroid. Figure 6.8 shows the clustering instances obtained for six clusters, clusters 0 to 5, with the following parameters.

There are 10 seed points and 44 iterations. In addition, the missing values were globally replaced with a mean/mode. Figure 6.8 shows six clusters. The number of samples are divided into these six clusters. Cluster 0 has 206 samples, cluster 1 has 77 samples, cluster 2 has 72 samples, and so on.

	Clustered Instances
0	206 (3%)
1	77 (1%)
2	72 (1%)
3	8 (0%)
4	1193 (17%)
5	5573 (78%)

Figure 6.8 Six clusters (0 to 5) with number of samples in each cluster.

6.6 Hierarchical Clustering

In hierarchical clustering, a series of steps are performed to partition the data into a cluster. Based on the similarity/dissimilarity of the objects, this technique may begin with one cluster that contains n objects, or with n clusters that each contain one object.

Hierarchical clustering consists of two methods: agglomerative or bottom-up hierarchical clustering techniques and divisive or top-down hierarchical clustering techniques.

6.6.1 Agglomerative Hierarchical Clustering

For agglomerative or bottom-up hierarchical clustering techniques, as shown in Figure 6.9, each object begins as a cluster. At each successive step, objects are

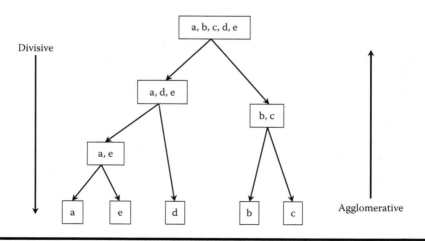

Figure 6.9 Hierarchical clustering.

combined into larger clusters based on their similarity, until a terminating point is reached or all the objects have been reassigned into one cluster. Divisive or top-down hierarchical clustering, as shown in Figure 6.9, is used as a reverse approach of the agglomerative technique. In this method, all objects are first assigned to one cluster, and then are divided into subclusters at each successive step until each point becomes a cluster, or a terminating point is reached (e.g., the required number of clusters is attained).

6.6.2 Cluster Splitting and Merging

Cluster splitting and cluster merging in hierarchical clustering, as discussed in Section 6.6.1, consist of two steps, a min-max cut algorithm and cluster merging using a Gaussian mixture, respectively (Ding and He 2001). These steps, which include merging nodes in agglomerative hierarchical clustering and splitting nodes in divisive hierarchical clustering, are described below.

The min-max cut algorithm is based on the min-max clustering principle that data should be assigned to clusters in such a way that intercluster similarity is minimized while intracluster similarity is maximized. For example, assume there are n data objects and the pairwise similarity matrix is $W = w_{ij}$, where w_{ij} is the similarity between i and j.

By using the min-max clustering principle, we divide the n data objects into two clusters C_1 and C_2. The similarity between C_1 and C_2 is defined as $s(C_1, C_2) \equiv \sum_{i \in C_1} \sum_{j \in C_2} w_{ij}$, which is also called the overlap between C_1 and C_2. The similarity within a cluster C_1 is the sum of pairwise similarities within $C_1 : s(C_1, C_1)$. Using the clustering principle, $s(C_1, C_2)$ is minimized while $s(C_1, C_1)$ and $s(C_2, C_2)$ are maximized.

Some data points are located near the boundaries of multiple clusters. These points are assigned to different clusters using probabilistic models. Based on the membership values, points are assigned to their natural cluster using Gaussian mixtures.

Splitting nodes in divisive hierarchical clustering is a three-step process in which cluster labeling, size priority cluster splitting, and average similarity must be performed. Cluster labeling is performed so that users can collect information about the clusters. Feature selection methods, such as mutual information (MI), information gain, and the chi-square method, can be used to identify cluster labels that characterize one cluster in contrast to other clusters. Size priority cluster split is the process in which the cluster with the largest split is selected. However, this approach is not optimal, as clusters are not of similar sizes. Average similarity is based on the min-max clustering principle, which requires that $s(C_1, C_1)$ be maximized. Therefore, clusters with high average similarity imply that data points in a cluster are similar. If we assume similarity is inversely proportional to distance, then data points in a cluster are similar to each other in Euclidean space. Therefore, the goal is to split clusters to increase the average similarity for all clusters.

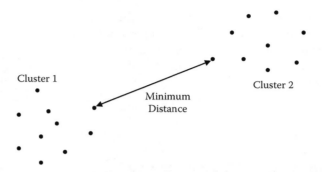

Figure 6.10   Single Link distance.

6.6.3 Calculate Distance between Clusters

Four measures, single link distance, complete link distance, centroid distance, and medoid distance, are commonly used to compute the distance between clusters.

The first measure is single link distance. It is the smallest or minimum distance between an object in one cluster and an object in the other cluster, i.e.,

$$dis(k_i, k_j) = \min(t_{ip}, t_{jq}), \tag{6.25}$$

where $\min(t_{ip}, t_{jq})$ is the minimum distance between two objects or points p in cluster i and q in cluster j. Single link distance is shown in Figure 6.10. In this figure, the minimum distance between elements of clusters 1 and 2 is shown. In Figure 6.10, dots are the data points within the clusters and the arrow indicates the distance between two data points. On left side of the arrow, the set of data points is called cluster 1, and similarly, on the right side of the arrow, the set of data points is called cluster 2.

The second measure is complete link distance. It is the largest distance between an object in one cluster and an object in the other cluster, i.e.,

$$dis(k_i, k_j) = \max(t_{ip}, t_{jq}), \tag{6.26}$$

where $\min(t_{ip}, t_{jq})$ is the distance between two objects or points, p in cluster i and q in cluster j. Figure 6.11 shows the complete link distance between cluster 1 and cluster 2. The figure also shows the maximum distance between elements of these clusters.

The third measure is centroid distance. It is the distance between the centroids of two clusters, i.e.,

$$dis(k_i, k_j) = dis(c_i, c_j) \tag{6.27}$$

where c_i is the centroid for cluster k_i, and c_j is the centroid for cluster k_j. Figure 6.12 shows a centroid distance.

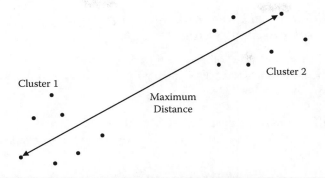

Figure 6.11 Complete link distance.

The fourth measure is medoid distance. It is the distance between the medoids of two clusters, i.e.,

$$dis(k_i, k_j) = dis(m_i, m_j) \quad (6.28)$$

where m_i is the medoid for cluster k_i and m_j is the medoid for cluster k_j.

6.6.4 Applications of Hierarchical Clustering Techniques in Bioinformatics

Researchers use hierarchical clustering techniques in bioinformatics to find the appropriate number of clusters or cluster stability estimation for microarray data. Three applications of distance-based clustering in bioinformatics are described below:

1. Hierarchical clustering based on partially overlapping and irregular data
2. Cluster stability estimation for microarray data
3. Comparison of gene expression sequences using pairwise average linking

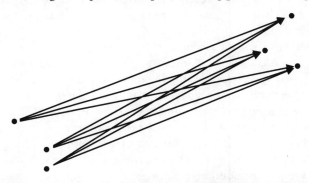

Figure 6.12 Centroid distance.

6.6.4.1 Hierarchical Clustering Based on Partially Overlapping and Irregular Data

Hierarchical clustering based on partially overlapping and irregular data is used to overcome the limitations of a clustering algorithm. The most important of these limitations are listed below:

1. It is difficult to select the appropriate number of clusters.
2. It is essential to distinguish partially overlapping and irregular data (Qu et al. 2007).

Similarity measures between subclusters can perform two roles:

1. They can control the merger process of hierarchical clustering.
2. Based on overlap similarity measure, these types of algorithms stop clustering automatically and cluster the overlapping data.

Hierarchical clustering algorithms perform well with data that are irregular or have overlapping partitions. However, these algorithms are usually unable to interpret the structures of partially overlapping data, for, e.g., those found in the IRIS dataset. Generally, for overlap, a similarity threshold value is set to control the number of clusters. However, it is not easy to select a global threshold for a dataset.

Qu et al. proposed the HCOSM clustering algorithm to merge overlapped subclusters. There are two phases to this algorithm: initialization into subclusters and merging pairs of subclusters. Each of the phases contains steps. These phases and their corresponding steps are described below.

In the first phase, data are partitioned into subclusters, following steps 1 and 2 below. In the second phase, these subclusters are merged, following steps 3–6.

Phase I: Initialization into subclusters
  Step 1: Use the k-means algorithm to partition the data into clusters.
  Step 2: Find pairs of subclusters that satisfy the conditions, so they can be merged in phase II.
Phase II: Merging pairs of subclusters
  Step 3: Use the COSM algorithm to determine overlap similarity between each pair of clusters and find the maximum overlap similarity measure.
  Step 4: Select and merge all possible candidate cluster pairs, and select a candidate that satisfies a certain threshold for merging. Then, combine subclusters into one subset and calculate their mean.
  Step 5: Update the number of clusters, and if the number of clusters is less than the maximum number of clusters, repeat steps 3–5; otherwise, go to step 6.
  Step 6: Output the number of clusters.

6.6.4.2 Cluster Stability Estimation for Microarray Data

Clustering output will depend on a number of factors, such as the number and stability of clusters in a dataset. However, most of the clustering is performed by analyzing or finding the number of clusters. This problem can be addressed with cluster stability scores using the subsampling technique, which can work for both known and unknown clusters (Smolkin and Ghosh 2003).

Clustering is performed by calculating the number of clusters using the Ben-Hur method and computing the random subspace measures (Smolkin and Ghosh 2003; Ben-Hur et al. 2002). The number of clusters can be determined using four steps, as described below:

Step 1: Estimate the number of clusters.
    1a: Partition samples into k clusters.
    1b: At each iteration, select samples and group subsamples into k clusters.
    1c: For each subset, calculate pair correlation between the clusters using the Jaccard coefficient.
    1d: If n correlations are computed for each cluster, and if the distributions of correlation coefficients are mapped, then distributions obtained from the correlation coefficients help determine the number of clusters (Fowlks and Mallows 1983).

The random subspace-based sensitivity measures can be computed as described below.

Step 2: Compute random subspace-based sensitivity measures.
    2.1: Perform the random subspace method if the number of clusters is known.
        2.1a: Partition the samples into k sets.
        2.1b: Randomly choose a subset (for example, 65% samples).
        2.1c: Create a dissimilarity matrix and follow the hierarchical clustering procedure.
        2.1d: Get k clusters.
        2.1e: Determine whether $Ai \subset Aj$, then randomly select a subspace and repeat B times.
        2.1f: Determine the sensitivity measure by calculating the proportion of B samples in which a set appears for each of the original sets $A1$, $A2$, ..., Ak.
        2.1g: If the value of the sensitivity measure is close to 1, then the cluster is more stable than sensitivity measure that is not close to 1.
    2.2: Randomly subspace for an unknown number of clusters.

2.2a: Estimate the number of clusters using the Ben-Hur method (Ben-Hur et al. 2002).
2.2b: Once the number of clusters in the above step is estimated, use the above-mentioned random subspace method to calculate sensitivity measures of the clusters. Follow the above steps.

Finally, to estimate the reliability of individual clusters: the R-index and the cluster scoring method can be used (Ray and Bandyopadhyay 2007; Tsai et al. 2004).

6.6.4.3 Comparing Gene Expression Sequences Using Pairwise Average Linking

Sokal and Michener (1958) applied a pairwise average linking cluster analysis to gene expression to compare the sequences. The gene similarity measure is based on a correlation coefficient that is found by computing a similarity score. A similarity score can be calculated for any two genes X and Y observed over a series of n conditions. For example, let G_i equal the (log-transformed) primary data for gene G in condition i. For any two genes X and Y, a similarity score can be calculated over a series of N conditions (Eisen et al. 1998), as

$$S(X,Y) = \frac{1}{N} \sum_{i=1,N} \left(\frac{X_i - X_{offset}}{\varphi_x} \right) \left(\frac{Y_i - Y_{offset}}{\varphi_y} \right) \quad (6.29)$$

where $\varphi_g = \sqrt{\sum_{i=1,N} \frac{(G_i - G_{offset})^2}{N}}$.

Step 1: Follow the agglomerative approach.
Step 2: Compute the similarity matrix using the metric above.
Step 3: Identify similar pairs of genes based on their highest value.
Step 4: Create a node by joining the two most similar genes; compute the gene expression profile for the node, and update the similarity matrix by replacing these two elements.
Step 5: Repeat the process $n - 1$ times.

6.7 Implementation of Hierarchical Clustering

We have implemented hierarchical clustering, using the GenePattern tool, on the all/aml dataset to run our experiments (as explained in Section 6.5). The algorithm is based on agglomerative hierarchical clustering and groups all elements into a cluster according to their pairwise distance, with the closest item pairs being merged first (Eisen et al. 1998). The comparison of gene expression sequences using pairwise average linking is discussed in Section 6.6.3.3.

We have selected the following parameters to implement hierarchical clustering in the GenePattern tool:

1. **Input file name:** Already explained in Section 6.5.
2. **Measure column distance:** Pearson correlation is used to measure column distance.
3. **Measure row distance: No row clustering**

Normalize row, center column, and column normalize are selected default values. Figure 6.13 shows hierarchical clustering results.

6.8 Self-Organizing Maps Clustering

Self-organizing maps (SOMs) are used to make the visualization of data easier by mapping or transforming the n-dimensional data into one-dimensional (1D) or 2D data (Germano 1999). For example, in Figure 6.14, if we need to map 40 dimensions each having 100 data points into two dimensions, then we will divide the 100 data points into a 10 × 10 matrix, and each block of the matrix will represent the data found in 40 dimensions.

There are two components of SOM: data and weights (Germano 1999). Data can be any number of data points in n-dimensional space, e.g., 100 data points in 40 dimensions. Weights are further divided into two parts: data and natural location. This data are different from the data mentioned above and should have the same dimensions in the above example. A natural location is the location of the data points in the matrix, such as (1, 1), (2, 2).

6.8.1 SOM Algorithm

The SOM algorithm is initiated by randomly selecting a data point (Ultsch and Siemon 1990). Once the data point is selected, use the best matching unit to search the data points that are similar or best represent the selected data point. In this step, distance is calculated between the selected data point and every other data point. The data point that has the shortest distance will be selected as the one most similar to the selected data point. The most common way of calculating distance (e.g., to find the distance between two data points p and q) is the Euclidean distance, which is given as

$$d(p,q) = \sqrt{\sum_{i=1}^{n}(p_i - q_i)^2}. \qquad (6.30)$$

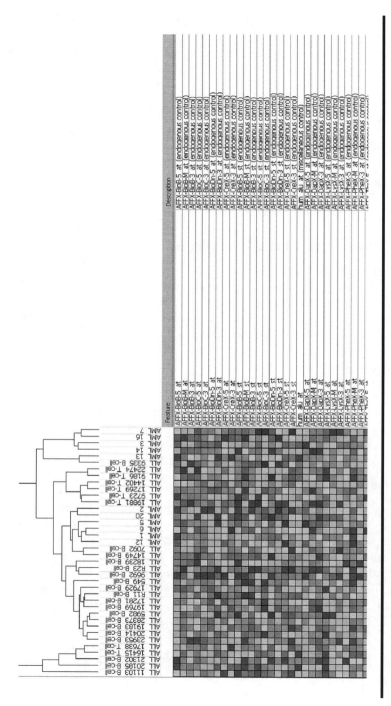

Figure 6.13 Hierarchical clustering results.

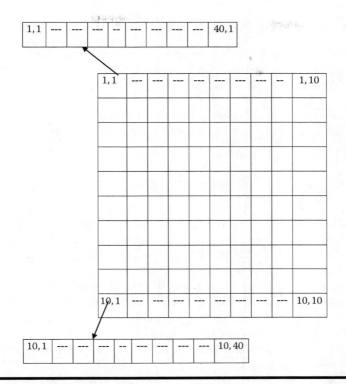

Figure 6.14 Self-organizing map.

Let us assume a randomly selected data point is p (0,6,2) and two other data points are q (0,3,4) and r (2,3,5). By using Euclidean distance, we will find which of the two data points is closer to the selected data point. This equation can be visualized as

$$d(p,q) = \sqrt{(0-0)^2 + (6-3)^2 + (2-4)^2} = \sqrt{13} = 3.6$$
$$d(p,r) = \sqrt{(0-2)^2 + (6-3)^2 + (5-2)^2} = \sqrt{18} = 4.24. \quad (6.31)$$

Thus, data point q (0, 3, 4) is the best matching unit because it has a shorter distance than data points p (0, 6, 2) and r (2, 3,5).

Once the most similar data point is selected, we check to see if its neighbors are similar. The neighbors can be found using methods such as concentric squares, hexagons, or Gaussian functions. If a neighbor is similar, it is selected, and the above steps are repeated iteratively. In this way, scaling of neighbors occurs, so that similar data points are grouped together (Fukunaga 1990).

6.8.2 Application of SOM in Bioinformatics

Researchers use SOM techniques for bioinformatics applications (Wang et al. 2001). To illustrate the usefulness of SOM techniques in bioinformatics, two such applications of SOM are described below:

1. Identifying distinct gene expression patterns using SOM
2. SOTA: Combining SOM and hierarchical clustering for convenient representation of genes

6.8.2.1 Identifying Distinct Gene Expression Patterns Using SOM

Due to the high complexity and dimensionality of microarray gene expression profiles, dimensional reduction and the feature selection of raw expression data are necessary (Wang et al. 2002). To solve this problem, Wang et al. have proposed a two-step analysis.

The first step of this analysis is to use a self-organizing map (SOM) to reduce the dimensionality of the original data and help visualize the data more effectively and efficiently in a SOM component plane. The second step is hierarchical and k-means clustering is used to identify gene expression patterns to classify samples.

From the training set, sample data are chosen at each training step, and the distances between sample data and all prototype vectors are calculated (Kohonen 1997). During training, data points are moved toward the dense area based on similarity using neighborhood data points, which leads to prototype vectors of neighboring units resembling each other (Vesanto 1999). The SOM component plane inspects the cluster structure, by comparing the spread of values in a component plane. Correlations can then be revealed between similar patterns in identical positions. To find an interesting group or cluster of map units, k-means clustering is used to further cluster trained prototype vectors m_i of SOM, which are combined to form clusters (Vesanto and Alhoniemi 2000).

Vesanto used validity indexes such as Davies-Bouldin to validate the best clusters by minimizing intercluster similarity and maximizing intracluster similarity. This validation scheme provided good clustering results for spherical clusters (Vesanto 1999). The algorithm has some limitations; clusters with nonspherical shapes are not recognized as one cluster. Moreover, as the cluster increases, the algorithm becomes sensitive to outliers and the number of samples in clusters decreases.

6.8.2.2 SOTA: Combining SOM and Hierarchical Clustering for Representation of Genes

SOTA combines the advantages of hierarchical clustering and SOM for convenient representation of genes. The biggest advantage of hierarchical clustering methods is that they help researchers visualize and represent genes more conveniently.

However, these methods are neither robust nor efficient, whereas SOM is insensitive to noise, but require that the number of clusters be known before implementation (Longde et al. 2006).

SOM is a neural network with a number of nodes that have the same length of the input data and are assigned random values when the process begins (Kohonen 1998; Tamayo et al. 1999). The reference vectors are grouped together based on their closeness or the similarity of genes with respect to reference vectors. The advantage of SOM is that the input of other genes can counterbalance and correct the effects of outliers.

The self-organizing tree algorithm is a divisive (top-down) clustering method, which starts with a node called a root, and there are two leaves, each representing one cluster (Dopazo and Carazo 1997; Herrero and Dopazo 2002; Tamames et al. 2002). The tree grows when the mean value between the cluster and the genes associated with it merge into a node. The growth of a tree can be stopped by customizing a specific number of loops. The SOTA algorithm is nondeterministic and is not sensitive to noise or outliers. It is more flexible than the HC method and SOM.

6.9 Fuzzy Clustering

In hard clustering each object is assigned to only one cluster when cluster boundaries are well defined. However, in many cases cluster boundaries are ambiguous; hence fuzzy clustering can help in overcoming this limitation. In fuzzy clustering, the object can be assigned to more than one cluster based on degree of membership associated with each object when the cluster boundaries are ambiguous (Dave 1992; Eschrich et al. 2003). The degree of membership indicates the strength of association between an object and a particular cluster (Wunsch and Xu 2005; Zadeh 1965). For example, when a coin is tossed, as explained in Section 6.3.1.5, there is an uncertainty as to whether the output will be heads or tails. This type of uncertainty is called fuzziness.

Figure 6.15 shows two clusters. In the first cluster, samples are denoted by ○. In the second cluster samples are denoted by □. In hard clustering, each object is

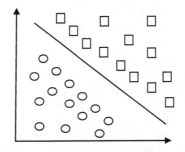

Figure 6.15 Hard clustering.

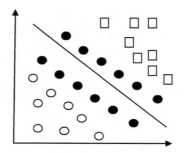

Figure 6.16 Fuzzy clustering.

assigned to only one cluster, whereas in Figure 6.16, an object can be assigned to more than one cluster. For example, object ● is assigned to more than one cluster based on degree of membership.

The objective function measures the overall dissimilarity within clusters. This dissimilarity needs to be minimized to obtain optimal partitioning where membership values determine how much fuzziness a fuzzy set contains. Memberships can determine important relations between a given object and the disclosed clusters.

The mountain function for a vertex is defined as

$$M(v_i) = \sum_{j=1}^{N} e^{-aD(x_j, x_i)},\qquad(6.32)$$

where $D(x_j, v_i)$ is the distance between the jth data object and the ith node, and is a positive constant. Therefore, if the data object is closer to a vertex, it contributes more to the mountain function. First, the center is selected based on the vertex v_{ml}, which has the maximum value of mountain function M_{vml} and removal of selected center mountain destruction is performed. Mountain destruction can be performed by subtracting the value of the mountain function for each of the remaining vertices, which depends on two factors:

1. Current maximum mountain function value
2. Distance between the vertex and the center

The algorithm stops when a terminating point is reached. The terminating point is the ratio between the current maximum and M_{vml}, which should be below a threshold value (Dave and Krishnapuram 1997).

6.9.1 Fuzzy c-Means (FCM)

Fuzzy c-means (FCM) attempts to find a partition (c fuzzy clusters) for a set of data points $x_j \in \Re^d$, $j = 1,...,N$, while minimizing the cost function, as denoted in (Hoppner et al. 1999)

$$J(U,M) = \sum_{i=1}^{c}\sum_{j=1}^{N}(u_i,j)^m D_{ij} \quad (6.33)$$

where $U = [u_{i,j}]_{c \times N}$ is the fuzzy partition matrix and $u_{i,j} \in [0,1]$ is the membership coefficient of the jth object in the ith cluster. $M = [m_i,...,m_c]$, then, is the cluster prototype (mean or center) matrix, $m \in [1,\infty)$ is the fuzzification parameter and is usually set to 2, and $D_{ij} = D(x_j, m_i)$ is the distance measure between x_j and m_j (Hathaway and Bezdek 2001).

Wunsch and Xu (2005) have summarized the standard FCM in the four-step algorithm that follows:

1. Select appropriate values for m, c, and a small positive number ϵ. Initialize the prototype matrix M randomly. Set step variable $t = 0$.
2. Calculate (at $t = 0$) or update (at $t > 0$) the membership matrix U by

$$u^{ij(j+1)} = \frac{1}{\left(\sum_{l=1}^{c}(D_{ij}/D_{ij})^{1/(1=m)}\right)} \quad (6.34)$$

   for $i = 1, ..., c$ and $j = 1, ..., N$.
3. Update the prototype matrix by

$$m_i^{(t+1)} = \left(\sum_{j=1}^{N}\left(u_{ij}^{(t+1)}\right)^m x_j\right) / \left(\sum_{j=1}^{N}\left(u_{ij}^{(t+1)}\right)^m\right) \quad (6.35)$$

   for $i = 1, ..., c$.
4. Repeat steps 2 and 3 until $\| M^{(t+1)} - M^{(t)} \| \leq \epsilon$, in which the Euclidean or L^2 norm distance function is used.

The fuzzy c-means (FCM) method measures the cluster centroid as the mean of all points, which are weighted by their location in the cluster. The weighting is inversely related to the distance from the centroid to the cluster. Since cluster centers and membership grades are updated in each iteration, the accuracy of FCM depends on selection of initial centroids.

FCM suffers from four potential problems: the presence of noise and outliers, identifying initial partitions, unknown locations of clusters (centers) *a priori*, and the number of points to be handled due to a large variability in cluster shape or density.

6.9.2 Application of Fuzzy Clustering in Bioinformatics

Researchers use fuzzy clustering techniques for bioinformatics applications (Dembele and Kastner 2003; Horimoto and Toh 2001). To illustrate the usefulness of fuzzy clustering techniques in bioinformatics, three such applications of fuzzy clustering are described below:

1. Clustering genes using fuzzy *J*-means and VNS methods
2. Fuzzy *k*-means clustering on gene expression
3. Comparison of fuzzy clustering algorithms

6.9.2.1 Clustering Genes Using Fuzzy J-Means and VNS Methods

The aim of fuzzy clustering is to assign a membership value ranging from 0 to 1 to a gene that can be in more than one cluster. In this algorithm, a value of 0 indicates a weak association with the cluster, and a value of 1 indicates a strong association with the cluster (Belacel et al. 2004).

The fuzzy *c*-means (FCM) method is an extension of the *k*-means method (Bezdek 1981; Dunn 1974; Ruspini 1969).

For a chosen number of clusters, c, and for an $n \times c$ matrix, $W = [w_{ik}]$, where w_{ik} is the membership degree for gene $i, i = 1, 2, \ldots, n$, to cluster $k, k = 1, 2, \ldots, n$, the FCM clustering problem can be represented as (Belacel et al. 2004)

$$\left(\begin{array}{c} \min \\ W,V \end{array} \right) J_m(W,V) = \sum_{i=1}^{n} \sum_{k=1}^{c} w_{ik}^m \, \|x_i - v_k\|^2 \qquad (6.36)$$

where $J_m(W,V)$ is the objectivity function that defines the quality of the result obtained for centroids V and memberships, W_m is the fuzzy parameter, and for $m = 1$ the partition is crisp, leading to the problem of minimum sum of squares clustering (Belacel et al. 2004).

$$V = [v_1, v_2, \ldots, v_c] = \begin{bmatrix} v_{11} & v_{12} & v_{1c} \\ v_{21} & v_{22} & v_{2c} \\ v_{31} & v_{32} & v_{3c} \end{bmatrix} \qquad (6.37)$$

gives a set of c centroids or prototypes, i.e., positions of cluster centers. In this instance,

$$||x_i - v_k||^2 = \langle x_i - v_k / x_j - v_k \rangle = \sum_{j=1}^{N} |x_{ij} - v_{ij}|^2 \qquad (6.38)$$

is the Euclidean norm determining distances between expression-level vectors and centroids, whereas membership degrees w_{ij} are defined such that $0 \le w_{ik} \le 1$ and $\Sigma_{k=1}^{c} w_{ik} = 1, \forall i = 1, 2, ..., n$.

The FCM algorithm is as follows (Belacel et al. 2004):

Step 1: Find the initial centroid.
Step 2: Calculate the membership and initial centroid.
Step 3: Calculate the new centroid.
Step 4: If the centroid is improved, go to step 5. Otherwise, go to step 2.
Step 5: Calculate the membership and the objectivity function.

The value of fuzzy parameter m has to be greater than 1, as $m = 1$ represents crisp clustering (Belacel et al. 2004).

Equation 6.38 can therefore be reformulated to (Hathaway and Bezdek 2001)

$$\left(\min_{v} \right) R^m(V) = \sum_{i=1}^{n} \left[\sum_{k=1}^{c} ||x_i - v_k||^{2(1-m)} \right]^{(1-m)} \qquad (6.39)$$

where $R_m(V)$ is the new objectivity function that depends on the centroid positions, which can be found by minimizing Equation 6.38, and therefore it will be used to compute membership values.

The FJM method, which was introduced by Belacel et al. (2002), is described below. To form defined neighborhoods, FJM uses all possible centroids-to-pattern relocations where membership values and centroids are calculated in the same way as in FCM (Belacel et al. 2002).

Step 1: Find the initial centroid.
Step 2: Calculate the objective function.
Step 3: Drop the least useful centroids.
Step 4: Once the centroid has been deleted, add the most useful pattern.
Step 5: Find membership values that help in changing the crisp solution to a fuzzy one and find the new centroid set using given memberships.
Step 6: Calculate the objective function R.
Step 7: If R improves, return to step 3. Otherwise, go to step 8.
Step 8: Calculate the memberships and the objectivity function.

Both FCM and FJM depend on starting centroid values (Belacel et al. 2004). Therefore, these values cannot guarantee an optimal clustering solution. The steps to the algorithm are as follows:

Step 1: Set the centroids and objective function to be optimal: set stopping conditions and k_{max}.
Step 2: If stopping conditions are satisfied, go to step 3; otherwise, go to step 8.
Step 3: Decrease the value of k by 1.
Step 4: If $k > k_{max}$, go to step 2. Otherwise, go to step 5.
Step 5: Generate at random new centroids, V, as the initial set of centroids.
Step 6: If R has improved, go to step 2. Otherwise, go to step 7.
Step 7: Increase the value of k by 1.
Step 8: Calculate memberships and objective functions.

6.9.2.2 Fuzzy k-Means Clustering on Gene Expression

Gasch and Eisen (2002) modified the fuzzy k-means clustering in two ways to determine overlapping clusters:

1. They ran fuzzy k-means clustering three times, and for the second and third runs, subsets of the data were used for clustering.
2. They used random initialization.

Genes representing rows and conditions representing columns in an expression value table were input for the algorithm. The steps for this process are outlined below.

Step 1: Clustering was begun by defining $k/3$ prototype centroids, where k and 3 are clusters and clustering cycles, respectively. Gasch and Eisen (2002) used PCA to identify these $k/3$ Eigen vectors.
Step 2: Find the correlation between the gene expression pattern and the centroid using the Pearson correlation method, which assigns a membership score to each gene for each prototype centroid.
Step 3: Calculate each centroid pattern again.
Step 4: Iterate the calculation of gene-centroid memberships and update the centroids until the required condition is met, for example, if centroid patterns become fixed or the termination criterion is met.

This method not only gives the unique centroid but also gives a matrix that provides membership scores for each gene for each centroid. Therefore, genes that belong to multiple clusters can be identified through their membership value.

6.9.2.3 Comparison of Fuzzy Clustering Algorithms

The comparison of fuzzy clustering algorithms method is a clustering algorithm that is used to classify fuzzy models (Almeida and Sousa 2001). This action is performed by comparing the computational efficiency and accuracy of the algorithm.

Selecting fuzzy models for classification is a complex task, and this task becomes more complex due to the large number of features in a large dataset; therefore it becomes necessary to select only the relevant features. The fuzzy clustering algorithm can be used to clustered data using different fuzzy clustering algorithms, such as possibilistic c-means, fuzzy possibilistic c-means, or possibilistic fuzzy c-means. These methods are all limited by the difficulty of determining which fuzzy clustering algorithm should be used for classification. In this section, different fuzzy clustering algorithms are compared for computational efficiency and accuracy.

Classification of systems using fuzzy clustering includes the following steps (Sousa and Kaymak 2002):

Step 1: Gather data by computing or constructing system-relevant features.
Step 2: Preprocess data to remove incomplete, noisy, and inconsistent data.
Step 3: Select and identify relevant features.
Step 4: Select a clustering algorithm and its parameters.
Step 5: Select the number of required clusters.
Step 6: Cluster data using the selected clustering algorithm.
Step 7: Determine the membership functions from clusters by projection.
Step 8: Determine the fuzzy rule from each cluster by using membership functions that are obtained in the previous step.
Step 9: Validate the model.

The possibilistic c-means is based on the FCM algorithm, which uses a condition that the sum of membership degrees must equal 1. Due to the presence of outliers, this condition is difficult to achieve. A possibilistic objective function is proposed to overcome this limitation (Krishnapuram and Keller 1993) and is given by

$$J(Z,\gamma,T,V) = \sum_{i=1}^{c}\sum_{k=1}^{N}(\mu_{ik})^m D_{ikA}^2 + \sum_{i=1}^{c}\gamma_i \sum_{k=1}^{N}(1-\mu_{ik})^m, \quad (6.40)$$

where γ_i are positive constants and D are $_{ikA}^{2}$, which is the squared inner-product norm defined in Equation 6.39.

The first term in the above equation is similar to the FCM objective function, shown in Equation 6.39. The distances between the feature vectors and the prototypes should be as small as possible, whereas the second term forces μ_{ij} to be as large as possible, whereby assigning all memberships to zero and minimizing the criterion function.

The difference between fuzzy-possibilistic c-means (FPCM) and other fuzzy clustering algorithms, such as FCM and PCM, is that FPCM produces both memberships and possibilities, along with the centers for each cluster. Both memberships and possibilities help visualize the correct interpretation of data. Membership helps classify a data object that has the representative vector closest to the data point. Possibility helps find the centroids to avoid the effect of outliers or noise. Moreover, FPCM overcomes the limitations of noise sensitivity defects and the coincident clusters problem of FCM and PCM, respectively (Pal et al. 1997). For example, in Equation 6.40,

$$J(U,T,V;Z) = \sum_{i=1}^{c}\sum_{k=1}^{N}(\mu_{ik})^m + (t_{ik})D_{ikA}^2, \qquad (6.41)$$

where D_{ikA}^2 is the squared inner-product norm defined in Equation 6.40 and $m > 1, n > 1, 0 \leq \mu_{ik}, t_{ik} \leq 1$.

FPCM is limited in that it constrains the typicality values. By removing the constraint on the typicality values and retaining the column constraint on the membership values, the objective function can be determined as (Hathaway and Bezdek 2001)

$$J(U,T,V;Z) = \sum_{i=1}^{c}\sum_{k=1}^{N}\left[(a\mu_{ik})^m + (bt_{ik})^n\right]D_{ikA}^2 + \sum_{i=1}^{c}\gamma_i\sum_{k=1}^{N}(1-t_{ik})^n \qquad (6.42)$$

where D_{ikA}^2 is the squared inner-product norm defined in Equation 6.40 and $m > 1$, $\eta > 1, 0 \leq \mu_{ik}, t_{ik} \leq 1$ and $\Sigma_{i=1}^{c}u_i = 1$.

Outliers can be reduced by using a bigger value for b than for a. Thus, similar effects can also be achieved by controlling the value of m. For example, to reduce the effect of outliers on the centroid, the large value of m and the small value of a can be used. However, in order to reduce the membership effects on the prototypes, a large value of m should be used, so that the model will behave more like the PCM model.

The fast fuzzy clustering algorithm (FFCA) is based on the self-organizing Kohonen network, in which the number of clusters is not known before application (Herrero et al. 2001; Qin et al. 2003). FFCA was introduced to select the input for nonlinear regression models. First, all datasets must be normalized so that all the data points lie in one particular range. Gaussian membership functions represent the clusters, and this algorithm uses the input patterns one by one. Initially, the first cluster center is defined by the input sample, and the initial cluster width is set to a default value, σ_{init}. For each pattern, it checks whether the sample belongs to a cluster c or not. If the sample does belong to cluster c, the pattern and cluster center v_i are added to that cluster. Otherwise, a new cluster is created.

6.10 Implementation of Expectation Maximization Algorithm

We have implemented the expectation maximization (EM) algorithm using the IRIS dataset, which is available in WEKA. In this dataset, there are 150 samples and 4 attributes. The four attributes are sepallength, sepalwidth, petallength, and petalwidth.

The EM algorithm works on the concept of probability distribution, where it indicates the probability of the instance belonging to each cluster. EM uses cross-validation, which helps in finding the number of clusters. However, users can also specify the number of clusters in the beginning that need to be generated.

Working of cross-validation is done to determine the number of clusters:

1. For 10-fold cross-validation, training data are split randomly into 10-fold and the number of clusters is set to 1.
2. EM is performed 10 times using 10-fold cross-validation, and the loglikelihood is averaged over all 10 results.
3. If loglikelihood has increased the number of clusters by 1, then go to step 2.

Figure 6.17 shows four clusters (0, 1, 2, 3) obtained from the expectation maximization (EM) algorithm by using a 10-fold cross-validation method. The mean, standard deviation, and distribution of points are given for each cluster of an attribute. There are four attributes in the dataset: sepallength, sepalwidth, petallength, and petalwidth. Moreover, it has three classes: Iris-setosa, Iris-versicolor, and Iris-virginica. As shown, cluster 0 contains 48 points, cluster 1 contains 50 points, cluster 2 contains 29 points, and cluster 3 contains 23 points.

6.11 Conclusion

Cluster analysis finds potential classes in the dataset, by either using a hierarchical structure or partitioning the data according to a prespecified number that includes steps ranging from preprocessing to cluster discovery and also provides great challenges to scientists. Although these algorithms solve several problems, each has limitations. Clustering algorithms usually follow certain assumptions and biases; therefore none of the clustering algorithms can solve all problems. Thus, selection of clustering algorithm depends on the application and needs of a researcher.

```
EM
==

Number of clusters selected by cross validation: 4

                            Cluster
Attribute                   0          1          2          3
                           (0.32)     (0.33)     (0.2)      (0.14)
==============================================================
sepallength
    mean                    5.897      5.006      6.9426     6.1304
    std. dev.               0.5279     0.3489     0.498      0.2943

sepalwidth
    mean                    2.7519     3.418      3.1103     2.8088
    std. dev.               0.3103     0.3772     0.2952     0.2361

petallength
    mean                    4.2267     1.464      5.8559     5.0993
    std. dev.               0.445      0.1718     0.4626     0.2462

petalwidth
    mean                    1.3134     0.244      2.1495     1.8254
    std. dev.               0.1864     0.1061     0.232      0.2152

class
    Iris-setosa             1          51         1          1
    Iris-versicolor         48.1125    1          1.0182     3.8693
    Iris-virginica          2.0983     1          31.0375    19.8641
    [total]                 51.2108    53         33.0557    24.7335
Clustered Instances

0       48 ( 32%)
1       50 ( 33%)
2       29 ( 19%)
3       23 ( 15%)

Log likelihood: -2.03504
```

Figure 6.17 Four clusters obtained using the expectation maximization algorithm.

References

Almeida, R.J., and J.M.C. Sousa. "Comparison of fuzzy clustering algorithms for classification". 2006 International symposium on Evolving Fuzzy Systems. Ambelside, U.K.: IEEE, 2006, 112–117.

Balasubramaniyan, R., E. Hullermeier, N. Weskamp, and J. Kamper. Clustering of gene expression data using a local shape-based similarity measure. *Bioinformatics* (2005).

Belacel, N., M. Cuperlovic-Culf, M. Laflamme, and R. Ouellette. Fuzzy J-means and VNS methods for clustering genes from microarray data. *Bioinformatics* (2004).

Belacel, N., P. Hansen, and N. Mladenovic. Fuzzy J-means: A new heuristic for fuzzy clustering. *Pattern Recog* 35 (2002): 2193–2200.

Ben-Hur, A., A. Elisseeff, and I. Guyon. "A stability-based method for discovering structure in clustered data." BIOCOMPUTING 2002. *Proceedings of the Pacific Symposium*, Kauai, HI: World Scientific Press, 2001, 6–17.

Bezdek, J.C. *Pattern recognition with fuzzy objective function algorithms.* New York: Plenum Press, 1981.

Chiang, C.-S., S.C. Chu, Y.-C. Hsin, and H.-M. Wang. Genetic distance measure for K-modes algorithm. *Int J Innovative Comput Inf Control* 2, no. 1 (2006): 33–40.

Newman, A.M and J.B. Cooper, and. AutoSOME: A clustering method for identifying gene expression modules without prior knowledge of cluster number. *BMC Bioinformatics* (2010).

Dave, R. Adaptive fuzzy c-shells clustering and detection of ellipses. *IEEE Trans Neural Networks* 3, no. 5 (1992): 643–662.

Dave, R., and R. Krishnapuram. Robust clustering methods: A unified view. *IEEE Trans Fuzzy Syst* 5, no. 2 (1997): 270–293.

Dembele, D., and P. Kastner. Fuzzy C-means method for clustering microarray data. *Bioinformatics* 19 (2003): 973–980.

Ding, C., and X. He. Cluster merging and splitting in hierarchical clustering algorithms. *Proceedings of the 2002 IEEE International Conference on Data Mining.* Washington, DC, USA: IEEE Computer Society, 2002, 139–146.

Dopazo, J., and J.M. Carazo. Phylogenetic reconstruction using an unsupervised growing neural network that adopts the topology of a phylogenetic tree. *J Mol Evol* 44 (1997): 226–233.

Dunn, J.C. A fuzzy relative ISODATA process and its use in detecting compact well-separated clusters. *J Cybernetics* 3 (1974): 32–57.

Eisen, M.B., P.T. Spellman, P.O. Brown, and D. Botstein. Cluster analysis and display of genome-wide expressions. *Proc Natl Acad Sci USA* 95, no. 25 (1998): 14863–14868.

Eschrich, S., J. Ke, L. Hall, and D. Goldgof. Fast accurate fuzzy clustering through data reduction. *IEEE Trans Fuzzy Syst* 31, no. 5 (2003): 262–270.

Fowlks, E.B., and C.L. Mallows. A method for comparing two hierarchical clusterings. *J Am Stat Assoc* 78 (1983): 553–569.

Fukunaga, K. *Introduction to statistical pattern recognition.* Boston: Academic Press, 1990.

Gasch, A.P., and M.B. Eisen. Exploring the conditional coregulation of yeast gene expression through fuzzy k-means clustering. *Genome Biol* 3, no. 11(2002): 1–22.

Germano, T. "Self organizing Maps" March 23, 1999. http://davis.wpi.edu/matt/courses/soms/(accessed September 15, 2011)

Hathaway, R., and J. Bezdek. Fuzzy c-means clustering of incomplete data. *IEEE Trans Systems Man Cybernetics* 31, no. 5 (2001): 735–744.

Herrero, J., and J. Dopazo. Combining hierarchical clustering and self-organizing maps for exploratory analysis of gene expression patterns. *J Proteome Res* 1, no. 5 (2002): 467–470.

Herrero, J., A. Valencia, and J. Dopazo. A hierarchical unsupervised growing neural network for clustering gene expression patterns. *Bioinformatics* 17, no. 2 (2001): 126–136.

Herzel, H., and I. Grosse. Measuring correlations in symbols sequences. *Physica A* 216, no. 4 (1995): 518–542.

Hoppner, F., F. Klawonn, and R. Kruse. *Fuzzy cluster analysis: Methods for classification, data analysis, and image recognition.* New York: Wiley, 1999.

Horimoto, K., and H. Toh. Statistical estimation of cluster boundaries in gene expression profile data. *Bioinformatics* 17, no. 12 (2001): 1143–1151.

Huang, D., and W. Pan. Incorporating biological knowledge into distance-based clustering analysis of microarray gene expression data. *Bioinformatics* 22, no. 10 (2006): 1239–1268

Kohonen, T. *Self-organized maps.* Vol. 117. Berlin: Springer, 1997.

Kohonen, T. The self-organizing map. *Neurocomputing* 21 (1998): 1–6.

Krishnapuram, R., and J. Keller. A possibilistic approach to clustering. *IEEE Trans Fuzzy Syst* 1, no. 2 (1993): 98–110.

Pal, N., K. Pal, and J. Bezdek. A mixed C-Means clustering model. In *Proceedings of the Sixth IEEE International Conference on Fuzzy Systems*. Barcelona: IEEE, 1997, pp. 11–21.

Priness, I., O. Maimon, and I. Ben-Gal. Evaluation of gene-expression clustering via mutual information distance measure. *BMC Bioinformatics* 8, no. 111 (2007): 1–12.

Qin, J., D.P. Lewis, and W.S. Noble. Kernel hierachical gene clustering from microarray expression data. *Bioinformatics* 9, no. 16 (2003): 2097–2104.

Qu, J., Q. Jiang, F. Weng, and Z. Hong. A hierarchical clustering based on overlap similarity measure. In *Eighth ACIS International Conference on Software Engineering, Artificial Intelligence, Networking, and Parallel/Distributed Computing*. Washington, DC: IEEE Computer Society, 2007.

Ray, S., and S. Bandyopadhyay. Dynamic range-based distance measure for microarray expressions and a fast gene-ordering algorithm. *IEEE Trans Syst Man Cybernetics* 37, no. 3 (2007): 742–749.

Ruspini, E.H. A new approach to clustering. *Int Control* 15 (1969): 22–32.

Shyu, M-L., K. Sarinnapakorn, I. Kuruppu-Appuhamilage, S-C. Chen, L. W. Chang, and T. Goldring. Handling Nominal Features in Anomaly Intrusion Detection Problems. *Proceedings of the 15th International Workshop on Research Issues in Data Engineering: Stream Data Mining and Applications*. Washington, DC, USA: IEEE Computer Society, 2005, 55-62.

Smolkin, M., and D. Ghosh. Cluster stability scores for microarray data in cancer studies. *BMC Bioinformatics* 4 (2003): 36.

Sokal, R.R., and C.D. Michener. A statistical method for evaluating systematic relationship. In *University of Kansas Bulletin*. Vol. 38. 1958, pp. 1409–1438.

Sousa, J., and U. Kaymak. *Fuzzy decision making in modeling and control*. Singapore: World Scientific Publication Company, 2002.

Tamames, J., et al. Bioinformatics methods for the analysis of expression arrays: Data clustering and information extraction. *J Biotechnol* 98, no. 2–3 (2002): 269–283.

Tamayo, P., et al. Interpreting patterns of gene expression with self-organizing maps: Methods and application to hematopoietic differentiation. *Proc Natl Acad Sci USA* 96, no. 6 (1999): 2907–2912.

Tsai, H.K., J.M. Yang, Y.F. Tsai, and C.Y. Kao. An evolutionary approach for gene expression patterns. *IEEE Trans Inf Technol Biomed* 8, no. 2 (2004): 69–78.

Ultsch, A., and H.P. Siemon. Kohonen's self-organizing feature maps for exploratory data analysis. In *Proceedings of the International Neural Network Conference*, Kluwer, Dordrecht, 1990, pp. 305–308.

Vesanto, J. SOM-based data visualization methods. *Intell Data Anal J* 3 (1999): 111–126.

Vesanto, J., and E. Alhoniemi. Clustering of the self-organizing map. *IEEE Trans Neural Networks* 11, no. 3 (2000): 586–600.

Wang, H.C., J. Badger, P. Kearney, and M. Li. Analysis of codon patterns using bacterial genomes using the self-organizing map. *Mol Biol Evol* 18 (2001): 792–800.

Wang, J., J. Delabie, H. Aasheim, E. Smeland, and O. Myklebost. Clustering of the SOM easily reveals distinct gene expression patterns: Results of a reanalysis of lymphoma study. *BMC Bioinformatics* 3, no. 36 (2002): 1–9.

Wunsch, D., and R. Xu. Survey of clustering algorithms. *IEEE Trans Neural Networks* 16, no. 3 2005): 645–678.

Yin, L., C.-H. Huang, and J. Ni. Clustering of gene expression data: Performance and similarity analysis. *BMC Bioinformatics* 7, no. 519 (2006): 1–11.

Zadeh, L. Fuzzy sets. *Inf Control* 8 (1965): 338–353.

Chapter 7
Advanced Clustering Techniques

This chapter is an extension to the clustering techniques in bioinformatics. The clustering algorithms in this chapter are aptly titled advanced clustering techniques because they are based on the clustering techniques described in Chapter 6, but with natural extensions. In this chapter we provide descriptions of how these advanced clustering techniques are applied to different areas of bioinformatics (Bader and Hogue 2003).

7.1 Graph-Based Clustering

Graph-based clustering is used to group similar vertices into one cluster such that the maximum number of edges connect within the cluster, and the minimum number of edges connect between the clusters (Bader and Hogue 2003).

For clustering, a graph is represented by $G(V, E)$, where V represents vertices of a graph, and E indicates edges that connect two vertices (Figure 7.1).

7.1.1 Graph-Based Cluster Properties

Graph-based cluster properties can be determined using the two-step process explained below (Schaeffer 2007).

1. If one vertex cannot be connected to another vertex through an edge, then these vertices will not be in the same cluster.
2. The edge connecting two vertices should be located within the cluster, not between clusters.

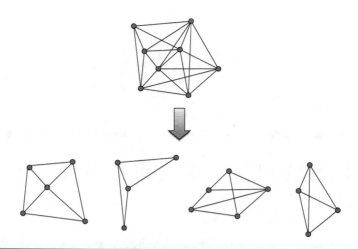

Figure 7.1 A partitioned graph. The graph (top) has been partitioned into four clusters based on data point similarity.

Edges are classified into two groups: internal edges and external edges.

Internal edges connect one vertex to another vertex within the cluster. Therefore, if the vertex has an internal degree of 0, then this vertex is not connected to any other vertex within a cluster. Hence, it should not be a part of the cluster (Figure 7.2).

External edges connect vertices between clusters. Therefore, if the vertex has an external degree of 0, then this vertex should be a part of a cluster, as it has no connections outside a cluster (Figure 7.3).

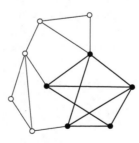

Figure 7.2 Edges in black are part of one cluster, as their internal edges are connected to one other. Hence, they have more internal edges than other edges, which are not part of this cluster.

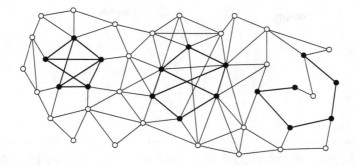

Figure 7.3 Three clusters (shown in black): The cluster on the left is of good quality due to the high connectivity of internal edges. The cluster in the middle is not as well defined, as its edges are more connected to external edges than to internal edges. The cluster on the right has fewer connections both outside and inside. Hence, it is not a good cluster.

7.1.2 Cut in a Graph

Cut S is defined as the division of the vertices V into two nonempty sets $(S, V/S)$ of a graph $G = (V, E)$, and cut size is defined as the number of edges that connect vertices in one set S to vertices in another set 'V/S'. The cut size is shown in Equation 7.1.

$$c(S, V/S) = \left| \{ \{v, u\} \in E \mid u \in S, v \in V/S \} \right|. \tag{7.1}$$

Cut size helps identify the sparseness of connections in the cluster, rather than with the rest of the graph (Goldberg and Tarjan 1986; Gomory and Hu 1961). This measurement can be taken by computing graph density where smaller cut sizes isolate the cluster better. Graph density is defined as the ratio of the number of edges present to the maximum possible edges.

A *minimum cut* (mincut) separates a graph so that the end product has a small number of edges (Figure 7.4).

7.1.3 Intracluster and Intercluster Density

Internal or intracluster density is defined as the density of the subgraph induced by the cluster (Davies and Bouldin 1979). For good clustering, internal density should be higher than the density of the graph $\delta(G)$,

$$\delta_{int}(e) = \frac{|\{\{v, u\} \mid v \in e, u \in e\}|}{|e|(|e|-1)}, \tag{7.2}$$

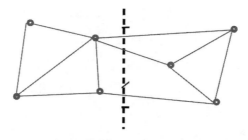

Figure 7.4 A cut in the graph is shown by a dotted line that cuts the graph into two partitions.

where e is a cluster, v and u are vertices of an edge, and $\delta_{int}(e)$ is the intracluster density of a cluster (e).

The external or intercluster density is defined as the ratio of intercluster edges to the maximum number of intercluster edges possible. For good clustering, the intercluster density of the clustering should be lower than the density of the graph,

$$\delta_{ext}(G \mid e_1,....,e_2) = \frac{|\{\{v,u\} \in e_i, u \in e_j, i \neq j\}|}{n(n-1) - \sum_{l=1}^{k}(|e_l|(|e_l|-1))} \quad (7.3)$$

where $e_1,....,e_k$ are clusters, v and u are vertices of an edge.

7.2 Measures for Identifying Clusters

Clusters can be determined using one of two methods. They can be determined by computing values for the vertices, and assigning them to clusters based on these values. Clusters can also be determined by calculating a fitness measure for the set of possible clusters (Schaeffer 2007) (Figure 7.5) (Maulik and Bandopadhyay 2002).

7.2.1 Identifying Clusters by Computing Values for the Vertices or Vertex Similarity

Some of the clustering algorithms compute similarity between the vertices by setting a threshold similarity. If similarity between vertices is higher than the threshold is, then the vertices are clustered together. This method is computationally more expensive than clustering a graph once the similarities are known. The cluster should contain only those vertices that are similar and remove other vertices.

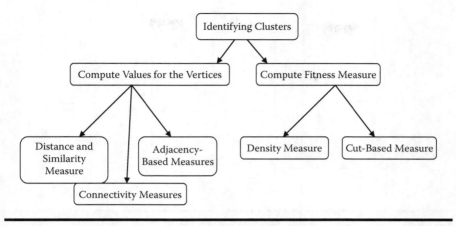

Figure 7.5 Measures to identify clusters.

A distance measure can be computed instead of similarity if the cluster boundary is located in a way that includes most of the outside vertices and significantly increases the intracluster distances by using distance and similarity measure, adjacency-based measures, and connectivity measures. We will explain each of these measures below.

7.2.1.1 Distance and Similarity Measure

Distance measures, such as Euclidean distance, can be used to calculate the distance between two vertices. That distance is compared with the threshold distance. If a calculated distance is less than the threshold distance, then an edge will be assigned between the two vertices, and the vertices are considered to be similar and sorted into one cluster.

7.2.1.2 Adjacency-Based Measures

Adjacency-based measures can be performed by measuring the similarity between vertices using an adjacency matrix, which determines whether two vertices are similar or not by analyzing the overlap of their neighbors. If a value is 0, then there is no common neighbor between the two vertices. If the value is 1, then the neighbors are identical. In Figure 7.6(a) there are six vertices in a graph that are labeled and their relationship is explained by using the adjacency matrix in Figure 7.6(b). If one vertex is connected to another vertex by an edge, then 1 is assigned in the adjacency matrix; otherwise, 0 is assigned. For example, vertex 1 is connected to vertex 2, and hence 1 is assigned in the adjacency matrix, whereas vertex 4 is not connected to vertex 2 by an edge, and hence 0 is assigned in the adjacency matrix.

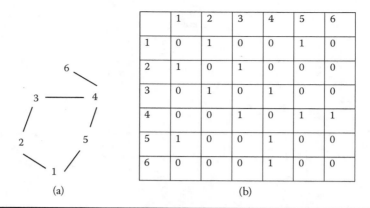

Figure 7.6 (a) Labeled graph. (b) Adjacency matrix.

7.2.1.3 Connectivity Measures

Connectivity measures can be used to find the similarity between the vertices whether the vertices are in the same cluster or in different clusters. Vertices will be in the same cluster. If many paths exist between a pair of vertices, the number of paths can be compared by a default or a threshold value; i.e., they should be highly connected with each other.

7.2.2 Computing the Fitness Measure

The cluster fitness measure is used to measure the quality of a given cluster (Rand, 1971; Rousseeuw 1987). Using vertex similarity, cluster fitness functions help classify vertices into the clusters. Cluster fitness can be determined using two approaches: density measure and cut-based measure (Figure 7.7).

7.2.2.1 Density Measure

Several algorithms that have a density higher than a threshold value have been proposed to search for subgraphs. These algorithms work because a cluster is a subgraph that is dense with respect to a given threshold density.

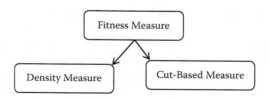

Figure 7.7 Identifying clusters by a fitness measure.

7.2.2.2 Cut-Based Measures

Cut-based measures find the subgraph that is not dependent or has no relationship with the remaining part of the graph by computing connectivity between the subgraph and the rest of the graph. Hence, high-quality clusters can be found.

7.3 Determining a Split in the Graph

Factors such as cuts, spectral methods, and betweenness determine graph splits (Figure 7.8).

7.3.1 Cuts

A cut in a graph is quite important, as a well-selected cut can divide a graph or separate two or more dense clusters. There are two limitations with such a division:

1. Simply removing single vertices will not help compute a cluster.
2. It is difficult to determine the terminating point of splitting the graph.

7.3.2 Spectral Methods

The second factor that helps determine a graph split is the use of spectral methods that are based on spectral clustering and computing eigenvectors. In a spectral method, eigenvectors are computed to correspond to the *second smallest* eigenvalue of the normalized Laplacian operator. Hence, the resulting eigenvector components are used to measure the similarity between the vertices for determining clusters. Although the spectral method is computationally expensive, this limitation can be overcome using a distributed algorithm to reduce the computational load (Kempe 2004) (Prodromidis et al. 2000).

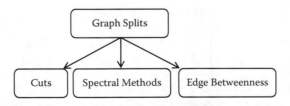

Figure 7.8 Graph can be split by using techniques such as cuts, spectral methods, and edge-betweenness.

7.3.3 Edge-Betweenness

Edge-betweenness is another way to determine graph split. Newman and Girvan (2003) assign numerical weights on the edges in an unweighted graph to determine clustering. These weights are called edge-betweenness, which is the number of shortest paths between two vertices that pass through an edge.

The algorithm can be performed in four steps:

1. Compute the edge-betweenness of all edges in the graph.
2. Eliminate those edges that have the highest edge-betweenness.
3. Calculate the edge-betweenness again for all the edges, as removal of the edges may change the edge-betweenness of other existing edges in the graph.
4. Repeat steps 2 and 3 until no edges remain.

Finally, we obtain a clustering algorithm.

7.4 Graph-Based Algorithms

Graph theory helps describe clustering problems that arise in graphs where the nodes of a graph represent data points and edges represent proximity between nodes or data points. A dissimilarity matrix is defined as

$$D_{ij} = \begin{cases} 1 \; if \; D(x_i, x_j) < d_0 \\ else \; 0 \end{cases}, \qquad (7.4)$$

where d_0 is the threshold distance and $D(x_i, x_j)$ is the distance between two points (x_i, x_j).

Applications of graph theory include the Chameleon and CLICK algorithms.

7.4.1 Chameleon Algorithm

The Chameleon algorithm is an agglomerative, hierarchical clustering algorithm, based on a nearest-neighbor graph, where an edge will be eliminated if the distance between the vertices is less than a defined threshold distance (Xu and Wunsch 2005). Chameleon is performed as follows (Figure 7.9):

1. Use minimal edge cut to divide the connectivity graph into a set of subclusters.
2. Ensure there are enough nodes in each subgraph so that there is an effective similarity computation.
3. Combine both relative interconnectivity and closeness to help find potential clusters, and merge these small subsets to obtain ultimate clustering solutions.
4. Normalize the average weight of the edges that are connected based on the closeness of the clusters.

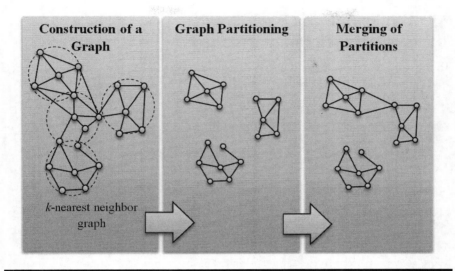

Figure 7.9 Steps of the Chameleon algorithm are explained in Section 7.4.1. (From Karypis, G., and E.H. Han, *IEEE Comput Soc* (1999): 68–75.)

7.4.2 CLICK Algorithm

The CLICK algorithm is based on computing the minimum weight cut to form clusters (Xu and Wunsch 2005). Probability and graph theory help assign weight to the edges, which are defined as shown in Equation 7.5:

$$e_{ij} = \log \frac{\Pr ob(i, j\, belong\, to\, the\, same\, cluster | S_{ij})}{\Pr ob(i, j\, does\, not\, belong\, to\, the\, same\, cluster | S_{ij})}. \tag{7.5}$$

Equation 7.5 can be replaced using the Bayes' theorem as shown in Equation 7.6, because CLICK further assumes that the similarity values within clusters and between clusters follow Gaussian distributions with different means and variances. For example,

$$e_{ij} = \log \frac{p_0 \sigma_B}{(1-p_0)\sigma_w} + \frac{(S_{ij}-\mu_B)^2}{2\sigma_B^2} - \frac{(S_{ij}-\mu_w)^2}{2\sigma_w^2}, \tag{7.6}$$

where p_0 is the prior probability that two objects belong to the same cluster, μ_B is the mean between cluster similarities, σ_B^2 is the variance between cluster similarities, μ_w is the mean within cluster similarities, and σ_w^2 is the variance within cluster similarities.

7.5 Application of Graph-Based Clustering in Bioinformatics

To illustrate the usefulness of graph-based clustering in bioinformatics, three such applications are described below:

1. Analysis of gene expression data using the shortest path (SP)
2. Construction of genetic linkage maps using the minimum spanning tree of a graph
3. Finding isolated groups in a random graph process

Each of these applications is described below.

7.5.1 Analysis of Gene Expression Data Using Shortest Path (SP)

Some gene pairs exist in the same biological pathway but do not show high expression similarity (Zhou et al. 2002). Therefore, transitive expression similarity is an important measure for finding transitive genes. These genes can be found using shortest path (SP) analysis. By using this method, genes are identified based on whether they are functionally related. Moreover, by using the functionality of known genes, the function for unknown genes that are on the same shortest path as the known genes can be predicted (Zhou et al. 2002).

To find the number of connected gene pairs in the graph, a threshold will be selected to help construct a graph and compute the shortest path by assigning an edge when the absolute expression correlation $C_{a,b}$ is higher than the predefined threshold. The edge length between vertices a and b is $d_{a,b} = f(C_{a,b}) = (1 - C_{a,b})^k$, where increasing the value of k in the gene provides more power to reveal transitive coexpression. To stabilize the number of genes, Zhou et al. set a value of k equal to or greater than 6. Moreover, to include only significant SPs, the authors set a threshold for a path length equal to 0.008. Using this threshold, only SPs that are less than 0.008 will be included.

Zhou et al. (2002) find the shortest path for all pairs of known genes and connect them. By using the functionality of known genes, Zhou et al. have predicted the functionality of unknown genes by finding a gene-specific function. They scanned the tree from the root to the lowest level of the tree. If the difference between the root and lowest level is less than four levels, then that function is a specific gene function, and hence this functionality can be assigned to an unknown gene on the SP.

7.5.2 Construction of Genetic Linkage Maps Using Minimum Spanning Tree of a Graph

This method is useful when the data are noisy and incomplete. Therefore by computing the minimum spanning tree, the correct order of markers can be determined (Wu et al. 2008; Matsuda et al. 1999).

Building a genetic map is a three-step process (Wu et al. 2008), as shown below.

> Step 1: Divide the markers into linkage groups, known as a group of loci that are connected. These markers act as a single group, and clustering is needed to assign markers into linkage groups.
> Step 2: Determine the correct order given to a set of markers in the same linkage group.
> Step 3: Calculate the distance between the adjacent markers.

Let $d_{i,j}$ be the Hamming distance between two markers, l_i and l_j, belonging to two linkage groups (Wu et al. 2008). The graph $G(M, E)$ is drawn so that the weight is given to an edge $(l_i, l_j) \in E$ of a pairwise distance $d_{i,j}$ between l_i and l_j to cluster the markers into linkage groups between all sets of markers. Markers will be assigned to the linkage groups when the distance between them is larger than or equal to δ. Then, that edge is removed from $G(M, E)$, and the resulting graph divides into connected components.

7.5.3 Finding Isolated Groups in a Random Graph Process

Brumm et al. used two approaches to find different representations of the relationships. First, they used a graph-based approach, in which a global threshold was set to classify all pairs that were above a predefined threshold, and the threshold graph was obtained. Second, they generated a dendrogram (or tree) using a clustering algorithm, whereby tree pruning was performed to obtain gene groups (Brumm et al. 2008).

Both of these methods are limited by the necessity to set a global threshold that is extremely sensitive. Therefore, it is difficult to discover whether two genes are related within a module (internal cohesion) or how two genes are unrelated to each other within a module (external cohesion) (Handl et al. 2005; Hubert and Arabie 1985). Moreover, in a heterogeneous biological system, it is difficult to reveal all the modules by using either one threshold graph or tree pruning.

To overcome the above limitations, a new method has been developed to detect modules in relational genomic data by giving ranks to the relationships between genes and threshold graphs by moving the global threshold from stringent to permissive (Brumm et al. 2008). Sequences of graphs having modules persist as cohesive subgraphs, which are identified as groups, which allows modules to identify with internal cohesion and find the statistical significance of each candidate module (Brumm et al. 2008).

The graph approach can find relationships across a range of thresholds (Brumm et al. 2008). The first step of the graph approach is to find the graph that has all genes and no edges. The second step is to place the edge that has the strongest relationship score, i.e., rank 1, between the pair of genes and obtain the next graph. Therefore, subsequent edges are added based on ranks.

In this way, a graph sequence is obtained. This graph sequence begins with an empty graph and ends with a complete graph. Hence, the entire analysis is based on a single graph.

7.5.4 Implementation in Cytoscape

We have implemented graph clustering by using a tool called Cytoscape. The input or dataset for Cytoscape is galFiltered.cys, which indicates a Cytoscape session file that contains interaction network and expression data. The files contain an interaction network of 331 genes that were significantly differentially expressed in at least 1 of the 20 experimental conditions (KeiichiroOno 2010).

To implement graph clustering we used a plug-in called ClusterONE in Cytoscape (Nepusz et al. n.d.). ClusterONE is based on a graph approach. This method is performed by "growing" dense regions using one or two vertices called seeds based on their cohesiveness (Nepusz et al. 2012). Cohesiveness is a quality measure in which a well-defined group should have more internal edges and fewer boundary edges. Weights, which define how reliable that edge is, are assigned to the edges. Whenever reliable edges are found, they are selected in a numeric edge attribute in Cytoscape that helps drive the cluster growth process.

The ClusterONE algorithm searches for high cohesiveness of clusters, which begins from a small set of vertices *that are strongly bound together*. This group can be increased by adding new vertices as long as cohesiveness increases. A vertex can be removed if it increases the cohesiveness of a group (Nepusz et al. 2012).

The termination condition, or the stopping condition, occurs when the cohesiveness of a group fails to increase. Subgroups that are less than a given threshold are discarded. Finally, subgroups that are cohesive and overlap are combined to form larger subgroups so that results can be interpreted more easily.

Once the clustering process is performed successfully, nodes will be colored based on the number of clusters. If a node is only in a single cluster, it will turn red. Similarly, if a node is in more than one cluster, it will turn yellow. Finally, if a node is an outlier, then it will turn gray (Figure 7.10).

We have selected the following default parameters to implement the ClusterONE algorithm in Cytoscape: **minimum size**, **minimum density**, **edge weights** and **merging method,** and **overlap threshold.** Two clusters are merged if overlap is larger than a given threshold.

7.5.4.1 Seeding Method

ClusterONE will start producing clusters from a single node or a single edge called initial seeds. There are three ways in which ClusterONE can select seeds:

1. *Every node* will be used as a seed.

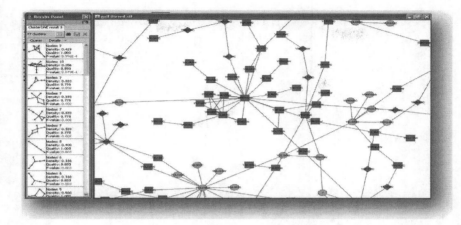

Figure 7.10 Implementation of ClusterONE in Cytoscape.

2. *From unused nodes,* in which the node that does *not* participate and also has the largest weight of the discovered clusters will be selected as the next seed.
3. *From every edge,* a measure will be taken once, each having a seed consisting of the two endpoints of the edge.

7.6 Kernel-Based Clustering

Most clustering algorithms are limited in that they can detect and cluster data only into spherical or elliptical shapes. This inability to recognize clusters in other shapes, such as nonspherical or arbitrary shapes, limits the applicability of the algorithm (Vapnik 1999).

A kernel is a nonnegative real-valued integral function K, which satisfies the following two requirements:

$$\int_{-\infty}^{+\infty} K(u)\,du = 1 \text{ and } K(-u) = K(u) \text{ for all values of } u. \tag{7.7}$$

If K is a kernel, then the function K^* is defined by $K^*(u) = \lambda^{-1} K(\lambda^{-1} u)$, where $\lambda > 0$.

Kernel-based learning algorithms are based on Cover's theorem (Xu and Wunsch 2005). Complex and nonlinear, separable patterns can be transformed nonlinearly into a higher-dimensional feature space. In this way, it is possible to separate these patterns linearly (Xu and Wunsch 2005). An inner-product kernel can be calculated to help compute the corresponding points in the transformed space.

For example, suppose we have a set of patterns $x_j \in \Re^d$ and a nonlinear map $\Phi : \Re^d \to F$ (Xu and Wunsch 2005). Here, F represents a feature space with arbitrarily high dimensionality. The objective is to find K centers so that we can minimize the distance between the mapped patterns and their closest center as

$$\| \Phi(x) - m_l \|^2 = \| \Phi(x) - \sum_{j=1}^{N} T_{lj} \Phi(x_j) \|^2$$
$$= k(x,x) - 2\sum_{j=1}^{N} T_{lj} k(x,x_j) + \sum_{i,j=1}^{N} T_{li} T_{lj} K(x_i, x_j), \quad (7.8)$$

where m_i is the center for the lth cluster and lies in a span of $\Phi(x_1),....,\Phi(x_N)$, and $k(x,x_j) = \Phi(x).\Phi(x_j)$ is the inner product kernel.

They define the cluster assignment variable as

$$C_{jl} = \begin{cases} 1, & \text{if } x_j \text{ belongs to cluster} \\ 0, & \text{otherwise}. \end{cases} \quad (7.9)$$

7.6.1 Kernel Functions

Kernel function transforms the data from the original space into a high-dimensional space nonlinearly, which may result in a better performance, as data would be more linearly separable (Scholkopf et al. 2001).

Most common kernel functions are uniform, triangle, Epanechnikov, quartic (biweight), tricube (triweight), Gaussian, and cosine. The most popular of these choices is the Gaussian function.

7.6.2 Gaussian Function

The Gaussian function can be read as

$$f(x) = ae^{\frac{-(x-b)^2}{2c^2}}. \quad (7.10)$$

Equation 7.10 shows a classic Gaussian function, in which a is the height of the curve peaks, b is the position of the center of the peak, and c controls the width of the curve. In signal processing, Gaussian functions are used as Gaussian filters, and in image processing, they are used as Gaussian blurs (Figure 7.11).

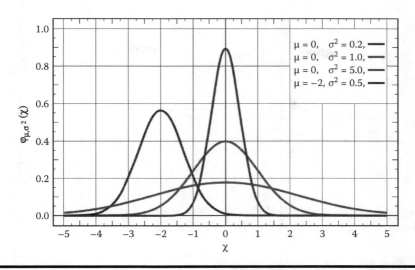

Figure 7.11 Gaussian function. (From http://upload.wikimedia.org/wikipedia/commons/thumb/7/74/Normal_Distribution_PDF.svg/720px-Normal_Distribution_PDF.svg.png.)

7.7 Application of Kernel Clustering in Bioinformatics

7.7.1 Kernel Clustering

Kernel clustering can find the number of clusters. In this method, the kernel matrix eigenvectors define the mapping based on the underlying nature of the data. The features of the data are presented for partitioning, and the sum-of-squares cost is used to evaluate the clustering method.

The sum-of-squares method will not work if the boundaries separating the clusters are nonlinear. This limitation can be overcome by transforming the data into a high-dimensional feature space nonlinearly and clustering within this feature space. However, this approach will only work if the feature space is not high. Hence, the kernel principal component analysis (KPCA) method should not be applied on the transformed variables. Kernel function can be used in the original data space to compute inner products between points. This ability may aid the users working on infinite feature spaces.

Next, the kernel, k-means algorithm, can be formulated as the following:

1. Initialize the centers m_l with the first i, $(i \geq K)$ observation patterns.
2. Take a new pattern x_{i+1}, and calculate $C_{(i+1)h}$, as shown in Equation 7.11.
3. Update the mean vector m_h, which has a corresponding $C_{(i+1)h}$ of 1, as

$$m_h^{new} = m_h^{odd} + \xi(\Phi(x_i + 1) - m_h^{old}), \text{ where } \xi = C_{(i+1)h} / \sum_{j+1}^{i+1} C_{jh}. \quad (7.11)$$

4. Adapt the coefficients T^{hj} for each $\varphi(x_j)$ as

$$T_{hj}^{new} = \{T_{hj}^{old}(1-\xi), \text{ for } j \neq i+1,$$
$$\xi, \text{ for } j = i+1. \quad (7.12)$$

5. Repeat steps 2–4 until convergence is achieved.

Kernel-based clustering algorithms have many advantages. For example, in high-dimensional or infinite feature space, it is possible to obtain a linearly separable hyperplane using kernel-based clustering algorithms. These algorithms can form arbitrary clustering shapes, except for hyperellipsoids and hyperspheres. Support vector clustering (SVC) is a form of kernel-based clustering algorithm and can address noise and outliers, and no prior knowledge is needed to determine the same topological structure.

7.7.2 Kernel-Based Support Vector Clustering

Support vector clustering is based on the kernel method that uses the kernel function for data clustering (Ben-Hur et al. 2001). This method is unable to detect non-convex-shaped clusters. To overcome this limitation, Yeh and Lee have proposed a two-step v-SVC method to aid in clustering the data into different groups. In the first step, a sphere centroid is calculated for each cluster. In the second step, the cluster results are improved iteratively using the k-means algorithm.

To make the method robust, excessive distances should be removed in order to avoid the excessive distance slack variables $\xi, \xi_i \geq 0, \forall \xi_i$ have used. Moreover, distances should not be smaller than R. Therefore,

$$\text{Min} R^2 + \frac{1}{vn}\sum_{i=1}^{n}\xi_i \geq 0, \forall_i \text{ s.t.} ||\phi(x_i) - a||^2 \leq R^2 + \xi_i, \xi_i \geq 0, \forall_i, \quad (7.13)$$

where v is a trade-off between sphere radius and the excessive distance of outliers, which helps determine the number of outliers and support vectors. Thus, slack variables are also called v-SVC. In addition, $\phi: X \mapsto F$ can be used as a nonlinear mapping function that connects an input space to a feature space F. In this equation, a and R are the center and radius of a sphere in the feature space, respectively. If the value of v is small, then only a few outliers will be found in a large sphere due to substantial penalty on the excessive distance. Similarly, if the value of v is large, then the radius will be small and there will be many outliers with large excessive distances.

SVC is time-consuming, as it can produce a large number of clusters, each containing a single sample. In addition, sometimes it can produce only one cluster having all the samples (Ben-Hur et al. 2001).

To overcome these limitations, Satish and Sekhar cluster the data by using a two-step v-SVC method and then use k-means to improve the cluster results (Ben-Hur et al. 2001). To remove the effect of noise or outliers they changed the parameters by setting the value of v to be large and provide a trade-off between the radius of the sphere and the excessive distance, and then improved the cluster results iteratively by using k-means algorithm.

7.7.3 Analyzing Gene Expression Data Using SOM and Kernel-Based Clustering

Kotani and Sugiyama (2002) examined and classified the gene expression data using self-organizing maps (SOMs) and a kernel-based clustering obtained from the DN.4 microarray. Thus, gene expression data are input to the SOM, and prototype vectors will be generated as input for kernel-based clustering. Hence, kernel-based clustering will be applied on the data (Kotani and Sugiyama 2002). Therefore, kernel-based clustering categorizes the units of the SOM.

The application of SOM is limited because the results are difficult to visualize or understand. Hence, it is hard to cluster boundaries, whereas data can be partitioned nonlinearly by using kernel-based clustering. Thus, Kotani and Sugiyama find cluster boundaries by using kernel-based clustering, making the results of SOM easy to visualize and understand.

Self-organizing maps (SOMs) mapped or transformed the n-dimensional data into one-dimensional (1D) or 2D data where mapping is defined by associated D-dimensional prototype vectors, p_i, for the ith unit. The unit chosen as the closest prototype vector to the nth input vector, x_n, is defined as (Kohonen 1995)

$$\min(x_{n_i} - p_i)^T (x_n - p_i). \tag{7.14}$$

The chosen unit updates its prototype vector according to

$$p_i(t+1) = p_i(t) + h_{ci}(t)[p_i(t) - x_n]. \tag{7.15}$$

where t is the learning iteration and h_{ci} is the neighborhood function that is a decreasing function. The authors have used a radial symmetric Gaussian function as the neighborhood (Kotani and Sugiyama 2002).

To increase the linear dispersion in feature space, kernel-based clustering (Girolami 2002) transforms the nonlinear data into higher-dimensional feature space. Φ is a smooth and continuous mapping from the data space to the feature space, F, and is defined as

$$\Phi : R^D \to F. \tag{7.16}$$

The trace of the within-group scatter matrix in F, S_W^Φ, is given by

$$\operatorname{tr}\left(S_W^\Phi\right) = \frac{1}{N} \sum_{k=1}^{N_0} \sum_{n=1}^{N} z_{kn} \left(\Phi(x_n) - m_k^\Phi\right)^T \left(\Phi(x_n) - m_k^\Phi\right), \quad (7.17)$$

where N_c is the number of clusters and N is the number of input vectors. The mean of each cluster, m_k^Φ, is defined as

$$m_k^\Phi = \sum_{n=1}^{N} z_{kn} \Phi(x_n)/N_k, \quad (7.18)$$

where $N_k = \sum_{n=1}^{N} z_{kn}$, z_{kn} is an indicator of whether the nth input vector belongs to the kth cluster, namely, if x_n belongs to the kth cluster, $z_{kn} = 1$, and otherwise $z_{kn} = 0$.

$N \times N$ kernel matrix, K, is defined as

$$K_{ij} = K(x_i, x_j) = \Phi(x_i).\Phi(x_j). \quad (7.19)$$

Kotani and Sugiyama obtain the following equation:

$$\operatorname{tr}\left(S_W^\Phi\right) = 1 - \sum_{k=1}^{K} \gamma_k \Re(x/C_k), \quad (7.20)$$

where

$$\Re(x/C_k) = \frac{1}{N_k^2} \sum_{i=1}^{N} \sum_{j=1}^{N} z_{ki} z_{kj} K_{ij}, \quad (7.21)$$

and

$$\gamma_k = \frac{N_k}{N}. \quad (7.22)$$

Girolami (2002) has used a radial basis function as the kernel function,

$$k(x_i, x_j) = \exp\left[-\|x_i - x_j\|^2\right]. \quad (7.23)$$

Finally, the optimal clustering of input vectors is obtained as

$$Z = \arg\min_{Z} tr\left(S_w^{\Phi}\right) = \arg\max_{Z} \sum_{k=1}^{K} \gamma_k \Re\left(x/C_k\right). \tag{7.24}$$

7.8 Model-Based Clustering for Gene Expression Data

Model-based clustering is based on the assumption that each component or group of data is generated by underlying probability distribution, which helps determine a relevant or good clustering method (Azuaje and Bolshakova 2002; Yeung et al. 2001). Yeung et al. have considered six such models: the Gaussian mixture model (GMM), the equal volume spherical model, the unequal volume spherical model, the unconstrained model, the elliptical model, and the diagonal model. The most common such models are Gaussian mixture and the diagonal model.

7.8.1 Gaussian Mixtures

Gaussian mixtures are the most statistically mature models for clustering areas in which each component is spherically symmetric. In these models, there are few parameters, and the spherical model is of equal volume (Ouyang and Welsh 2004). In GMM, each component is modeled by multivariate normal distribution parameters, μ_k (mean vector) and Σ_k (covariance matrix), which helps determine geometric features such as shape, volume, and orientation for each component of k (Banfield 1993). For example,

$$f_k\left(y_i/\mu_k, \Sigma_k\right) = \frac{\exp\left[-\frac{1}{2}(y_i - \mu_k)^T \Sigma_k^{-1}(y_i - \mu_k)\right]}{\sqrt{\det(2\Pi\Sigma_k)}}. \tag{7.25}$$

7.8.2 Diagonal Model

The diagonal model can find the number of clusters and model parameters using the EM algorithm in both the MCLUST and diagonal model implementation. In the EM algorithm, the expectation (E) steps determine the probability whether a sample belongs to a particular cluster or not, and in the maximization (M) step, the model parameters are determined based on the current given group of membership probabilities. When the EM algorithms are combined, observations are assigned to their group based on maximum conditional probability

7.8.3 Model Selection

Model-based clustering is a probabilistic approach that helps users to select the best clustering algorithm and the correct number of clusters (Yeung et al. 2001). There is a trade-off between the probability model and the number of clusters. For example, a complex model requires a small number of clusters, while a simple model requires a large number of clusters. Let us assume D is an observed data, and $M1$ and $M2$ are two models with parameters of $a1$ and $a2$, respectively. The integrated likelihood represents the probability that D is observed, given that the underlying model is Mk (Yeung et al. 2001).

To choose between models $M1$ and $M2$, (*Kass and Raftery, 1995*) suggested the use of Bayes factor. The Bayes factor is defined as the ratio of the integrated likelihoods of the two models $B12 = p(D/M1)/p(D/M2)$. If $B12 > 1$, model $M1$ is preferred over model $M2$. The limitation of using the Bayes factor is the estimation of integrated likelihood.

Schwarz et al. have used a Bayesian information criterion (BIC) to compare models and find the BIC score of differences greater than a threshold. Models that contain this difference can be a strong reason for preferring one model over the other.

7.9 Relevant Number of Genes

The clustering process partitions data into a number of clusters or groups that are denoted by K (Yuan and Li 2008; Tseng and Wong 2005). Their quality depends on the number of clusters. Some clustering algorithms need the value of K as an input. Sometimes K can be difficult to understand and evaluate if there are many clusters. Similarly, if there are fewer, clusters K cause loss of information (Youness and Saporta, 2010). Research in this area is ongoing, and it is difficult to find the appropriate value of K (Tseng and Wong 2005). Tseng and Wong have proposed a method that does not need to assign all the points into the clusters and produces tight and stable clusters (Yuan and Li 2008). Some representative examples are illustrated in the following.

7.9.1 A Resampling-Based Approach for Identifying Stable and Tight Patterns

To identify stable and tight patterns (Tseng and Wong 2005), cluster the samples based on similar expression patterns. In microarray experiments, there are many genes that are not related to any biological process. Therefore, there are no correlation variations within clusters of genes. Hence, these genes should not be clustered, and are thus called scattered genes. Due to scattered genes, estimating the appropriate number of clusters is not easy. As a result, we get distorted clusters that are difficult to visualize and analyze because these scattered genes divide the algorithm into clusters at all points.

For simplicity, Tseng and Wong (2005) used k-means and Euclidean distance as a dissimilarity measure by assuming data are in Euclidean space.

7.9.2 Overcoming the Local Minimum Problem in k-Means Clustering

k-Means clustering will be used in the tight clustering algorithm. However, because the k-means method is the local minimum, its applications are limited, although it can be overcome by minimizing within-cluster dispersion (sum of squares) (Tseng and Wong 2005). However, it is computationally expensive to search for the global minimum. Therefore, to stabilize within-cluster dispersion, the algorithm performs iterative reallocation. Poor selection of input values can give inaccurate results, as minimization falls in a local minimum quickly, and it becomes more prominent when scattered points exist.

7.9.3 Tight Clustering

The subsampling procedure is used to create variability so that it is easy to distinguish between points that are stably clustered and those that are clustered by chance. From the original data, X takes a random subsample X', for example, 60% of the original sample size, and applies k-means with the prior knowledge of k on X' to obtain the cluster centers $C(X', k) = (C1, C2, ..., Ck)$, which can be used to cluster the original data X based on the distances from each point to the cluster centers. Following the convention of Tibshirani et al. (2001), the resulting clustering is represented by a comembership matrix $D[C(X', k), X]$, where $D[C(X', k), X]_{ij}$.

Repeat independent random subsampling B times to obtain subsamples $X^{(1)}$, $X^{(2)}$, ..., $X^{(B)}$. The average comembership matrix is defined as $\bar{D} = \text{mean}(D[C(X^{(1)}, k), X], ..., D[C(X^{(B)}, k), X])$. Search for a set of points $V = \{v_1, v_2, ..., v_m\} \subset \{1, ..., n\}$ such that $\bar{D}_{v_i v_j} \geq 1 - \alpha, \forall i, j$, where α is a constant close to 0. Order sets with this property by size to obtain V_{k1}, V_{k2}, etc. These V sets are candidates of tight clusters.

7.9.4 Tight Clustering of Gene Expression Time Courses

Tight clusters are the most informative clusters and are obtained as small clusters. Such clusters usually include 20 to 60 genes in genomic signal processing (Yuan and Li 2008; Roddick and Spiliopoulou 2002). Moreover, tight clusters are more interpretable than existing partitions because they find core patterns. The k-means method is used to find the initial partition, and helps reveal more information. For example, a new function can be discovered when genes belonging to the same functional category are assigned into different tight clusters. Tight clusters can be obtained, by classifying some genes as scattered genes, but such classification can disturb biologically relevant patterns. Hence, Yuan and Li have proved that some scattered genes can be of biological importance and should not be removed as outliers (Yuan and Li 2008).

7.10 Higher-Order Mining

The knowledge, information, or patterns obtained from large raw data are widely acknowledged, but many a times these raw data are not available due to several reasons. First, agencies do not want to share their data. Second, streaming data is only available temporarily, which will be in some other form. Finally, it is difficult to achieve the required computational speed, which is dependent on hardware technologies (Wijsen and Meersman 1998). Therefore, there is a strong need to define methods that can extract knowledge or information even if there is no accessibility of primary or raw data. Hence, to overcome the above limitation, higher-order mining is defined. Higher-order mining is a data mining form in which derived data, statistical information, or patterns are the input instead of raw data (Roddick et al. 2008). More formally, let $\Xi = \{\varepsilon_i \mid i = 1....n, n \geq 1\}$ be a set of models or patterns, such that ε_i has been extracted from a dataset D_i. Higher-order mining discovers new pattern or model $\hat{\varepsilon}$ from the set Ξ through the use of data mining methods (Roddick et al. 2008).

7.10.1 Clustering for Association Rule Discovery

Due to the lack of discreteness in the nonprimary data, it is difficult to apply association rule mining on it (Tuzhilin and Adomavicius 2002). This limitation can be overcome by using clustering as a preprocessing step that helps formulate discrete intervals to obtain association rules of adequate frequency. Yang and Miller find numerical ranges for ordinal values by creating distance-based association rules by using the BIRCH algorithm (Zhang et al. 1996), which can be obtained from the generated clusters (Yang and Miller 1997). Moreover, Yang and Miller used the distance measure to find the distance between two clusters instead of finding support and confidence. This distance helps in determine how strong a rule is; i.e., if a distance between two clusters C_x, C_y, is large, then the rule is weak, $C_x \rightarrow C_y$ (Yang and Miller 1997).

7.10.2 Clustering of Association Rules

There can be some cases when there are large numbers of rules and these large numbers can make their interpretation difficult to understand (Toivonen et al. 1995). A set of rules of the form

ID[1] $\rightarrow$ Insurance[yes]
ID[2] $\rightarrow$ Insurance[yes]
ID[3] $\rightarrow$ Insurance[yes]

might be better described as

ID[1–3] → Insurance[yes]

Lent et al. (1997) cluster the association rules by using a concept similar to the binning method used by Agrawal and Srikant (Lent et al. 1997; Agarwal and Srikant 1995), where each bin represents an association rule. Gupta et al. (1999) used the concept of finding distances to cluster the association rules, and Denton and Perrizo (2003) combined various forms of data mining algorithms and made a framework based on partitions.

7.10.3 Clustering Clusters

Partition-based clustering algorithms are iterative, as the center continues to move until it reaches a stopping or threshold condition based on some criterion. Therefore, initial points are computed to find the mean and allow the iterative process to begin. According to Bradley and Fayyad, if the starting point is suboptimal, then clustering algorithms will reach suboptimal solutions (Bradley and Fayyad 1998). Because of this problem, Fayyad suggested that clustering subsamples can aid in the discovery of an improved local minimum, and that the combination of solutions through clustering can reach an improved starting point. However, due to a suboptimal starting point, this method will be computationally expensive. A smoothing process can improve the chance that a researcher will reach a good solution (Bradley and Fayyad 1998).

7.11 Conclusion

In conclusion, Chapters 6 and 7 contain a detailed list of clustering (unsupervised) techniques of data mining. In these two chapters we have also provided insights into their application in bioinformatics and the challenges they pose.

References

Agarwal, R., and R. Srikant. Mining sequential patterns. *Proceedings of the Eleventh International Conference on Data Engineering.* Washington, DC: IEEE Computer Society, 1995, 3–14.

Azuaje, F., and N. Bolshakova. Clustering genome expression data: Design and evaluation principles. In D. Berrar, W. Dubitzky, and M. Granzow (Eds.), *Understanding and using microarray analysis techniques: A practical guide.* London: Springer Verlag, 2002.

Bader, G.D., et al. An automated method for finding molecular complexes in large protein interaction networks. *BMC Bioinformatics* 4, no. 2 (2003): 1–27.

Banfield, J.A. Model-based Gaussian and non-Gaussian clustering. *Biometrics* 49, no. 3 (1993): 803–821.

Ben-Hur, A., D. Horn, and H. Siegelmann. Support vector clustering. *J Machine Learn* 2 (2001): 125–137.

Bradley, P.S., and P.B. Fayyad Refining initial points for k-means clustering. In J. Shavlik (Ed.), *15th International Conference on Machine Learning (ICML'98)*. San Francisco: Morgan Kauffmann, 1998, pp. 91–99.

Brumm, J., E. Conibear, and W. Wasserman. Discovery and expansion of gene module by seeking isolated groups in a random graph process. *PLoS One* 3, no. 10 (2008): 1–9.

Davies, D.L., and D.W. Bouldin. A cluster separation measure. *IEEE Trans Pattern Anal Machine Intell* 1 (1979): 224–227.

Denton, A., and W. Perrizo. Framework unifying association rule mining, clustering and classification. In *International Conference on Computer Science, Software Engineering, Information Technology, e-Business and Applications (CSITeA03)*, Rio de Janeiro, Brazil, 2003.

Goldberg, A.V., and R.E. Tarjan. A new approach to the maximum flow problem. *Proceedings of the Eighteenth Annual ACM Symposium on Theory of Computing*. New York: ACM, 1986, 136–146.

Gomory, R.E., and T.C. Hu. Multiterminal network flows. *SIAM J* 9 (1961): 551–570.

Gupta, G.K., et al. Distance based clustering of association rules. In *Intelligent Engineering Systems through Artificial Neural Networks, ANNIE*. St. Louis, MO: ASME, 1999, pp. 759–764.

Girolami, M. Mercer kernel based clustering in feature space. *IEEE Trans Neural Networks* (2002).

Handl, J., J. Knowles, and D. Kell. Computational cluster validation in post-denomic data analysis. *Bioinformatics* 21, no. 15 (2005): 3201–3212.

Hubert, L. Comparing partitions. *J Classification* 2 (1985): 193–218.

Karypis, G., and E.H. Han. Chameleon: Hierarchical clustering using dynamic modelling. *IEEE Comput Soc* (1999): 68–75.

Kass, R.E., and A.E. Raftery. Bayes factors. *J Am Assoc* 90 (1995): 773–795.

KeiichiroOno. (2010, May 28). Cytoscape wiki. Retrieved October 16, 2010, from http://cytoscape.wodaklab.org/wiki/Data_Sets?action=AttachFileanddo=viewandtarget=Readme_may_txt.

Kempe, D., et al. A decentralized algorithm for spectral analysis. In *Proceedings of the Thirty-Sixth Annual Symposium on Theory of Computing*. New York: ACM Press, 2004.

Kohonen, T. *Self-organizing maps*. Heidelberg: Springer-Verlag, 1995.

Kotani, M., and A. Sugiyama. Analysis of DNA microarray data using self organizing map and kernel based clustering. In *Proceedings of the 9th International Conference on Neural Information Processing (ICONIP'02)*, Vol. 2, Singapore, IEEE. 2002, pp. 755–759.

Lent, B., et al. Clustering association rules. In A. Gray and P.-A. Larson (Eds.), *13th International Conference on Data Engineering*. Birmingham, UK: IEEE Computer Society Press, 1997, pp. 220–231.

Maulik, U., and S. Bandopadhyay. Performance evaluation of some clustering algorithms and validity indices. *IEEE Trans Pattern Anal Machine Intell* 24, no. 12 (2002), 1650–1654.

Matsuda, H., T. Ishihari, and A. Hashimoto. Classifying molecular sequences using a linkage graph with their pairwise similarities. *Theor Comput Sci* 210, no. 2 (1999): 305–325.

Meersman, R., et al. On the complexity of mining quantitative association rules. *Data Mining Knowledge Discovery* 2, no. 3 (1998): 263–281.

Nepusz, T., and H. Yu. (n.d.). Detecting overlapping protein complexes in protein-protein interaction networks. In preparation.

Nepusz, T., H. Yu, and A. Paccanaro. Detecting overlapping protein complexes in protein-protein interaction networks. *Nature Methods* 8 (2012): 417–472.
Nepusz, T., H. Yu, and A. Paccanaro. (n.d.). *Cluster ONE Cytoscape plugin*. Retrieved October 15, 2010, from http://www.cs.rhul.ac.uk/home/tamas/assets/files/cl1/cl1-cytoscape-0.1.html.
Newman, M., and M. Girvan. Mixing patterns and community structure in networks. In *Statistical mechanics of complex networks: Proceedings of the XVIII Sitges Conference on Statistical Mechanics*. Lecture Notes in Physics, vol. 625. Berlin: Springer Verlag, 2003.
Ouyang, M., and W.J. Welsh. Gaussian mixture clustering and imputation of microarray data. *Bioinformatics* 20, no. 6 (2004): 917–923.
Prodromidis, A., et al. Meta-learning in distributed data mining systems. In H. Kargupta and P. Chan (Eds.), *Advances in distributed and parallel knowledge discovery*. Ann Arbor, MI: AAAI Press, 2000.
Rand, W.M. Objective criteria for the evaluation of clustering methods. *J Am Stat Assoc* 66 (1971): 846–850.
Roddick, J.F., M. Spiliopoulou, D. Lister, and Ceglar, A. Higher order mining. *SIGKDD Explorations* 10, no. 1 (2008), 5–17.
Rousseeuw, P.J. Silhouettes: A graphical aid to the interpretation and validation of cluster analysis. *J Comput Appl Math* 20 (1987): 53–65.
Schaeffer, S.E. Graph clustering. *Comput Sci Rev* (Elsevier) 1, no. 1 (2007): 27–64.
Scholkopf, B., et al. Estimating the support of a high-dimensional distribution. *Neural Comput* 13 (2001): 1443–1471.
Schwarz, G. Estimating the dimension of a model. *Ann Stat* 6 (1978): 461–464.
Spiliopoulou, M., et al. A survey of temporal knowledge discovery paradigms and methods. *IEEE Trans Knowledge Data Eng* 14, no. 4 (2002): 750–767.
Srikant, R., et al. (1995). Mining sequential patterns. In P. Yu and A. Cheneditors, *11th International Conference on Data Engineering (ICDE'95)*. Taipei, Taiwan: IEEE Computer Society Press, pp. 3–14.
Tibshirani, R., and G. Walther. Cluster validation by prediction strength. *J Comput Graphical Stat* 14, no. 3 (2005): 511-528.
Toivonen, H., et al. Pruning and grouping of discovered association rules. In *ECML-95 Workshop on Statistics, Machine Learning and Knowledge Discovery in Databases*, Heraklion, Greece, 1995, pp. 47–52.
Tseng, G.C., and W.H. Wong. Tight clustering: A resampling-based approach for identifying stable and tight patterns in data. *Biometrics* 61, no.1 (2005): 10–16.
Tuzhilin, A., and G. Adomavicius. Handling very large numbers of association rules in the analysis of microarray data. In *8th ACM SIGKDD International Conference on Knowledge Discovery and Data Mining*. Edmonton, Alberta, Canada: ACM, 2002, pp. 396–404.
Vapnik, V. An overview of statistical learning theory. *IEEE Trans Neural Networks* 10, no. 5(1999): 988–999.
Wu, Y., P.R. Bhat, T.J. Close, and S. Lonardi. Efficient and accurate construction of genetic linkage maps from the minimum spanning tree of a graph. *PLoS Genet* (2008): 1–11.
Xu, R., and D. Wunsch. Survey of clustering algorithms. *IEEE Trans Neural Networks* 16, no. 3 (2005): 645–678.
Yang, Y., and R. Miller. Association rules over interval data. In J. Peckham (Ed.), *ACM SIGMOD International Conference on the Management of the Data*. Tucson, AZ: ACM Press, 1997, pp. 452–461.

Yeh, C.-Y, and S.-J. Lee. A kernel-based two-stage nu-support vector clustering algorithm. *2007 International Conference on Machine Learning and Cybernetics*. Hong Kong, China: IEEE, 2007, 2551–2256.

Yeung, K.Y., C. Fraley, A. Murua, and A.E. Raftery. Model-based Gaussian and non-Gaussian clustering for gene expression data. *Bioinformatics* 17, no. 10 (2001): 977–987.

Youness, G., and G. Saporta. Comparing partitions of two sets of units based on the same variables. *Adv Data Anal Classification* 4 (2010): 53–64.

Yuan, Y., and C.T. Li. Partial mixture model for tight clustering of gene expression time-course. *BMC Bioinformatics* 9, no. 287(2008): 1–17.

Zhang, T., et al. BIRCH: An efficient clustering method for very large databases. In *ACM SIGMOD Workshop on Research Issues on Data Mining and Knowledge Discovery*. Montreal, Canada: ACM, 1996, pp. 103–114.

Zhou, X., M.C. Kao, and W.H. Wong. Transitive functional annotation by shortest-path analysis of gene expression data. *Proc Natl Acad Sci USA* 90, no. 20 (2002): 12783–12788.

SUPERVISED LEARNING AND VALIDATION

Chapter 8

Classification Techniques in Bioinformatics

Supervised learning, like unsupervised learning, is one of the data mining tasks introduced in the knowledge discovery in databases (KDD) process and consists of two phases, training and testing. In the training phase, we build a model using samples that is representative of the hypothesis (or real-world use of the function) and connects (learns from) the input parameters to achieve a learning objective, such that the samples can accurately and efficiently predict the learning outcome. We then extract features of interest from the samples. In this step, we ensure that the features are not too large in order to avoid the curse of dimensionality. Once we have completed the training phase, we begin the test phase. We test the trained model using random samples of data. Typically, the test phase includes evaluation routines such as holdout and k-fold cross-validation techniques. In this chapter we provide an overview of the various supervised learning techniques, better known as classification techniques, and their application in the field of bioinformatics.

8.1 Introduction

There is a wide variety of classification techniques that one can choose from, and they perform differently under different data and learning domains. It is important to understand how these algorithms work in order to understand how the performances and results will differ. Before we explain the working principle of each algorithm, we will highlight the intricacies of supervised learning.

8.1.1 Bias-Variance Trade-Off in Supervised Learning

When trying to understand how a supervised learning algorithm works, it is imperative to understand two key terms, *bias* and *variance*. To elucidate the effects of bias and variance and help define their prominence in supervised learning, let us first consider a situation in which we have two train sets. Let us assume that these two train sets do not share any samples but have the same number of classes and the same number of samples in each class. Let us consider a random test sample x that is used to test models built by a supervised learning algorithm using both training sets independently. If the sample x is incorrectly classified across both train sets, the model has a high degree of bias. If however, sample x is assigned a different class for different train sets, then the model has a high degree of variance. These two variables have direct implications on the prediction error of the model, as it is directly proportional to the sum of the bias and variance.

When we create a train set with low bias, the result is often a "flexible classifier." However, such a classifier may be too flexible and will fit differently in different datasets. Thus, there is a natural trade-off between bias and variance.

8.1.2 Linear and Nonlinear Classifiers

In this section we draw the distinction between linear and nonlinear classifiers. The distinction is drawn by how the input object's characteristics are modeled for decision making.

A linear classifier decides class membership of a sample by comparing a linear combination of the features to a fixed threshold. For example, let us consider a set of points that belong to two classes represented in a two-dimensional (2D) space as shown in Figure 8.1. A linear classifier attempts to fit a line $c_1f_1 + c_2f_2 = H$ so that

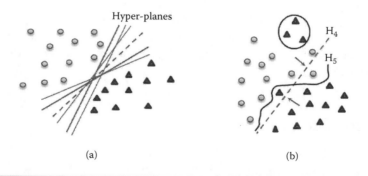

Figure 8.1 The triangles and dots can be separated by multiple linear classifiers in (a). In some cases, the separation of data by a linear function can lead to false alarms (and dismissals) in discrimination, but a nonlinear function can achieve better separation, as shown in (b).

the line separates or divides the points into two corresponding classes. This step is best described in Figure 8.1(a). Since we consider only two features f_1 and f_2 for analysis, the resultant rules for classification are a linear combination of these two features, in which a sample is assigned to the first class if it satisfies $c_1 f_1 + c_2 f_2 > H$. Otherwise, the feature is assigned to the second class.

In our example, $(f_1, f_2)^T$ is the 2D vector representation of a data point. Both the parameter vector $(c_1, c_2)^T$ and the constant H play a vital role in defining the decision boundary. The resultant 2D representation of the decision boundary is a straight line that is a plane when viewed in three dimensions. When the number of dimensions is greater than 3, the resultant decision boundary is generalized to what is referred to as a hyperplane. If a hyperplane perfectly separates two classes, then the two classes are linearly separable. It should be noted that if the property of linear separability is maintained, then there are an infinite number of linear separators. Figure 8.1(a) is pictorial representation of a scenario in which the number of possible hyperplanes can be infinite. In reality, data are plagued by noise. While dealing with a linearly separable problem using noisy data for training, the challenge of choosing the best hyperplane is questioned, requiring a stringent criterion for selecting among all decision hyperplanes that perfectly separate the training data. In general, some hyperplane will perform well on new data and some will not. Thus, linear classifiers may not be as simple to use as they are to conceive due to a difficulty in determining the optimal set of parameters of $\bar{c}$ and H from a given train set.

The nonlinearity of a nonlinear classifier is intuitively clear when the decision boundaries of the classifier are locally linear segments. However, we generally have a complex shape that is not equivalent to a line in two dimensions or a hyperplane in higher dimensions.

Figure 8.1(b) shows one such example of a nonlinear problem. In this figure, there is no one good linear separator between the two classes, as an isolated cluster of points that belong to a different class is surrounded by points of another class. This lack of clear boundaries would make accurate classification using linear methods almost impossible.

Since nonlinear classifiers can capture complex decision boundaries, they produce better classification accuracies. Since we know that they perform better for complex classification problems, we must determine whether they perform well enough in other areas to justify using nonlinear classifiers for all classification problems. To answer this question, we look into bias and variance and their roles in both linear and nonlinear classification. It is imperative to describe a means to estimate the error associated with the classifier. A commonly used error estimate is the mean squared error (MSE), which is the difference between the predicted output of a classifier Υ given a test sample x and the probability of x belonging to a class C (represented as $P(C|x)$), represented by the following notation:

$$MSE(\Upsilon) = E_x[\Upsilon(x) - P(C|x)] \qquad (8.1)$$

where is E_x the expectation with respect to $P(d)$. Our objective is thus to minimize the MSE that is averaged over train sets. To achieve this objective, we reexamine the concept of bias and variance with respect to linear and nonlinear classifiers. We then define bias as the squared difference between $P(C|x)$ (the actual probability of a sample x belonging to a class C) and the predicted outcome $\Gamma_D(x)$ of the learned classifier averaged across train sets. The bias is large if the classifier is consistently inaccurate across different train sets. On the contrary, the bias may be small for several reasons: (1) if the classifier performs consistently well across different datasets, (2) if different train sets cause errors on different test samples, or (3) if different train sets result in positive and negative outcomes on the same test sample, but average out to near zero.

Linear models are considered to have a high bias for nonlinear problems, as they can only be employed to model a linear hyperplane. If one of the input train sets has a class that is a nonlinear class boundary, then the resultant bias is high. This high bias is a result of a large number of data points in the train set that would be consistently misclassified by the linear classifier. They therefore require intuitive knowledge of the problem for fitting a linear classifier with data that are believed to exhibit linear characteristics, yielding lower error estimates and more correctly classified instances. On the contrary, if true class boundaries are not linear and we incorrectly "force" the classifier to be linear, then the classification accuracy will drop.

Nonlinear models are considered to have a low bias. As discussed, the decision boundaries generated by these classifiers vary greatly and are dependent on the distribution of the data points of a class in the train set. The variability offered by the nonlinear classifiers provides flexibility in classifying different classes with different degrees of accuracies as per the application needs.

We define variance as the variation of the prediction of a learned classifier across different datasets. It is the average squared difference between the predicted outcome and its averaged prediction across different train sets.

The variance is large if different train sets D give rise to very different predictions for a given test sample x. It is small if the train sets have a minor effect on classifier prediction, be it correct or incorrect. Thus, variance is a measure of inconsistency with the decisions and does not take into consideration whether they are correct or incorrect.

Linear models are considered to have low variance, as most train sets that are randomly generated produce similar decision hyperplanes. The decision lines produced by linear learning methods will deviate slightly from the main class boundaries, depending on the train set, but the class assignment for the vast majority of samples (with the exception of those close to the main boundary) will not be affected. The circular enclave will be consistently misclassified.

Nonlinear models have high variance. It is apparent that these models create complex boundaries between classes and are thus sensitive to noise. As a result, the variance is high. For instance, if the test sample is close to noisy samples in the

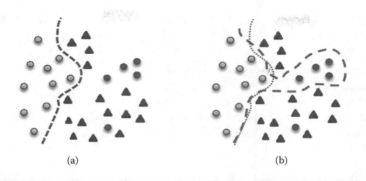

Figure 8.2 A schematic representation of noise in data and its effects on the performance of a nonlinear classifier. (a) A nonlinear classifier that does not overfit the data and (b) a nonlinear model that overfits to suit the distribution of noise in the data.

train set, it can get misclassified by default. This noise results in a high variation across train sets.

High variances in learning methods are prone to overfitting training data, which may prevent us from capturing true properties of the underlying distribution (as shown in Figure 8.2). In overfitting the learning model learns from noise as well as from features. Overfitting increases the MSE and frequently is attributed to high variance learning methods.

8.1.3 Model Complexity and Size of Training Data

With the large increase in bioinformatics data, a large volume of research has been published, creating a growing importance of using data mining to utilize, understand, and discover patterns of interest. In this section we try to explain the method for selecting a subset of data and the method's effects on variance and bias of a classifier used.

The work performed by Brian and Webb (1999) illustrates the effects of increasing the size of the dataset on the variance and bias of different learning methods. They attempted to explore the possibility of designing algorithms specifically for large datasets. Their analysis was aimed at proving that if the number of samples in the dataset was increased, they could decrease the variance to develop algorithms that obtain more interesting results.

As discussed in Section 8.1.2, variance measures the degree to which the predictions of the classifiers developed by a learning algorithm differ among datasets. If the train sets are small, we assume that the relative impact of the train sets cannot be a sufficient representation of the population, and thereby expect a large variance for the learning algorithm. We can solve this problem using the method described by Brian and Webb (1999). First, we assume that 50% of the samples in a population (large database) have a common characteristic. If a randomly selected train set is of size 10, there is a probability of 0.38 that 40% or less of the samples in the train set will retain the common characteristic from the population. Furthermore, there is a negligible 0.17% probability that 30% or less of the samples in the train set will retain the characteristic. Thereby, if a smaller train set is taken from a large population, it has a slimmer chance of being a good representation of the population. To negate the effect of this, we use a train set of size 1,000,000 if 50% of the population exhibits a common characteristic. The probability that 40% or less of the samples in the train set exhibits this characteristic is less than 10–22, and the probability that 30% or less of the samples in the train set exhibits it is less than 10–26.

Furthermore, if 1% of the population exhibits a common characteristic and we randomly choose a train set of size 100, there is a probability of 0.37% that this characteristic will not be found in the samples of the train set and a probability of 10–17 that this characteristic will not be captured in a larger train set of 1,000,000. In contrast, in a random sample of size 1,000,000, if 50% of the population exhibits the characteristic, then the probability that 40% or less of the sample will exhibit the characteristic is less than 10–22. The probability that 30% or less of the sample will exhibit the characteristic is less than 10–26.

From this description, it can be expected that classifiers learned from many small train sets will differ more significantly than classifiers learned from larger train sets. This hypothesis was tested and verified using both linear and nonlinear models. It was observed that in both linear and nonlinear models, the mean square error dropped as the size of the train set increased. However, when both independent variances and biases were compared, it was observed that nonlinear models exhibited a reduction in variance and a similar reduction in bias. While we expect a similar trend in linear models, it was observed that though there is a consistent decrease in variance with the increase in train set sizes, the bias fluctuated.

The statistical conclusions drawn from these experiments reinforced the hypothesis that both linear and nonlinear classifiers perform better as train data are increased. This result is evident with the decrease in variance values obtained with increased train set sizes. The various trends in bias show that the computational complexity of learning affects the performance of the models generated as the size of train sets increases. Thus, if the presented results are extrapolated to millions of train samples, then the complexity of the learned models can be expected to be orders of magnitude higher than that for the small train sizes from which models are normally developed. This change in learning complexity may also be attributed to the decrease in variance, leading to the next important feature of reducing the complexity of these algorithms to strike a balance between both bias and variance. Various forms of classification schemes are available based on the number of phenotypes or classes available in the train sets.

Binary classification: In this kind of classification scheme, there are two phenotypes considered in the train set. The objective of the learning scheme is to discriminate between samples that belong to either of the two classes. Better known as binary class classification, this scenario is derived from a single hypothesis H, leading to a single conclusion, either positive or negative.

Multiclass classification: On the other hand, we call learning multiclass classification if there are more than two phenotypes or classes associated with the instances in the training and testing sets. In this form of classification, the objective is to classify a single sample into one of the many classes. It is more complicated than the binary class classification, as comparing multiple hypotheses makes the decision.

Multiclass classification can thus be viewed as a collection of binary class classification strategies. Several commonly used strategies can make this possible, namely, the one versus all (*OvA*) and the all versus all (*AvA*). The *OvA* strategy works on the pretext that a single hypothesis separates one class from the rest of the classes. This strategy equates a multiclass classification approach into a standard binary class problem, whereas the *AvA* strategy employs multiple hypotheses, where independent hypotheses exist between each pair of classes. Thus, decisions are performed based on the cumulative results of various underlying hypotheses being satisfied.

8.1.4 Dimensionality of Input Space

It is apparent that the complexity of the learning method is connected to the size of the train set. For simplicity, let us assume a train set X is a collection of data samples $\{x_1, x_2,...,x_n\}$. Each data sample x_i is described by a set of features $\{f_1, f_2,...,f_m\}$, also referred to as dimensions. In our discussion on linear and nonlinear models, the complexity of a classifier is closely tied to the number of features or dimensions used to describe each x_i.

The complexity of the learning algorithm grows exponentially larger when the number of features (m) exceeds the number of samples (n) in the train set (i.e., when $m \gg n$). This growth in complexity is owed to the curse of dimensionality as discussed in Chapter 2. Furthermore, the distance (Euclidean distance) between the points in an m-dimensional space increases as the number of dimensions increases, thereby introducing sparseness in the distribution of data. With the increase in sparseness of data, it becomes a computational challenge to determine the boundaries of the data, and it is therefore difficult to determine a single hypothesis.

This problem is increasingly prevalent in bioinformatics, as many datasets are considered to be high-dimensional (Ma and Huang 2008). For instance, cancer classification using gene expression analysis (Golub et al. 1999; West et al. 2001), epigenetics (Zukiel et al. 2004; Piyathilake and Johannig 2002), and proteomics using mass spectrometry (Leslie et al. 2004) all contain a large number of features that far exceed the number of samples in the train set. It is believed that not all features are useful in describing the samples and as a solution require a data preprocessing step of feature selection to reduce the number of features by filtering out relevant or redundant features.

8.2 Supervised Learning in Bioinformatics

Supervised learning finds its application in many facets of bioinformatics, especially in genomics and proteomics. With the prevalence of high-throughput techniques in biology, it has become increasingly difficult to analyze data of large magnitudes. For example, microarray technology makes it possible to view the expression of thousands of genes under a variety of experimental conditions. Microarray gene expression experiments have been conducted to identify biomarkers in the different manifestations of cancer (Ramaswamy et al. 2001), including breast cancer (Lukes et al. 2009), head cancer, neck cancer, lung cancer (Vachani et al. 2007), and lymphoma (Golub et al. 1999). Researchers analyze the regulation (up- or downregulation) of subsets of genes to draw associations between genes that will elucidate their role in cancer. As with any biological data, the data obtained from microarray studies have categorical phenotypes of interest that are hierarchical, such as cancer occurrence, stages, or subtypes. Moreover, the number of genes in these studies typically exceeds the number of samples available, making statistical inference about the genes difficult. This mismatched ratio of genes to samples mandates the use of supervised learning. Supervised learning can be used to reduce the number of genes.

Furthermore, in proteomics, supervised learning techniques have been utilized to analyze an array of problems of biological significance. For instance, one of the central problems of bioinformatics is the classification of protein sequences into functional and structural families based on sequence homology. It is easy to sequence proteins but difficult to obtain protein structures. Analytic solutions are required based on statistical techniques to classify protein sequences into families and superfamilies. These classification strategies usually rely upon extracted features that exploit structure and functional relationships between proteins and their constituents. Furthermore, supervised learning has been exploited to determine the subcellular location of proteins in a cell. The subcellular location of a protein is necessary in determining its functional characteristics, as the protein's location in the cell aids in inferring its biological functions.

The automatic prediction, using supervised learning techniques, of protein subcellular localization is an important component of bioinformatics. Thus the prediction of protein function is now an integral part of bioinformatics and can aid in identification of drug targets.

As discussed in Chapter 5, mass spectrometry is an analytic technique that measures the mass-to-change ratio of ions. It is generally used to find the proteomic/peptide composition of a physical sample. Some types of cancer affect the concentration of certain molecules in the blood, which allows early diagnosis by analyzing the blood mass spectrum. This is a data-rich facet of proteomics that can benefit from supervised learning. Each feature is measured with mass spectra, and often summary statistics of the peaks can be used to discriminate between individuals with different cancer phenotypes. Researchers have used mass spectra to commonly detect prostate, ovarian, breast, bladder, pancreatic, kidney, liver, and colon cancers.

Thus, the aim of supervised learning in bioinformatics is to broadly address two objectives: to build accurate classifiers or predictive tools and to derive inferences from the results obtained.

1. *To build accurate classifiers or predictive tools*, users can apply one of several learning methods that could be linear or nonlinear. Such linear models used include support vector machines (SVMs) and the naïve Bayes (NB). Nonlinear models include the *k*-nearest-neighbor (kNN) classifier and tree-based classifiers such as C4.5. The classifier used is determined by the nature of the dataset used for training and testing. Moreover, the classifier should enable reliable discrimination between different phenotypes under analysis.
2. *To derive inferences from the results obtained*, biologists survey the data for relevant information. Though building of accurate classifiers is important, biologists are not merely interested in accurate predictive tools. They also look for additional information that could be extracted from the data but that could not be derived from simple statistical analysis. For example, it is of growing importance for researchers to identify biomarkers of diseases from a set of microarray samples obtained from different biological states. These biomarkers refer to a small set of relevant genes that lead to the correct discrimination between different biological states, which are derivatives of patterns obtained from classification rules.

To provide a conceptual view of the data, we use microarray data as an example to introduce concepts and challenges in this data process. Microarray technology enables the measurement of the expression level of thousands of genes simultaneously in a cell mixture (Wang et al. 2005). A phenotype is the outward, physical manifestation of an organism, and phenotype classification is used to classify tissue samples into different classes of phenotypes, including cancer versus normal, using gene expression data (refer to Figure 8.3). These phenotypes are determined using the measured expression levels of thousands of genes in the samples as features.

Thus, to conceptualize these data, let us assume that from given N tissue samples and expression levels of M genes, we can store the data in a $N \times (M+1)$ matrix as shown below, where each vector (column) represents a sample and each element in the vector represents the expression value of the M genes. We introduce an

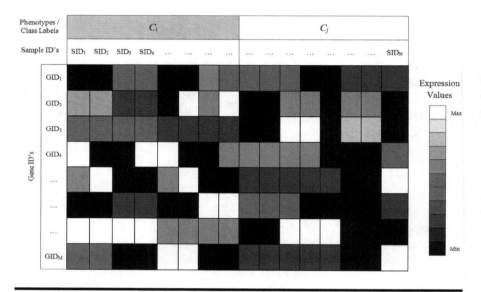

Figure 8.3 A schematic representation of gene expression data obtained from high-throughput microarrays.

additional (C_j) element into the vector that represents the phenotype from which the sample is drawn.

One of the challenges in using supervised learning in bioinformatics lies in the embedded challenges that the raw data possess. As mentioned, the success of using a supervised classification scheme can only be exemplified if the following data issues are addressed in the preprocessing stage of the KDD process: the removal of data inconsistency and missing values, the removal of noise, normalization, and the reduction of dimensionality.

1. *Removal of data inconsistency and missing values* must be performed to determine which method is best. Biological data typically consist of data generated by biological experiments. Legacy systems are typically plagued with manually curated data that have varied nomenclature and missing values. It is thus imperative that these issues be addressed before subjecting the data to any learning approach.
2. *Removal of noise* is performed to filter out samples that do not meet the standards for data. The inconstant recording of results from biological experiments plagues these systems. Not all biological experiments fail to systematically follow a set of standards; then these experiments, rather than increasing the volume of data, contribute to the noise of the system. It is thus imperative to filter out samples that do not confirm the standards by applying appropriate filtering approaches.

3. *Normalization* is a process that ensures that all the samples are treated equally. Typically data in bioinformatics are obtained from disparate sources, making normalization essential for effective comparison and learning.
4. *Dimensionality reduction* reduces the expense of computational systems in evaluating high-dimensional data. It is known that not all features/dimensions are important or at times redundant and can be removed, thereby reducing the computational load and decreasing the scarcity of the data. It is preferred to understand the nature of the data and use effective feature selection techniques.

In general, the analysis of biological data entails several hundred to thousands of features (as in the case of microarray data), with only a few dozen to hundreds of samples available. In such cases, the number of dimensions exceeds the number of samples ($M \gg N$). Most learning algorithms exploit chance patterns and elaborate models that perform well on training data but poorly on new data, leading to overfitting. The risk of overfitting must be reduced by selecting a set of features proportionate with the number of samples. Moreover, the selection of a reduced set of features requires fewer computational efforts for model learning and enables a better understanding of the process that underlies the data. Depending on how the selection process is combined with the classification process, attribute selection methods belong to one of the following three categories: filter methods, wrapper methods, or embedded methods. These methods are explained in Chapter 4. In the remainder of this chapter, we highlight key supervised learning approaches and their applications in the field of bioinformatics. There are several supervised learning strategies in existence, and we have logically separated them into the following categories: linear models, which include SVMs; naïve Bayes, nonlinear models, which include tree-based models and Bayesian networks; and ensemble approaches, which include bagging and boosting.

8.3 Support Vector Machines (SVMs)

Support vector machines (SVMs) are powerful classification algorithms. They are prominently used in computational biology and have been successfully applied to a gamut of problems, like protein homology detection (Melvin et al. 2007), functional classification of promoter regions (Holloway et al. 2005), and the prediction of protein-protein interactions (Chatterjee et al. 2011). SVMs are based on two key concepts, the margin of separation and kernel functions (Ben-Hur et al. 2008). The philosophy behind the use of the SVM is to fit a linear separating line or plane between the distributions of points. This philosophy is based on the performance methods of any linear model as described in previous sections. We refer to this separating line or plane as the hyperplane. In a 2D view, this hyperplane is as simple as drawing a line that separates

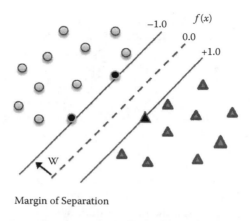

Figure 8.4 A 2D representation of a linear classifier separating two classes.

the points into two groups. Points that lie on one side of the hyperplane are considered to be positive, and the remaining points are considered to be negative.

Let us consider a situation with a set of points that belong to two classes that are well separated. We assume that the separation between these points is such that we could intuitively draw a separating hyperplane that is as far as possible from the points in both classes. Such a hyperplane is believed to possess a large margin of separation. This hyperplane is pictorially represented in Figure 8.4.

The decision boundary (or hyperplane) is represented using dashes and the function $f(x)$. The maximum margin boundary is computed by a linear SVM. The region between the two lines defines the margin area. The data points highlighted with black centers are support vectors. Thus, the first objective of the algorithm is to maximize this margin of separation. Since the fitting of the hyperplane is closely connected to the distribution of the data points, it becomes a challenge to fit a maximum margin of separation, when the data points overlap in their distributions. When the data point distributions overlap, the data are inherently believed to be nonlinear. We can extend the linear SVM to suit the nonlinearity of data using kernel functions. Thus, as the name suggests, kernel functions are transformation functions that transform the linear classifier into a nonlinear classifier. Such functions consist of mapping the nonlinear data to an abstract feature space where the maximum margin of separation exists. We discuss these concepts in depth in the following sections.

8.3.1 Hyperplanes

To introduce the first objective of the SVM algorithm of fitting a maximum separating hyperplane between data points that belong to different classes, let us represent each data point x by a vector of length N, i.e., x_j, where $j = 1,\ldots,N$, and j represents the features that describe the data point in an N-dimensional space. Using this nomenclature, a data matrix of M data points is represented as a matrix of the form $\{(x_i, y_i)\}_{i=1}^{M}$, where y_1 is the class label associated with the data point x_i.

SVMs use a linear discriminant function to fit a linear plane for the given data matrix that is represented as follows:

$$f(x) = \langle w, x \rangle + b. \tag{8.2}$$

Here, $\langle w, x \rangle$ represents the dot product between two vectors w and x. This dot product is also referred to as the scalar product between two vectors and is represented as follows

$$\langle w, x \rangle = \sum_{j=1}^{n} w_j x_j. \tag{8.3}$$

The purpose of the discriminant function $f(x)$ is to assign a score for a given data point x. This score is then used to decide how to classify x using the weight vector w and the bias b, a scalar value.

In a scenario where there are just two dimensions, the points satisfying the equation $\langle w, x \rangle = 0$ correspond to a straight line that passes through the origin. In the scenario where there are three dimensions, a plane and more generally a hyperplane pass through the origin. The bias b translates the hyperplane with respect to the origin.

The hyperplane divides the space into two half spaces according to the sign of $f(x)$, which indicates the side of the hyperplane a point is located on. If $f(x) > 0$, then the point is located in the positive class; if $f(x) < 0$, then the point is located in the negative class. The boundary between regions is classified as positive, and the decision boundary of the classifier is called negative. A classifier with a linear decision boundary, defined by a hyperplane, is called a linear classifier.

8.3.2 Large Margin of Separation

In a linearly separable dataset, a hyperplane correctly classifies all data points, and there may be many separating hyperplanes. We are thus faced with the question of which hyperplane to close, ensuring that not only the training data, but also feature examples, unseen by the classifier at training time, are classified correctly. Our intuition as well as statistical learning theory suggests that hyperplane classifiers are defined as the distance of the closest example to the decision boundary. Let us adjust b such that the hyperplane is halfway between the closest positive and negative samples. If we scale the discriminant function to take the values +/−1 for these samples, we find that the margin is $1/\|w\|$, where $\|w\|$ is the length of w, also known as its norm calculated using $\sqrt{\langle w,w\rangle}$.

The hard margin SVM, applicable to linearly separable data, is the classifier with maximum margin among all classifiers that correctly classify all the input examples. To compute w and b corresponding to the maximum margin hyperplane, one has to solve the following optimization problem:

$$\min_{w,b} \frac{1}{2}\|w\|^2$$
$$\text{subject to: } y_i(\langle w, x_i\rangle + b) \geq 1, \quad \text{for } i = 1,\ldots,n. \tag{8.4}$$

where the constraints ensure that each example is correctly classified, and minimizing $\|w\|^2$ is equivalent to maximizing the margin. The set of formulas above describes a quadratic optimization problem, in which the optimal solution (w, b) is described to satisfy the constraints $y_i(\langle w, x_i\rangle + b) \geq 1$, while the length of w is as small as possible. Such optimization problems can be solved using standard tools from convex optimization.

8.3.3 Soft Margin of Separation

Data are often not linearly separable; and even if they are, a greater margin can be achieved by allowing the classifier to misclassify some points. Theory and experimental results show that the resulting larger margin will generally provide better performance than the hard margin SVM. To allow errors we replace the inequality constraints in Equation 8.4 with $y_i(\langle w, x_i\rangle + b) \geq 1 - \xi_i$, for $i = 1,\ldots,n$, where $\xi_i \geq 0$ are slack variables that allow an example to be in the margin or misclassified. To discourage excess use of the slack variables, a term $C\sum_i \xi_i$ is added to the function to be optimized:

$$\min_{w,b} \frac{1}{2}\|w\|^2 + C\sum_i \xi_i$$
$$\text{subject to: } y_i(\langle w, x_i\rangle + b) \geq 1 - \xi_i, \quad \xi_i \geq 0. \tag{8.5}$$

The constant $C > 0$ sets the relative importance of maximizing the margin and minimizing the amount of slack. This formulation is called the soft margin SVM.

8.3.4 Kernel Functions

Instead of the abstract idea of data points in space, one can think of data points as representing objects using a set of features derived from measurements performed on each object. For large margin separation, the relative position or similarity of the points to each other is important, and the exact location is unimportant. In the simplest case of linear classification, the similarity of two objects is computed by the dot product (or scalar or inner product) between the corresponding feature vectors. To define different similarity measures leading to nonlinear classification boundaries, one can extend the idea of dot products between points with the help of kernel functions. Kernels compute the similarity of two points and are the second important concept of SVMs. The domain knowledge inherent in any classification task is captured by defining a suitable kernel (i.e., similarity) between objects.

A more straightforward way of turning a linear classifier nonlinear or making it applicable to nonvectorial data is mapping data to vector space, referred to as the feature space, using a mapping function ϕ. The use of this mapping function is represented as follows:

$$f(x) = \langle w, \phi(x) \rangle + b. \tag{8.6}$$

For example, if $f(x)$ is a nonlinear function in the original input space the mapping function ϕ maps each point to linearly separable feature space, as shown in Figure 8.5.

There are different forms of mapping functions, the simplest of which is one that considers all products of pairs of features in the input space. For example, let us

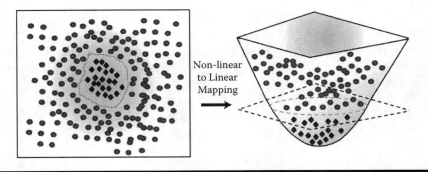

Figure 8.5 A schematic representation of the mapping of a nonlinear input space to a linear feature space where a simple hyperplane can separate between data points of different classes.

assume that we have three features, x_1, x_2, and x_3, in the input space. The resultant features in the feature space would be as follows: $x_1.x_2$, $x_1.x_3$, $x_2.x_3$, x_1^2 x_2^2, and x_3^2. This feature space is therefore quadratic in nature. Though simple to conceive, this approach of explicitly computing nonlinear features (i.e., product of features) does not scale well with a large number of features. Furthermore, if we use monomials of degree d rather than degree 2, as above, the dimensionality would be exponential in d, resulting in a substantial increase in memory usage and the time required to compute the discriminant function. If our data are high-dimensional to begin with, as in the case of gene expression data, this method will not provide acceptable results.

Kernel methods avoid this complexity by avoiding the set of explicitly mapping the data to a high-dimensional feature space.

It is known that (as discussed previously) the weight vector of a large margin separating a hyperplane can be expressed as a linear combination of training points, i.e., $w = \sum_{i=1}^{n} y_i \alpha_i x_i$. This expression can also be used for a large class of linear algorithms. Our discriminant function (Equation 8.6) then becomes

$$f(x) = \sum_{i=1}^{n} y_i \alpha_i \langle \phi(x_i), \phi(x) \rangle + b. \tag{8.7}$$

The representation in terms of the variable α_i is known as the dual representation. We observe that the dual representation of the discriminant function depends on the data only through dot products in feature space. Dual representation is also present for the dual optimization problem when we replace x_i with $\phi(x_i)$. If the kernel function $k(x, x')$ is defined as

$$k(x, x') = \langle \phi(x), \phi(x') \rangle \tag{8.8}$$

it can be computed efficiently. Once this function is defined, the dual formulation can solve the problem without carrying out the mapping ϕ into a potentially very high-dimensional space.

The two most commonly referred to kernel functions are the polynomial and Gaussian kernels. The polynomial kernel of degree d is defined as

$$k_{d,K}^{polynomial}(x, x') = (\langle x, x' \rangle + K)^d \tag{8.9}$$

where K is often chosen to be 0 (homogeneous) or 1 (heterogeneous). The feature space for the heterogeneous kernel consists of all monomials with a degree up to d. And yet, its computation time is linear in the dimensionality of the input space. The kernel with $d = 1$ and $K = 0$, denoted by k^{linear}, is linear, leading to a linear discriminant function.

The degree of the polynomial kernel controls the flexibility of the resulting classifier. The lowest-degree polynomial is the linear kernel, which is not sufficient when a nonlinear relationship between features exists. The second widely used kernel is the Gaussian kernel, defined as

$$k_\sigma^{Gaussian}(x, x') = \exp\left(-\frac{1}{\sigma} \|x - x'\|^2\right) \quad (8.10)$$

where $\sigma > 0$ is a parameter that controls the width of the Gaussian method. The Gaussian kernel plays a similar role, as the degree of the polynomial kernel controls the flexibility of the resulting classifier. The Gaussian kernel is essentially zero if the squared distance $\|x - x'\|^2$ is larger than σ; i.e., for a fixed x' there is a region around x' with high kernel values. The discriminant function is thus a sum of Gaussian bumps centered around each support vector. When σ is large, a given data point x has a nonzero kernel value relative to any sample in the set of samples. Therefore, the whole set of support vectors affects the value of the discriminant function at x, leading to a smooth decision boundary. As we decrease σ, the kernel becomes more local, leading to greater curvature of the decision surface. When σ is small, the value of the discriminant function is nonzero only in the close vicinity of each support vector, leading to a discriminant that is essentially constant outside the close proximity of the region where the data are concentrated.

8.3.5 Applications of SVM in Bioinformatics

SVMs are used for a variety of applications, such as splice site detection or recognition (Sonnenburg et al. 2007; Eichner et al. 2011; Degroeve et al. 2002), remote protein homology detection (Liao and Noble 2003), and gene expression data analysis (Brown et al. 2000). In this section, we describe the experimental design required to apply SVM in these areas of analysis and the nature of data and the modifications that are brought about to the algorithm.

8.3.5.1 Gene Expression Analysis

Here, we briefly describe the work performed by Brown et al. (2000), in which the SVM was used to analyze gene expression data.

8.3.5.1.1 Raw Data

Brown et al. (2000) used the gene expression data obtained from experiments conducted using budding yeast *Saccharomyces cerevisiae* (Eisen et al. 1998). The data consist of 79 samples, each consisting of 2,467 genes. As described in previous sections, we represent gene expression microarray data in the form of a gene expression matrix. Each element of the gene expression matrix contains values of an expression ratio, i.e., the expression levels of a specific gene with respect to two experimental conditions. Typically, the numerator of this expression ratio represents the expression level of the gene in the condition of interest. The denominator of the expression ratio represents the expression level of the same gene in a specific reference condition. Therefore, in a scenario where data are from a series of m experiments, each of the n genes is represented as an m-dimensional vector resulting in an $n \times m$ gene expression matrix. In this case we have $n = 2,467$ genes and $m = 79$ samples. The functional annotation information for these genes was obtained from the Munich Information Center for Protein Sequences (MIPS) (Mewes et al. 2000).

8.3.5.1.2 Data Preprocessing

The raw data are then subjected to the logarithm normalization scheme. The normalized logarithm of the gene expression value is the logarithm of the ratio of expression level E_i for gene X in experiment i, to the expression level R_i of the same gene X in the reference state. The logarithm of the ratio is further divided by the square root of the sum of logarithms such that the expression vector $\tilde{X} = (X_1, \ldots, X_{79})$ has the Euclidean length 1. This length is represented using the following relation:

$$X_i = \frac{\log(E_i/R_i)}{\sqrt{\sum_{j=1}^{79} \log^2(E_j/R_j)}}. \qquad (8.11)$$

Therefore, the expression value of gene X in experiment i is positive if the gene is upregulated with respect to the reference state and negative if it is downregulated.

8.3.5.1.3 Problem Illustration

The aim of this study is to use SVMs to create a model from a set of genes that have common functions and to discriminate between members and nonmembers of a given functional class based on expression data. Thus, with the expression features of the class learned, the SVM can recognize new genes as members or as nonmembers of the class based on its expression data. Second, the inferences drawn by the SVM can provide potential insight about the gene expression patterns that are characteristic for a functional group and whether a specific gene is likely to be a member of a functional group.

8.3.5.1.4 Methodology

The methodology employs the following steps:

1. Each vector X in the gene expression matrix is viewed as a point in an m-dimensional space.
2. Construct a hyperplane that separates samples from two phenotypes.
   a. The data are nonlinear; therefore, the authors used kernel functions. They use multiple kernels to obtain optimal results: (i) a simple kernel, (ii) a quadratic kernel, and (iii) the Gaussian kernel.
      i. Using a simple kernel $K(X,Y)$ that can measure the similarity between genes X and Y by using the dot product in the input space

      $$(X,Y) = (\vec{X}.\vec{Y} + 1) = \sum_{i=1}^{79} X_i Y_i.$$

      ii. Squaring the kernel $K(X,Y) = (\vec{X}.\vec{Y} + 1)^2$. This step yields a quadratic hyperplane. The corresponding separating hyperplane in the feature space includes features for all pairwise expression interactions $X_i X_j$, where $1 \leq i, j \leq 79$. Raising the kernel to higher powers yields polynomial separating surfaces of higher degrees in the input space.
         In general, the kernel of degree d is defined by $K(X,Y) = (\vec{X}.\vec{Y} + 1)^d$. In the feature space of this kernel any gene X features for all d-fold interactions between expression measurements are represented in terms of the form $X_{i_1}, X_{i_2}, \ldots, X_{i_d}$, where $1 \leq X_i, X_j \leq 79$.
      iii. Use the Gaussian kernel of the form $K(X,Y) = \exp(-\|\vec{X} - \vec{Y}\|^2/2\alpha^2)$, where α is the width of the Gaussian.
         As the data have been preprocessed such that the vectors follow the Euclidean distance, the value of α is set to be equal to the median of the Euclidean distances from each positive example to the nearest negative example.

The objective of this study was to test the ability of different kernels to distinguish between genes that belonged to two independent classes. Furthermore, the authors wanted to test the ability of the kernel functions to overfit the data being analyzed. They were successful in demonstrating that kernels of higher order were more successful in differentiating genes that belonged to different classes.

8.3.5.2 Remote Protein Homology Detection

We describe the work performed by Liao and Noble (2003). This work focuses on the use and modification of SVMs for the detection of remote homology in protein sequences.

8.3.5.2.1 Raw Data

The data used in this study consist of protein domains that belong to the different superfamilies defined by the Structural Classification of Proteins (SCOP) version 1.53 (Murzin et al. 1995). Sequences were selected using the Astral database (Brenner et al., 2000) by filtering similar sequences based on a threshold E value. In this example, the threshold was set at 10^{-25}, and the resulting 4,352 distinct protein sequences with known family and superfamilies were considered for analysis. For each family, the protein domains with the family were considered positive test samples, and the protein domains outside the family but within the superfamily were considered positive training samples. This designation resulted in sequences that were categorized into 54 families containing at least 10 samples (positive test) and 5 superfamily members outside of the family (positive train). Negative samples are taken from outside of the positive sequences' fold and are randomly split into training and testing sets in the same ratio as the positive samples.

With the exponentially growing number of protein sequences, finding protein sequence similarity is a major challenge of computational biology. It is a constant endeavor among bioinformatics researchers and practitioners to develop algorithms that can effectively detect the remotest sequence similarity between sequences of known families (classes) of proteins to recently generated sequences. Evolution controls the relation between families of proteins that is dictated by sequence, structure, and function. It is hypothesized that proteins that belong to a common family of proteins share a certain degree of similarity among each other. Traditional sequence similarity algorithms, such as the Smith-Waterman dynamic programming algorithm, basic local alignment search tool (BLAST), and FASTA, have been consistently performed and are used as benchmark techniques in the field of sequence similarity. However, they fail when the degree of similarity between proteins is less than 30%.

Data mining has played a vital role in this process, and several algorithms have been implemented. The SVM-Fisher (Jaakkola et al. 1999) and SVM-pairwise algorithms have been successfully employed. Both algorithms have been successful in integrating traditional sequence similarity techniques with supervised learning. The SVM-Fisher algorithm was successful in integrating the hidden Markov model (HMM) sequence profiling technique with the SVM algorithm. In this section, we describe how a pairwise sequence similarity technique can be integrated with the SVM using the SVM-pairwise algorithm (Liao and Noble 2003).

8.3.5.2.2 SVM-Pairwise Implementation

The SVM-pairwise implementation consists of two steps: (1) feature extraction, i.e., converting a protein sequence into fixed-length feature vectors, and (2) training an SVM using a train set that consists of protein sequences in the vectorized form.

Feature extraction: Converting a give protein sequence into fixed-length feature vectors is a feature extraction phase of the KDD process. Here we create a feature vector (for each protein sequence) using a list of pairwise sequence similarity scores, computed with respect to all of the sequences in the train set. Since the number of sequences in the train set is finite, the length of the feature vector for each sequence is fixed.

The pairwise sequence similarity score used for feature extraction has the following advantages: (a) The pairwise score representation is simpler. (b) It does not require multiple alignment of the train set. It allows for detection of motif or domain-sized similarities. (c) Pairwise score representation makes room for negative training samples, thereby allowing the SVM to leverage from the diversity of the training samples.

The vectorization step of SVM-pairwise uses the Smith-Waterman algorithm. The feature vector corresponding to a protein X is $Fx = fx_1, fx_2, \ldots, fx_n$, where n is the number of proteins in the train set fx_i and is the E value of the Smith-Waterman score between sequence X and the ith train set sequence. Default parameters, a gap opening penalty, and extension penalties of 11 and 1 are used, along with the BLOSUM 62 matrix.

Training of SVM: From the vectorized proteins this is performed to determine a similarity score between pairs of input vectors. At the heart of the SVM is a kernel that acts as a similarity score between pairs of input vectors. The base SVM kernel is normalized such that each vector has a length 1 in the feature space.

$$K(X,Y) = \frac{X.Y}{\sqrt{(X.X)(Y.Y)}}. \tag{8.12}$$

This kernel $K(X,Y)$ is then transformed into a radial basis kernel $\hat{K}(X,Y)$, as follows:

$$\hat{K}(X,Y) = e^{-\frac{K(X,X) - 2K(X,Y) + K(Y,Y)}{2\sigma^2}} + 1 \tag{8.13}$$

where the width σ is the median Euclidean distance (in feature space) from any positive training example to the nearest negative example. The constant 1 is added to the kernel to draw the data away from the origin. This translation is necessary because the SVM optimization algorithm we employ requires that the separating hyperplane pass through the origin.

An asymmetric soft margin is implemented by adding a value $0.02 \times \rho$, where ρ is the fraction of train set sequences that have the same label as the current sequence to the diagonal of the kernel matrix. The output of the SVM is a discriminant score that is used to rank the members of the test set. The same SVM parameters are used for the SVM-Fisher and SVM-pairwise tests. It was observed that the SVM-pairwise test performed well under these conditions.

8.4 Bayesian Approaches

The many forms of Bayesian approaches are derived from the Bayes' theorem. This section provides an overview of Bayesian approaches, which are discussed along with their applications in bioinformatics (Kelemen et al. 2003; Wilkinson 2007). In this chapter, we investigate two approaches: the naïve Bayes algorithm and the Bayesian network algorithm.

8.4.1 Bayes' Theorem

If we attempt to determine the probability density model $P(C|X)$, then we can determine the probability that a sample X, described by a set of features $\{x_1, x_2,\ldots, x_n\}$, belongs to the class C. In this example, let sample X be the evidence, and let X belonging to class C be the underlying hypothesis H. This example reduces the problem to the determination of the posterior probability of a hypothesis H provided that evidence X is true. The determination of the posterior probability is best defined using the Bayes' theorem. In its simplest form, the Bayes' theorem establishes this posterior probability using the following relation:

$$P(H|X) = \frac{P(H)P(X|H)}{P(H)}. \tag{8.14}$$

Thus, to determine the posterior probability the Bayes' theorem entails the computation of two prior probabilities, $P(X)$ and $P(H)$ and the posterior probability $P(X|H)$. To determine these probabilities, let us consider a train set T that has a set of m samples described by the same set of n features used to describe the sample X. $P(X)$ is the probability of the event occurring in train set T, and similarly $P(H)$ refers to the probability that the hypothesis H holds in train set T. The posterior probability $P(X|H)$ refers to the probability that the event X is conditioned on H. This posterior probability indicates that event X occurs if hypothesis H is true.

The Bayes' theorem is employed in classification in the forms of the naïve Bayes classifier and the Bayesian network classifier, which have been named under the category of Bayesian approaches. Extensions to these algorithms are prominently used in all fields of bioinformatics.

8.4.2 Naïve Bayes Classification

When the Bayes' theorem is extended into a classification algorithm, it becomes the naïve Bayes classifier. For example, if we retain the annotations used in the previous section, then train set T consists of m, n-dimensional vectors representing m samples. Let these samples belong to a fixed set of l classes $C = \{C_1, C_2, \ldots, C_l\}$.

Based on the computation of the posterior probability using the Bayes' theorem, the naïve Bayes classifier computes the posterior probability for given evidence. In this example, the test sample that does not have a class label for each hypothesis but refers to the different classes in the set C.

The objective of the naïve Bayes classifier is to assign a class label to the test sample. This label should exhibit the highest posterior probability. In other words, the evidence (test sample X) is assigned to class C_i if the following holds true for all values of j:

$$P(C_i|X) > P(C_j|X) \quad \text{for } 1 \leq j \leq l, \quad j \neq i. \tag{8.15}$$

This condition is called the maximum posteriori hypothesis and in general is represented as follows:

$$class(X) = argmax_i(P(C_i|X)) = argmax_i\left(\frac{P(C_i)P(X|C_i)}{P(X)}\right). \tag{8.16}$$

The implementation of the Bayes' theorem in the naïve Bayes classifier requires the following modifications: handling of prior probability and handling of posterior probability. These modifications are carried out to facilitate reduction in computational cost for determining the posterior probability $P(C_i|X)$.

8.4.2.1 Handling of Prior Probabilities

Since the prior probability $P(X)$ is marginal it can be deducted from the posterior probability $P(C_i|X)$ in computation; i.e., $P(X)$ is constant across all classes. This modification therefore reduces the Bayes' theorem to the form

$$P(C_i|X) \propto P(C_i)P(X|C_i). \tag{8.17}$$

The prior probability $P(C_i)$ is simple to compute. It is estimated as the ratio of the number of samples that belong to class C_i to the total number of samples in the train set T, i.e., $P(C_i) = |C_{i,T}|/|T|$. In situations where all classes in T have equal numbers of samples, the $P(C_i)$ can be treated as a constant. The Bayes' theorem is therefore reduced to the form

$$P(C_i|X) \propto P(X|C_i). \tag{8.18}$$

Thus, maximizing the posterior probability $P(C_i|X)$ relies heavily on the maximization of $P(X|C_i)$.

8.4.2.2 Handling of Posterior Probability

Since most of the datasets in data mining are high-dimensional and since the computational complexity of estimating the posterior probability is $P(X|C_i)$, the naïve Bayes algorithm assumes that all features of the evidence and train set T are independent of each other. This assumption, also known as the class conditional independence criterion (Keller et al. 2000), drastically reduces the computational complexity of the naïve Bayes algorithm by taking the product of probabilities of its attributes for a given class in the train set T.

$$P(X|C_i) = \prod_{k=1}^{n} P(x_k|C_i)$$
$$= P(x_1|C_i) \times P(x_2|C_i) \times \cdots \times P(x_k|C_i) \tag{8.19}$$

8.4.3 Bayesian Networks

Bayesian networks are useful for describing processes composed of locally interacting components; the value of each component depends on the values of a relatively small number of components. In addition, statistical foundations for learning Bayesian networks from observations, and the computational algorithms to do so, have been successfully tested in many applications. Below, we concentrate on the contributions of Friedman et al. (2000), which examine the dependence and conditional independence in data.

8.4.3.1 Methodology

Bayesian networks, though defined by probabilities and conditional independence statements, can derive connections using the direct causal influence of variables. The concept of Bayesian networks is best illustrated as follows. Let $P(X,Y)$ be a joint distribution over two variables X and Y. Further, let variables X and Y be independent if and only if $P(X,Y) = P(X)P(Y)$ for all values of X and Y, i.e., $P(X|Y) = P(X)$; otherwise, let the variables be considered dependent. If X and Y are dependent, then learning the value Y gives us information about X. Note that the correlation between variables implies dependence. However, dependent variables might be uncorrelated. For example, assume gene X is a transcriptional factor of gene Y. In such a case, we expect their levels of expression to be dependent. For example, when the expression level of X increases, we should see a similar increase in the expression level of Y. However, if gene X inhibits gene Y, then we see the reverse; when the expression level of X increases, the expression level of Y decreases.

These dependencies can be captured using graphs, with each gene represented as a node and the relation between nodes being represented using a directed edge.

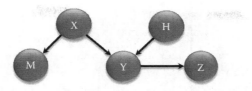

Figure 8.6 The representation of the parent relation between genes using graphs.

The direction of the edge between nodes represents the dependence between nodes. Since gene Y is dependent on gene X, we represent dependence with an edge that is directed from X to Y (X is the parent of Y). See Figure 8.6 for representation.

We now consider a slightly more complex scenario involving a system of three genes: X, Y, and Z. In this system, let us consider that gene X transcribes gene Y, and gene Y in turn transcribes gene Z. In such a situation, the expression levels of pairs of genes are dependent. However, gene X and gene Z do not share a direct relation, as they share only a common factor, gene Y. If gene Y is removed, then gene X and gene Z are independent of each other. In such a situation, gene Y is considered to be the mediator between gene X and gene Z and is represented as follows:

$$P(X|Y,Z) = P(X|Y), \qquad (8.20)$$

and we emphasize that genes X and Z are conditionally independent, given Y. This relation of conditional independence is represented as $I(X;Z|Y)$. The conditional independence between genes has no representation in the graph (see Figure 8.6).

In a more complex scenario of interaction between genes, let us assume a random gene M is regulated by gene X. As described above, the genes related to gene X are genes M and Y, whereas genes M and Y are conditionally independent of each other. This conditional independence is represented using the relation $I(M;Y|X)$, and gene X dictates the dependence between genes M and Y. We formalize the relation between genes M, Y, and X, as gene X is the common cause of genes Y and M. If gene X was nonexistent or not measured, then there would be dependence between genes M and Y, and in such an instance, we would refer to gene X as a hidden common cause (Friedman et al. 2000).

If, in another example, gene H transcribes gene Y, then gene Y is regulated by two genes X and H. We refer to genes X and H as the parent genes of gene Y, i.e., $pa(Y)$. Modeling the influence of two parent genes on a gene leads to an important parameter of Bayesian network models, where each node or variable is described as a conditional probabilistic function of its parents. Through this conditional probability function of Y, we specify the probability of gene Y to have the expression value y given the values of its parents $pa(Y)$ as $P(y|pa(Y))$.

8.4.3.2 Capturing Data Distributions Using Bayesian Networks

The next challenge in Bayesian network modeling is capturing the data distribution using a Bayesian network. To obtain such representations, we formalize the definition of Bayesian networks as follows.

Let us use a variable $X = \{X_1, \ldots X_n\}$, such that X_i represents a random variable whose value is x_i of finite domain. Similarly, we represent other variables Y and Z as vectors of random variables.

A Bayesian network is a representation of a joint probability distribution (JPD). This representation consists of two components. The first component, G, is a directed acyclic graph that has vertices that correspond to the random variables $X_1, \ldots X_n$. The second component describes a conditional distribution for each variable, given its parents in G. Together these two components specify a unique distribution on $X_1, \ldots X_n$.

Therefore, the graph G encodes the Markov assumption where each variable X_i is independent of its nondescendants given its parents in G. We formalize this Markov assumption as

$$\forall i, I(X_i; NonDescendants(X_i)|Pa(X_i)), \tag{8.21}$$

where $Pa(X_i)$ is the set of parents of X_i in G, and $NonDescendants(X_i)$ are the nondescendants of X_i in G.

By applying the chain rule of probabilities and properties of conditional independencies, any joint distribution that satisfies the above constraint can be decomposed in the product form

$$P(X_1, \ldots X_n) = \prod_{i=1}^{n} P(X_i | Pa(X_i)). \tag{8.22}$$

To specify a joint distribution, we also need to specify the conditional probabilities that appear in the product form. This component of the network describes distributions $P(x_i | pa(X_i))$ for each possible value x_i of X_i and of $pa(X_i)$. In the case of finite valued variables, we represent these conditional distributions as tables. Generally, Bayesian networks are flexible and can accommodate many forms of conditional distribution, including various continuous models. Given a Bayesian network, we might want to answer many types of questions that involve joint probability (for example, what is the probability of $X = x$ given the observation of some of the other variables?) or independencies in the domain (for example, are X and Y independent once we observe Z?).

8.4.3.3 Equivalence Classes of Bayesian Networks

A Bayesian network structure G implies a set of independence assumptions in addition to the independence statement. Let $Ind(G)$ be the set of independence statements (of the form Z is independent of Y given Z) that hold in all distributions satisfying these Markov assumptions. These can be derived as consequences, as shown in Equation 8.21.

More than one graph can imply exactly the same set of independencies. For example, consider graphs over two variables X and Y. The graphs $X \rightarrow Y$ and $X \leftarrow Y$ both imply the same set of independencies (i.e., $Ind(G) = \varnothing$). We say that two graphs G and G' are equivalent if $Ind(G) = Ind(G')$. This notation is crucial, since when we examine observations from a distribution, we often cannot distinguish between equivalent graphs.

8.4.3.4 Learning Bayesian Networks

With the modeling of data using Bayesian networks, the next challenge is learning from the modeled data. In simplistic terms, learning is achieved by identifying an optimal network that represents the complexities between variables, i.e., how they relate to each other in the training data. Moreover, it is important to quantify which is a challenge. Several methods have been proposed, of which the statistical scoring means determining the best network topology that captures inherent relationships between variables. These scoring functions have been motivated to select the optimal network based on the score obtained.

8.4.3.5 Bayesian Scoring Metric

Before we look into the Bayesian scoring metrics, we will formulate the definition of a Bayesian network as a graph to simplify the definition of the metrics.

Building on the concepts of a directed acyclic graph (DAG) and the definition of the joint probability distribution (JPD), $P(X_1,\ldots X_n)$, in the previous sections, for a set of variables $X_1,\ldots X_n$, a Bayesian network decomposes the JPD as

$$P(X_1,\ldots X_n) = P(X_{i1})P(X_{i2}|X_{i1})\ldots P(X_{in}|X_{i1},X_{i2},\ldots X_{in-1})$$

$$= P(X_{i1})P(X_{i2}|\pi_{i1})\ldots P(X_{in}|\pi_{in}) \qquad (8.23)$$

where $(i_1, i_2,\ldots, i_n)$ is a permutation of the variables index $(1,2,\ldots,n)$ and π_{ik} denotes the parent set of the variable x_k. It should be noted that there is no perfect representation of a Bayesian network, and the JPD can be represented in different forms depending on the order of each node.

8.4.3.5.1 Node Order and Acyclicity Constraint

Because the nodes' order can affect the JPD, it is important to define the order of nodes in the Bayesian network as it imposes parent-child relationships between nodes. If gene X precedes gene Y in the ordering, then gene Y cannot be a direct or indirect parent of gene X. This constraint of order ensures that the test of acyclic graphs need not be carried out whenever a new node or edge is added to the graph that violates the Bayesian network creation. Several methods are used to estimate the order of nodes in a Bayesian network. These methods include maximum a posteriori (MAP) and expectation maximization (EM) that can be used to estimate the conditional probabilities after the structure of a Bayesian network is determined.

8.4.3.5.2 Likelihood Equivalence

The likelihood equivalence assumption is as follows: if two structures are equivalent, their parameter joint probability density functions (PJPDFs) are identical, and thus coined the BDe score metric. The likelihood equivalence assumption measure implies that the Dirichlet distribution of the parameters and the resulting BDe score metric have a property of score equivalence, i.e., two equivalent structures have the same score. This score equivalence is advantageous in cases in which we do not want the data to distinguish the equivalent structures. However, it is disadvantageous in estimating the causal relationship between variables, as equivalent structures represent different causal relationships.

For example, if gene Y transcribes genes X and Z, then $X \leftarrow Y \rightarrow Z$. If we want to know the causal relationship, we require a scoring metric to differentiate between true structure and the equivalent $X \rightarrow Y \rightarrow Z$. On the other hand, if we simply want to learn a network to infer one gene Y given another gene X, or the probability $P(Y=k|X=j)$, either gene could fulfill this task. Thus, theoretically speaking, it is advantageous to have score equivalence. However, in network learning it is not clear that the score equivalence property is of any use, as when node order is specified there is only one resultant outcome and node order removes the need for score equivalence.

8.4.3.5.3 Score Metrics

There are various other score metrics in the literature and they have been described in brief as follows (Yang and Chang 2002):

Uniform prior score metric (UPSM): If the network parameters are assumed to have a uniform distribution (uniform priors), the score metric can be expressed as

$$P(B_s, D) = P(B_s) \prod_{i=1}^{n} \prod_{j=1}^{q_i} \frac{(r_i - 1)!}{(N_{ij} + r_i - 1)!} \prod_{k=1}^{r_i} (N_{ijk})! \qquad (8.24)$$

where N_{ijk} denotes the number of cases in the given database D in which the variable x_i took its kth value ($k=1,2,\ldots,r_i$), and its parent π_i was instantiated as its jth value ($j=1,2,\ldots,q_i$), and $N_{ij} = \sum_{k=1}^{r_i} N_{ijk}$.

Conditional uniform prior score metric (CUPSM): If the conditional uniform distribution is assumed, the score metric can be written as

$$P(B_s,D) = P(B_s)\prod_{i=1}^{n}\prod_{j=1}^{q_i}\prod_{k=1}^{r_i-1} B \times \left(N_{ijk}+1, 1+\sum_{m=0}^{r_i-k-1} N_{ij(r_i-m)}\right), \quad (8.25)$$

General Dirichlet prior score metric (DPSM): If the Dirichlet distribution is assumed, then the score metric can be written as

$$P(B_s,D) = P(B_s)\prod_{i=1}^{n}\prod_{j=1}^{q_i}\frac{\Gamma(N'_{ij})}{\Gamma(N'_{ij}+N_{ij})} \times \prod_{k=1}^{r_i-1}\frac{\Gamma(N'_{ijk}+N_{ijk})}{\Gamma(N'_{ijk})}, \quad (8.26)$$

where N'_{ijk} is the corresponding Dirichlet distribution orders for a set of parameters, which need to be assigned some values by users, and $N'_{ij} = \sum_{k=1}^{r_i} N'_{ijk}$. The uniform distribution can be considered a special case of $N'_{ijk}=1$.

BDe score metric (BDe): If the likelihood equivalence assumption is used instead of the Dirichlet distribution assumption, and the same formula as DPSM is derived. However, the user does not assign the orders arbitrarily. They are determined by the equivalent sample size N' and the assumed local joint probability. Specifically,

$$N'_{ijk} = N'\, p\left(x_i=k, \Pi_i = j \,|\, B_{SC}^h, \xi\right). \quad (8.27)$$

8.4.4 Application of Bayesian Classifiers in Bioinformatics

Using the BDe score metric of likelihood equivalence, various class models for the data, an example model M_i for class I, and a test sample vector $x = \{x_1, x_2, x_3, \ldots, x_n\}$ drawn from some probability distribution, one can classify x according to the model with maximum posterior probability (for a posterior probability), given the sample:

$$class(x) = argmax_i(\log p(M_i|x)), \quad (8.28)$$

where $p(M_i|x)$ is the Bayesian a posteriori probability that M_i is true given the test sample x. By the Bayes' theorem,

$$p(M_i|x)p(x) = p(x|M_i)p(M_i). \quad (8.29)$$

Assuming equal prior probabilities, $p(M_i)$, for each model, we obtain:

$$class(x) = argmax_i(\log p(x|M_i)) \tag{8.30}$$

i.e., the computed class of the sample is the model for which the sample has the greatest likelihood. Finally, the naïve Bayes method makes the additional assumption that, given the class model, values for each component of x are independent of one another, so that the above becomes

$$class(x) = argmax_i\left(\sum_g \log p(x_g|M_i)\right). \tag{8.31}$$

This assumption of class attribute independence greatly facilitates the computation of the likelihoods for the data given each model, since it is much easier to infer individual class attribute value probabilities from the training data than it is to infer joint class attribute value probabilities. This simplification has been used successfully in a number of domains, including some with known class attribute dependencies.

In the case of microarray data, we model each class as a set of Gaussian distributions, one for each gene computed from the training samples of that class:

$$M_i = \{M_i^1, M_i^2, \ldots, M_i^n\} \tag{8.32}$$

where M_i^g is the class I Gaussian distribution for gene g. The class of a test sample x is given by

$$class(x) = argmax_i\left(\sum_{gene\ g} \log p(x_g|M_i^g)\right) \tag{8.33}$$

which, when substituting M_i^g for a Gaussian distribution with sample mean μ_i^g and standard deviation σ_i^g, becomes

$$class(x) = argmax_i\left(\sum_{gene\ g}\left[-\log(\sigma_i^g) - 0.5\left((x_g - \mu_i^g)/\sigma_i^g\right)^2\right]\right). \tag{8.34}$$

Since $p(x_g|M_i^g)$ is proportional to $(1/\sigma_i^g)\exp\left(-0.5\left((x_g - \mu_i^g)/\sigma_i^g\right)^2\right)$, it can be interpreted as the probability that the gene g component of x is within some small nonzero interval centered at x_g. Furthermore, if one again assumes equal prior probabilities for all models, the relative log probabilities between any two models M_a and M_b with respect to x can be expressed simply as the difference between their log likelihoods:

$$logp(M_a|x) - logp(M_b|x) = logp(x|M_a) - logp(x|M_b) =$$
$$\sum_{gens\ g}\left[-\log(\sigma_a^g) - 0.5((x_g - \mu_a^g)/\sigma_a^g)^2 + \log(\sigma_b^g) - 0.5((x_g - \mu_b^g)/\sigma_a^g)^2\right] \quad (8.35)$$

Such a difference can be used as a confidence measure for choosing class a over class b.

8.4.4.1 Binary Classification

In this section, we cover the role of the NB classifier for the use of likelihood selection of genes. In binary classes, genes in the NB classifier each vote for the likelihood of alternative models, M_1^g and M_2^g, given the test sample vector component x_g. Intuitively, we want genes that can distinguish between samples of each class, finding M_1^g is more likely than M_2^g given a sample of class 1, and M_2^g is more likely than M_1^g given a sample of class 2. We define two relative log likelihood scores, $LIK_{1\to 2}$ and $LIK_{2\to 1}$, for gene g:

$LIK_{1\to 2} = logp(M_1^g|X_1) - logp(M_2^g|X_1)$ where X_1 are training samples of class 1, and $LIK_{2\to 1} = logp(M_2^g|X_2) - logp(M_1^g|X_2)$ where X_2 are training samples of class 2.

The ideal gene for the NB classifier should have both LIK scores much greater than zero, indicating that the gene, on average, votes for class 1 on training samples of class 1, and for class 2 on training samples of class 2. If a test sample is selected from the same probability distribution as the training data, then one can expect this gene to vote for class 1 for test samples of class 1, and for class 2 for test samples of class 2. The greater the values of the LIK scores above zero, the greater the contribution one expects the gene to make toward the correct classification of a test sample.

It is difficult to find genes for which both LIK scores are far greater than zero. Instead, one can select two sets of genes, $GENES_{1\to 2}$ and $GENES_{2\to 1}$, each maximizing one of the two LIK scores, while merely requiring the other to be greater than zero:

$$GENES_{2\to 1}:\ LIK_{1\to 2} > 0\ and\ LIK_{2\to 1} \gg 0. \quad (8.36)$$

Genes in each set are ranked according to their values of the LIK score maximized by that set. Combining the $n/2$ top-ranking genes from each set then produces an NB classifier with n genes.

8.4.4.2 Multiclass Classification

This method for using *LIK* scores to select genes for a naïve Bayes classifier extends beyond the case of two classes. In cases where the number of classes is c, we define $c(c - 1)$ different *LIK* scores:

$LIK_{j \to k} = logp\left(M_j^g | X_j\right) - logp\left(M_k^g | X_j\right)$ where X_j are training samples of class j and $1 \leq j, k \leq c, j \neq k$.

Similarly, we select $c(c - 1)$ distinct sets of genes, each maximizing one *LIK* score, while merely requiring all others to be greater than zero:

$$GENES_{j \to k}: \quad LIK_{j \to k} \gg 0$$
$$LIK_{j' \to k'} > 0 \quad j' \neq k', 1 \leq j', k' \leq c. \tag{8.37}$$

Genes in each $GENES_{j \to k}$ set should therefore distinguish test samples of class j with better accuracy than the alternative model M_k^g.

When equal numbers of genes from all $c(c-1)$ $GENES_{j \to k}$ sets are combined, the resulting NB classifier should again have the desired properties. Consider a test sample x of class j. Genes in the $(c - 1)$ different $GENES_{j \to k}$ sets, $1 \leq j', k' \leq c$, $j' \neq k'$, $j' \neq j$, will on average make a contribution to the log likelihood term of M_j^g at least as large as that of terms of the alternatives. As a result, the summed log likelihood term of M_j^g will on average be larger than that of all other models, so $argmax_i(\log p(x|M_i)) = j$ and the classifier votes for class j.

8.4.4.3 Computational Challenges for Gene Expression Analysis

Based on the above description of Bayesian networks, one can treat each gene in a microarray as a variable. In addition, other attributes that affect the system can be modeled as additional random variables. These attributes include temporal indicators, experimental conditions, and background variables such as exogenous cellular conditions. By using learning based on a Bayesian network based on statistical dependencies, one can answer a wide range of queries, such as whether there is dependence between expression levels of a gene and the experimental conditions under study. However, these inferences are connected to the statistical constraints and interpretation of results obtained. Moreover, the modeling of a complex system of genes entails a degree of algorithmic and processing complexity.

Most difficulties in this modeling process revolve around the curse of dimensionality, which exists due to the thousands of genes and few samples for analysis. On the positive side, it is believed that only a handful of genes affect the transcription of a gene. This sparcity of genes aids Bayesian network performance, as Bayesian networks perform best on these types of data. The implementation of Bayesian networks for gene sets was described by Bauer et al. (2010).

8.5 Decision Trees

In this section of the chapter we provide a brief description of the different supervised learning strategies inspired by decision trees. We start a discussion with the C4.5 algorithm, a natural extension of its predecessor, the ID3 algorithm, that can be used to construct a univariate decision tree. In the simplest terms, decision tree model generation can be viewed as a recursive splitting of the train set. Therefore, the train set in its entirety is found at the root of the tree. This train set is split into smaller chunks of data based on the values that each attribute possesses in the train set. This recursive data splitting is performed until the leaves of the tree result in individual records or a group of records that have the same phenotype. The decision tree algorithm for model generation has two major components, attributes selection and termination criteria.

The following are the guidelines for model generation using any form of decision tree:

1. The leaf of a tree could be a single sample or a group of samples that has a common phenotype.
2. Estimate the potential information content of each feature or attribute.
3. Based on a selection criterion find the best attribute to branch on.

Based on the above guidelines, the biggest challenge in constructing the tree model is estimating the potential information content of each feature or attribute that describes a sample (Figure 8.7).

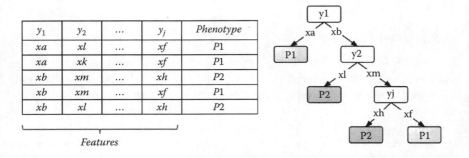

Figure 8.7 A schematic representation of the construction of a decision tree, where each node of the tree represents a feature and the values of the features determine the link between the nodes. The leaves of the decision tree are the associated phenotypes.

The C4.5 algorithm to construct the tree model uses the measure of entropy to measure the disorder of data using the following relation:

$$Entropy(\Upsilon) = -\sum_{i=1}^{m} \frac{|y_i|}{|\Upsilon|} \log\left(\frac{|y_i|}{|\Upsilon|}\right) \quad (8.38)$$

where Υ represents the train set and $Entropy(\Upsilon)$ represents the information content of the train set, iterating over all possible phenotypes that belong to Υ, and y_j represents a subset of samples that belong to a specific phenotype.

In order to estimate the conditional entropy of Υ for a given attribute j, we use the following relation:

$$Entropy(j|\Upsilon) = \sum_{j=1}^{n} \frac{|y_j|}{|\Upsilon|} \log \frac{|y_j|}{|\Upsilon|}, \quad (8.39)$$

The conditional entropy for a given attribute j allows us to estimate its relevance in the train set, where represents the partition of the train set Υ, where each partition is determined based on the n possible values possessed by attribute or feature j. We define the entropy gain of attribute j relative to the entropy possessed by the entire train set as follows:

$$Gain(\Upsilon, j) = Entropy(\Upsilon) - Entropy(j|\Upsilon). \quad (8.40)$$

The aim of using this definition is to maximize the gain, dividing by overall entropy due to split argument $\bar{y}$ by value j.

8.5.1 Tree Pruning

The problems associated with decision tree models stem from two issues: (1) The class that has the most number of samples (majority phenotype/class) would result in rules that overpower rules generated from minority phenotypes/classes. (2) It is difficult to determine a test set that could traverse all the nodes in a tree. This makes it difficult to actually determine an ideal test set.

However, apart from these two issues it should be noted that decision tree models are sensitive to outliers that are present in the train set. It is therefore common to see decision tree models that overfit the train set. Tree pruning is hence an important step that helps remove rules that are influenced by these outliers. From a bioinformatics perspective, it is importation to subject any decision tree model to pruning, as bioinformatics data are prone to noise and outliers.

The objective of tree pruning is twofold: (1) to reduce classification errors, caused by outlier instances embedded in the train set, and (2) make the decision tree more generalized to avoid overfitting (Esposito et al. 1997).

Considering the need for decision tree pruning there are several pruning approaches in the literature; for instance, some approaches proceed from the root of the decision tree and proceed down toward the leaves while examining the branches to prune. These are referred to as the top-down approaches. On the contrary, there exist approaches that traverse the decision tree in the opposite direction, known as the bottom-up approaches. Other approaches to tree pruning use the train set to evaluate the accuracy of a pruned decision tree, while others use an addition dataset called the pruning set to establish the performance of the pruned decision tree.

With the gamut of pruning approaches, they are commonly categorized into prepruning and postpruning. In the prepruning approach, the pruning step is integrated into the model building step. As part of the prepruning step, the data splitting that occurs at every node of the decision tree is terminated abruptly based on a predetermined threshold of the attribute evaluated. This abrupt termination of data splitting ensures that further splitting of data at a node does not take place at the next iteration. Here the node is treated as a leaf and assigned a phenotype label of the majority phenotype/class.

Unlike prepruning, there are several approaches to postpruning, such as reduced error pruning (REP), pessimistic error pruning (PEP), minimum error pruning (MEP), cost-complexity pruning (CCP), critical value pruning (CVP), and error-based pruning (EBP) (Esposito et al. 1997). For the purpose of brevity, we discuss the simplest form of postpruning, reduced error pruning (REP).

In the REP, postpruning is carried out using an independent pruning set. The iterative pruning process starts with the completed decision tree (T_{all}). For each node i of T_{all} the postpruning approach compares the number of classification errors made on the pruning set when the subtree T_i is kept with the number of classification errors made when i is turned into a leaf and associated with the best class. Sometimes, the simplified tree has a better performance than the original decision tree T_{all}. In such cases, T_i is pruned from T_{all}. This pruning operation is repeated on the simplified tree until further pruning increases the misclassification rate.

8.6 Ensemble Approaches

In this section we describe prominently used ensemble learning approaches. Ensemble learning is an effective technique that has increasingly been adopted to combine multiple learning approaches to improve overall classification accuracy. High dimension and relatively small number of samples typically characterizes biological data—frequently characterized as a small sample size problem. Moreover, these samples are typically plagued by noise and missing values. These ensemble techniques alleviate the small sample size problem by averaging classification results over multiple classifiers. It is believed that this philosophy of averaging the performance of multiple classifiers reduces the potential for overfitting the final classification results. Furthermore, through the use of an ensemble of classifiers, the train set may be used in a more efficient way, which is critical to many biological applications with small sample size (Yang et al. 2010; Webb and Zheng 2004). Thus, an ensemble of classifiers is designed to boost classification accuracy and enhance generalization. By the term *boosting* we refer to enhancing the classifier performance specifically in scenarios of high-dimensional data where the number of samples m are far lower than the number of features n ($m \ll n$) in the train set. The term *generalization* refers to the ability of the ensemble classifiers to classify samples of unknown classes after the training is performed. Both boosting of accuracy and generalization of classification are closely tied to the bias and variance of the ensemble of classifiers. It is shown that the ensemble of classifiers can control the variance and bias using boosting, bagging, and averaging strategies. However, the time and space complexity of these techniques are believed to be high. Thus, ensemble classifiers are applied in scenarios where accuracy is important.

Several ensemble approaches are prevalent in bioinformatics. In this section, we introduce three such techniques and describe their workings. These techniques include bagging, boosting, and random forests ensemble methods. But before we describe the characteristic differences between each of these techniques, we will illustrate the workings of an ensemble classifier. Throughout this chapter thus far, we focus on learning techniques. The significance of these techniques lies in choosing a single hypothesis from a set of hypotheses that best discriminates between samples of the training data. Typically, we envision a scenario in which the train set is free of noise and missing values. We believe that a resultant hypothesis generated from such a train set best discriminates between the classes of the train set and refer to it as the best hypothesis (h_{best}). For a visual explanation, see Figure 8.8(a).

In small sample size problem scenarios, determining the best hypothesis (h_{best}) is a challenge considering the fact that there could be several optimal hypotheses, and choosing the best hypothesis that covers all of the several hypotheses presents a challenge of its own. Figure 8.8(b) best describes this scenario. In this case, a traditional learning approach would choose a single hypothesis that would not generalize well considering the disparities in the hypothesis space.

The philosophy for using an ensemble of classifiers is the intelligent manipulation of the train set to obtain different hypothesis spaces with different classifiers, i.e., ($H_1, H_2, H_3 \ldots H_L$), where L is the number of classifiers. By manipulating the train set, we can effectively narrow down a consensus of hypotheses space H_o, represented by the overlap of the hypotheses spaces (Figure 8.8(b)).

Theoretically, this H_o is obtained by combining the classification rules of multiple classifiers using an integration method that takes advantage of the overlapped region. The best classification rule is obtained by approximating multiple rules.

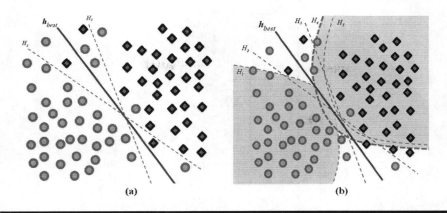

Figure 8.8 A schematic illustration of hypothesis space partitioning with the ensemble of classifiers as proposed by Yang et al. (2010).

This approximation yields to classifiers that are more accurate and provides effective generalization. As previously mentioned, there are several ensemble techniques that are prevalent in bioinformatics. The following section provides an overview of three such prominent techniques.

8.6.1 Bagging

Bagging, also known as bootstrap aggregation, is one of the first and simplest forms of ensemble-based techniques. This method was proposed by Breiman (1996). The working principle of the bagging technique is analogous to the following. Let us consider a panel of evaluators who have been chosen to help come up with the best possible decision given a compiling set of evidence. In the bagging technique, each evaluator in the panel is given equal importance, by dividing all the evidence into equal subsets of evidence across all the evaluators. It should be noted that the subset of evidence given to each evaluator is chosen at random to avoid biases. The decisions (votes) made by each of these evaluators are then tabulated and subjected to a voting scheme where decisions that are consistent across all evaluators are chosen as the best.

In a classification scenario using bagging, the train data (D) are first subjected to a bootstrap sampling strategy, where subsets of samples are chosen at random from D. Note that bootstrap sampling employs sampling with replacement, resulting in unbiased subsets that are subjected to independent classifiers (evaluators). Figure 8.9 provides an illustration of the bagging strategy in creation of an ensemble of classifiers. Each independent classifier that is part of the ensemble (referred to as a weak learner) generates rules from the independent subset of training data allocated to it through bootstrap sampling. These rules are subject to various voting

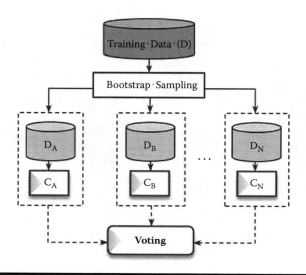

Figure 8.9 The bagging strategy in an ensemble of classifiers.

strategies to choose the most consistent rules that would ultimately be used for classification by the ensemble.

Various voting strategies are applied in bagging (Erp et al. 2002) and have been effectively classified into three categories, each derived from human voting strategies: unweighted voting methods, confidence voting methods, and ranked voting methods. Though these methods may seem complex, the voting strategies are simple to implement, as they are independent of the classifiers in the ensemble (Ho et al. 1994).

8.6.1.1 Unweighed Voting Methods

The unweighted voting methods consist of methods in which each vote carries equal weight. The only differentiation between the classes is the number of votes they have received. As a consequence, classifiers cannot express the degree of preference of one class over the other. Although this method removes relevant information, it also results in less complex methods to implement. Moreover, these methods do not perform well in the cases of ties.

> **Plurality:** The benefit of this voting strategy lies in the simplicity and ease of use. In this method, every classifier votes one class label for a given sample. Ultimately, the sample is assigned to that class that receives the highest number of votes. However, plurality voting may assign a sample to a wrong class due to erroneous assignments by the classifiers in the ensemble. There is a real possibility of the sample being assigned to a wrong class by a small number of wrong votes.

Majority voting: Majority voting builds on the problems of plurality voting and at times is confused with plurality voting. As in plurality voting, majority voting allows each classifier to vote one class label for a sample. The sample is ultimately assigned the class label that receives the highest vote with the constraint that receives a majority of more than half of the number of classifiers in the ensemble that have the same vote. Majority voting assigns a class label to a sample if and only if the majority constraint is satisfied. This majority constraint makes the voting strategy less error-prone if the ensemble has a large number of classifiers and is therefore a widely accepted technique. However, when a sample fails to satisfy the same majority constraint, the sample is rejected by the voting strategy and no class label is assigned.

Multiclass scenarios and variations in voting schemes require a hierarchy of steps; thus, the following methods are commonly referred to as multistep methods. These multistep methods of unweighted voting are difficult to implement as they rely on the preference of classifiers by taking pairs of classes into consideration.

8.6.1.2 Confidence Voting Methods

Unlike the unweighted voting methods, confidence voting methods rely on the classifiers in the ensemble to express their preference toward a class. The preference is therefore a scalar value called the confidence score of a classifier for a class. The higher the confidence score, the more the class is preferred by the classifier. The confidence scores of each classifier toward the classes are generated prior to the actual classification process.

Pandemonium: Every classifier is given one vote, which it can cast for any class. The classifier casts the vote by stating its confidence in the class. The class that receives the vote with the highest confidence of all votes wins. This method, known as Selfridge's pandemonium (Selfridge 1958), is one of the first examples of using separate experts/agents in computer science. It is very simple, but misses the possibility for a classifier to express differences of preference between classes. Only the classifier's top choice and its confidence are known. Furthermore, there is no limit to the amount of confidence that classifiers may adhere to. While limits are easily added to the method, a correct scale is difficult to implement. However, with well-scaled classifiers, this method could be sufficient.

Sum rule: When the sum rule is used each classifier has to give a confidence value for each class. Next all confidence values are added for each class, and the class with the highest sum wins the election.

Product rule: As with the sum rule, each classifier gives a confidence value for each class. Then, all confidence values are multiplied per class. The class with the highest confidence product wins. The product rule is highly subjective to low confidence values. A very low value can ruin a class's chance of winning the election no matter what its other confidence values are.

8.6.1.3 Ranked Voting Methods

In ranked voting methods the classifiers are asked for a prior preference ranking of the classes. In this way, more information on the classifier's preference is used than in the unweighted voting methods. However, unlike the confidence voting methods, the ranked voting methods reflect the degree of preference between two classes in the form of ranks. These ranks do not correspond to the confidence of the classifiers used in the ensemble classifier.

Borda count: This method only runs if a complete list of preference ranks is available from all classifiers over all classes. It then computes the mean rank of each class over all classifiers. The classes are reranked by their mean rank, and the top-ranked classes win the election. Note that the Borda count is the ranked variant of the sum rule.

Single transferable vote (STV): Also known as alternative voting (in case of one-winner solutions), each classifier gives a preference ranking of the classes. Incomplete ranks are possible, though such ranks may result in a classifier losing its vote. A majority vote is held based on the highest-ranked class of each classifier's ranking. If some class gains the majority, it wins the election. Otherwise, the class with the least number of votes in the majority voting is eliminated from further participation. This class is removed from all preference rankings. Now, the process repeats itself, starting with the majority vote, until one class gains the majority.

One low rank in an STV election has less effect on class selection than a low rank in the Borda count does. However, due to the elimination procedure, complex and illogical side effects may occur (for example, voting for a candidate may result in the candidate's loss of the election). Thus, in any ensemble based on bagging, there are three comprehensive components: the bootstrapping sampling, the classifiers, and the voting strategy. Bootstrap sampling divides the train set into unbiased subsets, which are provided to each classifier in the ensemble. The set of classifiers that composes the ensemble typically consists of a diverse set of classifiers. The most vital component of the bagging technique is the voting strategy used to combine the decisions derived from the classifiers. The variations of known bagging techniques are driven by the different voting strategies.

8.6.2 Boosting

As with bagging, the boosting technique is characterized by three components: sampling of the train set, a set of classifiers that form the ensemble, and a voting strategy. Boosting relies on strategic resampling that is geared toward providing the most informative training subset to each of the classifiers in the ensemble. The boosting strategy can be viewed as a cascade of classifiers—in which each classifier generates decisions based on a refined subset of the training subset as we iterate through the cascade. The refining of the training subset is the responsibility of each of the classifiers in the ensemble. A schematic representation of boosting is provided in Figure 8.10. It should be noted that the voting techniques used in bagging can be employed in boosting.

To formalize the procedure behind the boosting ensemble of classifiers strategy, let us consider an ensemble ξ that consists of a set of N classifiers, i.e., $\xi = \{C_1, C_2, \ldots, C_N\}$. Let us assume that each of the classifiers in ξ is binary, where C_i classifies a sample x_i to only two classes, i.e., $C_i(x_i) \in \{-1,1\}$. The final decision of ξ in classifying sample x_i is the weighted sum of the outputs of the classifiers in the ensemble represented by the following relation:

$$\xi(x_i) = \alpha_1 C_1(x_i) + \alpha_2 C_2(x_i) + \cdots + \alpha_N C_N(x_i). \tag{8.41}$$

In Equation 8.41, $\{\alpha_1, \alpha_2, \ldots, \alpha_N\}$ corresponds to the weight assigned to the decisions by each of the classifiers.

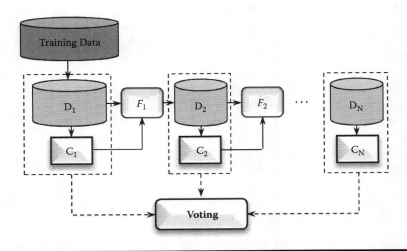

Figure 8.10 The schematic representation of the boosting strategy in an ensemble of classifiers.

The AdaBoost algorithm is targeted toward choosing the best set of classifiers from a pool of a diverse set of potential classifiers (Freund and Shapire 1995). The objective of the AdaBoost algorithm is to choose a set of classifiers that complement each other in an optimal manner. This objective is brought about by the following framework and reflected in Equation 8.41. It thus consists of (1) a set of diverse classifiers $\{C_1, C_2, \ldots, C_N\}$, (2) a weight ($\alpha_i$) associated with each classifier, which acts an indicator (or flag) of the classifier's ability to yield a decision given the train set, and (3) a function that sums the outputs of each of the classifiers to yield a common result.

AdaBoost follows an iterative framework. At each iteration, a classifier is chosen from a pool (or a pertinent set) of potential classifiers depending on their ability to classify samples that were previously misclassified by classifiers higher up the cascade. This process ensures that classifiers in the ensemble complement each other and at the same time boost the overall performance in the ensemble. The heart of the algorithm is realized through the choice of classifiers, and is composed of the following three steps: seeking prospective classifiers for the ensemble, choosing an optimal set of classifiers, and assigning weight to a chosen classifier.

8.6.2.1 Seeking Prospective Classifiers to Be Part of the Ensemble

The objective of this step is to select new classifiers from a pool of classifiers that can help with the classification of samples that are still misclassified by a classifier higher in the cascade. Let us assume that we start with our initial train set T, which consists of N-dimensional samples $x_i = \{x_i^1, x_i^2, \ldots, x_i^N\}$ of data having class labels $y_i \in \{-1, 1\}$. Let us further assume that we have a finite set of k classifiers to choose from and the AdaBoost algorithm is subject to M iterations. First, we set up an error criterion that is iteratively carried out. AdaBoost uses the exponential loss error criterion (Wyner 2002), where each classifier is assigned a cost $e^{-\beta}$ for every hit (correctly classified instance) and a weight e^{β} for a miss (misclassified instance). It should be noted that $\beta > 0$ such that misses are penalized more than hits.

The main idea in AdaBoost is to proceed systematically by extracting one classifier from the pool in each of the M iterations. The elements in the dataset are weighted according to their current relevance (or urgency) at each iteration. At the beginning, all elements are assigned the same weight. During each iteration, those samples that are misclassified are assigned higher weights. Thus, when a new classifier is selected, importance is given to the classifier that performs well with those samples that are weighed higher.

8.6.2.2 Choosing an Optimal Set of Classifiers

In each iteration of the AdaBoost algorithm the k classifiers in the pool are reranked, taking into consideration the new weights assigned to the samples of

T (as discussed above). In this step, we focus on determining the next C_m and its corresponding weight, α_m. Based on Equation 8.41, we obtain

$$\xi_{(m-1)}(x_i) = \alpha_1 C_1(x_i) + \alpha_2 C_2(x_i) + \cdots + \alpha_{m-1} C_{m-1}(x_i), \tag{8.42}$$

and we want to extend it to

$$\xi_m(x_i) = \xi_{(m-1)}(x_i) + \alpha_m C_m(x_i). \tag{8.43}$$

At the first iteration ($m = 1$), $\xi(m-1)$ is the zero function. We define the total cost, or total error, of the extended classifier as the exponential loss.

$$E = \sum_{i=1}^{N} e^{-y_i(\xi_{(m-1)}(x_i) + \alpha_m C_m(x_i))}, \tag{8.44}$$

where α_m and C_m are to be determined in an optimal manner. We rewrite the above Equation 8.44 as follows:

$$E = \sum_{i=1}^{N} w_i^{(m)} e^{-y_i \alpha_m C_m(x_i)} \tag{8.45}$$

where

$$w_i^{(m)} = e^{-y_i(\xi_{(m-1)}(x_i))}. \tag{8.46}$$

In the first iteration, we get $w_i^{(1)} = 1$, for $i = 1, \ldots, N$. During later iterations, the vector $w^{(m)}$ represents the weight assigned to each data point in the train set at iteration m. We then divide the sum into two numbers that reflect the weighted cost of all hits plus the weighted cost of all the misses.

$$E = \sum_{y_i = C_m(x_i)} w_i^{(m)} e^{-\alpha_m} + \sum_{y_i \neq C_m(x_i)} w_i^{(m)} e^{\alpha_m} \tag{8.47}$$

$$E = W_c e^{-\alpha_m} + W_e e^{\alpha_m}.$$

For selecting C_m the exact value of $\alpha_m > 0$ is irrelevant since minimizing E is equivalent to minimizing $e^{\alpha_m} E$ for a fixed α_m and $e^{\alpha_m} E = W_c + W_e e^{2\alpha_m}$. Further, since $e^{2\alpha_m} > 1$, we can rewrite the above expression as

$$e^{\alpha_m} E = (W_c + W_e) + W_e(e^{2\alpha_m} - 1). \tag{8.48}$$

$(W_c + W_e)$ is a constant; thus $e^{\alpha_m} E$ is minimized for the mth iteration if a classifier is picked that has the lowest weight W_e. Thus, the next choice of C_m should be the one with the lowest penalty given the current set of weights.

8.6.2.3 Assigning Weight to the Chosen Classifier

With the classifier chosen, C_m the immediate step is to determine its corresponding weight α_m.

Considering the error E represented as

$$E = W_c e^{-\alpha_m} + W_e e^{\alpha_m} \tag{8.49}$$

we differentiate both sides by the weight α_m:

$$\frac{\delta E}{\delta \alpha_m} = -W_c e^{-\alpha_m} + W_e e^{\alpha_m}. \tag{8.50}$$

On multiplying both sides by e^{α_m} and equating it to zero, we obtain

$$-W_c + W_e e^{2\alpha_m} = 0. \tag{8.51}$$

On simplification the optimal α_m is provided by the following relations:

$$\alpha_m = \frac{1}{2} \ln\left(\frac{W_c}{W_e}\right) \tag{8.52}$$

$$\alpha_m = \frac{1}{2} \ln\left(\frac{W - W_e}{W_e}\right) = \frac{1}{2} \ln\left(\frac{1 - e_m}{e_m}\right) \tag{8.53}$$

where $e_m = W_e / W$, the percentage rate of error given the weights of the data points.

The above steps are iteratively captured as follows:

For $m = 1$ to M,

1. Select and extract from the pool of classifiers the classifier C_m, which minimizes

$$W_e = \sum_{y_i \neq C_m(x_i)} w_i^{(m)},$$

2. Set the weight α_m of the classifier to

$$\alpha_m = \frac{1}{2} \ln\left(\frac{1-e_m}{e_m}\right)$$

where $e_m = \frac{W_e}{W}$, and
3. Update the weights of the data points for the next iteration. If $C_m(x_i)$ is a miss, set

$$w_i^{(m+1)} = w_i^{(m)} e^{\alpha_m} = w_i^{(m)} \sqrt{\frac{1-e_m}{e_m}}.$$

Otherwise, $w_i^{(m+1)} = w_i^{(m)} e^{-\alpha_m} = w_i^{(m)} \sqrt{\frac{e_m}{1-e_m}}$.

8.6.3 Random Forest

Random forest is an ensemble approach that is suited to handle high-dimensional data, as different models work on independent feature sets (subsets of the high-dimensional space). The results are assimilated to a single result. A random forest is a collection of individual decision tree classifiers, where each tree is a forest that has been trained using a bootstrap sample of instances from the data, and each split attribute in the tree is chosen from among a random subset of attributes. Classification of instances is based on aggregate voting over all trees in the forest.

Individual trees are constructed as follows from data having N samples and M explanatory attributes:

1. Choose a train set by selecting N samples, with replacement from the data.
2. At each noted in the tree, randomly select m attributes from the entire set of M attributes in the data (the magnitude of m is constant throughout the forest building).
3. Choose the best split at that node from among the m attributes.
4. Iterate the second and third steps until the tree is fully grown (no pruning).

During the first step of the process a subset of N samples is chosen from the initial train set using the bootstrap sampling with replacement, resulting in a split of the train set. The set of N samples is used for analysis and generation of a tree t. The remaining sets that are not part of set D_i are called out-of-bag sets. The samples of this set D_i are used for error prediction estimation.

The samples in set D_i are used to construct the tree without pruning. It should be noted that this is an iterative process, and each iteration results in the creation of a tree for analysis. Moreover, each tree is constructed differently as the number of randomly selected attributes would vary from each tree in the forest, as dictated in step 2.

To predict the class of an observation using a tree, the observation is assigned to a terminal node (i.e., a leaf) based on its predictor values. The class containing the majority of train set observations in the leaf is selected as the class prediction for the observation. With a forest of classification trees, each tree gets one vote for each out-of-bag observation, and for a given observation, the class receiving the most votes is the forest prediction. Again, ties are resolved by selecting the class with the lowest label. The probability of ties is very small if the number of trees is large. The random forest prediction for an observation is computed by averaging the tree predictions over trees for which the given observation is out of bag (Figure 8.11).

Repetition of these steps yields a forest of trees, each of which has been trained on bootstrap samples of instances. Thus, for a given tree, certain instances will have been left out during training. Prediction error is estimated from these out-of-bag instances. The out-of-bag instances are also used to estimate the importance of particular attributes via permutation testing. If randomly permuting values of a particular attribute do not affect the predictive ability of trees on out-of-bag samples, that attribute is assigned a low importance score.

8.6.4 Application of Ensemble Approaches in Bioinformatics

Association studies have become an integral part of bioinformatics over the past decade. Association studies can help determine individual susceptibility to various diseases as well as their responses to drugs based on their genetic variations. A widely used design for association study is to screen common single nucleotide polymorphisms (SNPs) and compare their variation between case and control samples for disease-associated gene identification at the genome-wide scale (termed as genome-wide association (GWA) studies). It is commonly accepted that complex diseases such as diabetes and cancer arise from a combination of multiple genes that often regulate and interact with each other to produce the traits. Therefore, the goal of these studies is to identify the complex interactions among multiple genes that, together with environmental factors, may substantially increase the risk of the development of diseases. Using SNPs as genetic markers, this problem is commonly formulated as the task of SNP-SNP and SNP-environment interaction identification.

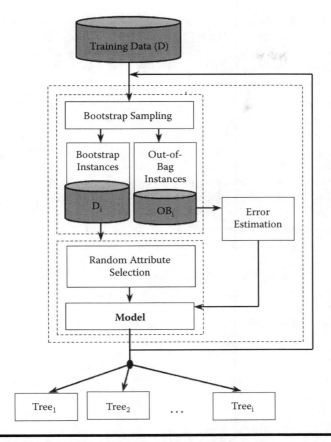

Figure 8.11 The schematic illustration of the random forest classifier.

Among many pattern recognition algorithms, the decision tree algorithm has long been recognized as a promising tool for SNP-SNP interaction identification. Initial attempts to identify gene-gene interaction using decision tree-based methods were investigated on relatively small datasets. For instance, Yang et al. (2010) explained the application of the CART algorithm with a multivariate adaptive regression spline model to explore the presence of genetic interactions from 92 SNPs.

With the increasing popularity of tree-based ensemble methods, such methods have become the focus of many recent studies under the context of SNP-SNP interaction identification for complex disease analysis. Although different ensemble methods have been proposed for identifying SNP-SNP interactions, random forests are the most popular. This popularity is largely due to the method's intrinsic ability to take multiple SNPs jointly into consideration in a nonlinear fashion. In addition, random forests can be used easily as an embedded feature evaluation algorithm, which is applicable for disease association studies.

The primary goal of a random forest analysis in the context of genetic association studies is to identify SNPs that may increase or decrease susceptibility to a disease (Bureau et al. 2005; Lunetta et al. 2004). This goal can be achieved by quantifying how much each SNP contributes to the predictive accuracy of a random forest by measuring its predictive importance. Finding that an SNP helps differentiate between cases and controls is an indication that the SNP either contributes to the phenotype or is linked to disequilibrium with SNPs, contributing to the phenotype.

We describe measures of predictive importance as a categorical response, such as the case or control status of individuals in a genetic study. For individual i, let X_i represent the vector of predictor variable values, y_i represent its true class, $V_j(X_i)$ represent the vote of tree j, and t_{ij} represent an indicator taking value 1 when individual i is out of bag for tree j and 0 otherwise. Let $T_i = \Sigma_{j=1}^{T} t_{ij}$ be the number of trees for which individual i is out of bag. The margin of votes $mg(X_i, y_i)$ is the difference between the proportion of votes for the true class and the largest proportion of votes among the other classes for a given individual. With only two classes, such as diseases and control, the margin becomes the difference between the proportion of votes for the true class and the proportion of votes for the wrong class. Letting $I(V_j(X_i) = y_i)$ denote the indicator function taking value 1 when $V_j(X_i) = y_i$ and 0 otherwise, the margin can be written:

$$mg(X_i, y_i) = \frac{1}{T_i} \sum_{j=1}^{T} I(V_j(X_i) = y_i) t_{ij} - \max_{k \neq y_i} \left\{ \frac{1}{T_i} \sum_{j=1}^{T} I(V_j(X_i) = k) t_{ij} \right\}. \quad (8.54)$$

With only two classes, 0 and 1, the margin simplifies to:

$$mg(X_i, y_i) = \frac{1}{T_i} \sum_{j=1}^{T} I(V_j(X_i) = y_i) t_{ij} - \frac{1}{T_i} \sum_{j=1}^{T} I(V_j(X_i) = 1 - y_i) t_{ij} \quad (8.55)$$

$$= \frac{2}{T_i} \sum_{j=1}^{T} I(V_j(X_i) = y_i) t_{ij} - 1. \quad (8.56)$$

The margin represents the level of confidence of the forest prediction. When most trees vote for the true class of an individual and the margin is close to 1, the pattern of predictor values for that individual unambiguously matches that of other individuals in the true class. When a large proportion of trees votes for another class and the margin is just above 0 or is negative, the pattern of predictor values has only weak similarity with other individuals in the same class and may point to another class.

Now, consider random permuting of the values of a predictor variable such as an SNP genotype among the individuals excluded from the bootstrap sample, such that the variable becomes independent of the response. If the variable is predictive of the response, it will be present in a large proportion of trees and near the roots of those trees. A large proportion of out-of-bag individuals have genotypes that will be directed to the wrong side of the tree. The margin is then expected to decrease from the original variable values. Conversely, if the variable is not related to the response, it will be present in few trees, and when it is present, it will be near the leaves.

8.7 Computational Challenges of Supervised Learning

The computational challenges of applying supervised learning on bioinformatics datasets are attributed to two aspects. The first is the very unbalanced nature of the datasets encountered in bioinformatics (Provost 2000). A dataset is believed to be unbalanced when one class contains more samples (majority class) than the other(s). For instance, in the case of splice site detection there are 100 times fewer positive samples than negative samples. Unbalanced datasets can present a challenge when training supervised learners. The standard approach to addressing this issue is to assign a different misclassification cost to each class. For SVMs, this cost is calculated by associating a different soft margin constant to each class according to the number of samples in the class. Often when data are unbalanced, the cost of misclassification is also unbalanced, where having a false negative proves more costly than having a false positive.

Another challenge in supervised learning is the problem of overfitting. A supervised model is considered to be overfit if the model is closely tied down to its train set. The results obtained from such models tend to be biased where the model fails to perform on random test samples. Overfitting stems from the fact when the model is overly trained to fit closely to the noise in the data. There are different model evaluation strategies that are discussed in Chapter 9, to gauge the effectiveness of the models.

8.8 Conclusion

In this chapter we described prominent classification or supervised learning strategies used in the field of bioinformatics. The chapter covered important concepts of supervised learning such as bias and variance and model complexity. The chapter also shed light on specific challenges that bioinformatics datasets pose to supervised learning. Apart form listing the different approaches to supervised learning, the chapter provided a description of the application of these approaches to different high-throughput data-rich areas of bioinformatics, such as gene expression data and protein structure prediction.

References

Bauer, S., J. Gagneur, and P.N. Robinson. GOing Bayesian: Model-based gene set analysis of genome-scale data. *Nucl Acids Res* 38, no. 11 (2010): 3523–3532.

Ben-Hur, A., C.S. Ong, S. Sonnenburg, B. Schölkopf, G. Rätsch. Support vector machines and kernels for computational biology. *PLoS Comput Biol* 4, no. 10 (2008): e1000173.

Breiman, L. Bagging predictors. *Machine Learn* 24, no. 2 (1996): 123–140.

Brenner, S.E., P. Koehl, and M. Levitt. The ASTRAL compendium for sequence and structure analysis. *Nucl Acids Res* 28 (2000): 254–256.

Brian, D., and G.I. Webb. On the effect of data set size on bias and variance in classification learning. In *The Fourth Australian Knowledge Acquisition Workshop (AKAW '99)*. Sydney, Australia: University of New South Wales.1999, pp. 117–128.

Brown, M.P.S., et al. Knowledge-based analysis of microarray gene expression data by using support vector machines. *Proc Natl Acad Sci USA* 97, no. 1 (2000): 262–267.

Bureau, A., et al. Identifying SNPs predictive of phenotype using random forests. *Genet Epidemiol* 28 (2005): 171–182.

Chatterjee, P., S. Basu, M. Kundu, M. Nasipuri, and D. Plewczynski. PPI_SVM: Prediction of protein-protein interactions using machine learning, domain-domain affinities and frequency tables. *Cell Mol Biol Lett* 16, no. 2 (2011): 264–278.

Degroeve, S., B. De Baets, Y. Van de Peer, and R. Rouze. Feature subset selection for splice site prediction. *Bioinformatics* 18, no. 2 (2002): S75–S83.

Eichner, J., G. Zeller, S. Laubinger, and G. Ratsch. Support vector machine-based identification of alternative splicing in *Arabidopsis thaliana* from whole genome tiling arrays. *BMC Bioinformatics* 12, no. 55 (2011): 1–17.

Eisen, M.B., P.T. Spellman, P.O. Brown, and D. Botstein. Cluster analysis and display of genome-wide expression patterns. *Proc Natl Acad Sci USA* 95 (1998): 14863–14868.

Erp, M.V., L. Vuurpijl, and L. Schomaker. An overview and comparison of voting methods for pattern recognition. In *International Workshop on Frontiers in Handwriting Recognition (IWFHR'02)*. Washington, DC: IEEE Computer Society, 2002, p. 195.

Esposito, F., D. Malerba, and G. Semeraro. A comparative analysis of methods for pruning decision trees. *IEEE Trans Pattern Anal Machine Intell* 19, no. 5 (1997): 476–491.

Freund, Y., and R. Shapire. A decision-theoretic generalization of on-line learning and an application to boosting. In *Proceedings of the Second European Conference on Computational Learning Theory*. London: Springer-Verlag, 1995, pp. 23–37.

Friedman, N., M. Linial, I. Nachman, and D. Pe'er. Using Bayesian networks to analyze expression data. *J Comput Biol* 7 (2000): 601–620.

Golub, T.R., et al. Molecular classification of cancer: Class discovery and class prediction by gene expression monitoring. *Science* 286 (1999): 531–537.

Ho, T.K., J.J. Hull, and S.N. Srihari. Decision combination in multiple classifier systems. *IEEE Trans Pattern Anal Machine Intell* 16, no. 1 (1994): 66–75.

Holloway, D.T., M. Kon, and C. DeLisi. Integrating genomic data to predict transcriptional factor binding. *Genome Inform* 16, no. 1 (2005): 83–94.

Jaakkola, T., M. Diekhans, and D. Haussler. Using the Fisher kernel method to detect remote protein homologies. In *Proceedings of the Seventh International Conference on Intelligent Systems for Molecular Biology*. Menlo Park, CA: AAAI Press, 1999, pp. 149–158.

Kelemen, A., H. Zhou, P. Lawhead, and Y. Liang. Naive Bayesian classifier for microarray data. In *Proceedings of the International Joint Conference on Neural Networks, 2003*. Portland, OR: IEEE, 2003, pp. 1769–1773.

Keller, A.D., M. Schummer, L. Hood, and W.L. Ruzzo. *Bayesian classification of DNA array expression data*. Seattle: Department of Computer Science and Engineering, University of Washington, 2000.

Leslie, C.S., A. Cohen, E. Eskin, J. Weston, and W.S. Noble. Mismatch string kernels for discriminative protein classification. *Bioinformatics* 20, no. 4 (2004): 467–476.

Liao, L., and W.S. Noble. Combining pairwise sequence similarity and support vector machines for detecting remote protein evolutionary and structural relationships. *J Comput Biol* 10, no. 6 (2003): 857–868.

Lukes, L., N.P.S. Crawford, R. Walker, and K.W. Hunter. The origins of breast cancer prognostic gene expression profiles. *Cancer Res* 69 (2009): 310–318.

Lunetta, K.L., L.B. Hayward, J. Segal, and P.V. Eerdewegh. Screening large-scale association study data: Exploiting interactions using random forests. *BMC Genet* 5, no. 32 (2004): 1–13.

Ma, S., and J. Huang. Penalized feature selection and classification in bioinformatics. *Briefings Bioinformatics* 9, no. 5 (2008): 392–403.

Melvin, I., E. Ie, R. Kuang, J. Weston, W.S. Noble, and C. Leslie. SVM-Fold: A tool for discriminative multi-class protein fold and superfamily recognition. *BMC Bioinformatics* 8, no. 4 (2007): 1–15.

Mewes, H.W., et al. MIPS: A database for genomes and protein sequences. *Nucl Acids Res* 28, no. 1 (2000): 37–40.

Murzin, A.G., S.E. Brenner, T. Hubbard, and C. Chothia. SCOP: A Structural Classification of Proteins database for the investigation of sequences and structures. *J Mol Biol* 247 (1995): 536–540.

Piyathilake, C., and G.L. Johannig. Cellular vitamins, DNA methylation and cancer risks. *Am Soc Nutr Sci* 2 (2002): 2340S–2344S.

Provost, F. Machine learning from imbalanced data sets 101. In *Proceedings of the AAAI 2000 Workshop on Imbalanced Data Sets*. Austin, TX: AAAI Press, 2000, pp. 1–3.

Ramaswamy, S., et al. Multiclass cancer diagnosis using tumor gene expression signatures. *Proc Natl Acad Sci USA* 98, no. 26 (2001): 15149–15154.

Selfridge, O. Pandemonium: A paradigm for learning in mechanisation of thought processes. In *Proceedings of a Symposium Held at the National Physical Laboratory*. London: HMSO, 1958, pp. 513–526.

Sonnenburg, S., G. Schweikert, P. Philips, J. Behr, and G. Ratsch. Accurate splice site prediction using support vector machines. *BMC Bioinformatics* 8, no. 10 (2007): S7.

Vachani, A., et al. A 10-gene classifier for distinguishing head and neck squamous cell carcinoma and lung squamous cell carcinoma. *Clin Cancer Res* 13 (2007): 2905–2915.

Wang, Y., F.S. Makedon, J.C. Ford, and J. Pearlman. HykGene: A hybrid approach for selection of marker genes for phenotype classification using microarray gene expression data. *Bioinformatics* 21, no. 8 (2005): 1530–1537.

Webb, G.I., and Z. Zheng. Multistrategy ensemble learning: Reducing error by combining ensemble learning techniques. *IEEE Trans Knowledge Data Eng* 16, no. 8 (2004): 980–991.

West, M., et al. Predicting the clinical status of human breast cancer by using gene expression profiles. *Proc Natl Acad Sci USA* 98, no. 20 (2001): 11462–11467.

Wilkinson, D.J. Bayesian methods in bioinformatics and computational systems biology. *Briefings Bioinformatics* 8, no. 2 (2007): 109–116.

Wyner, A.J. On boosting and the exponential loss. In *Proceedings of the Ninth Annual Conference on AI and Statistics*. Key West, FL: Society for Artificial Intelligence and Statistics, 2003.

Yang, P., Y.H. Yang, B.B. Zhou, and A.Y. Zomaya. A review of ensemble methods in bioinformatics. *Curr Bioinformatics* 5, no. 4 (2010): 296–308.

Yang, S., and K.-C. Chang. Comparison of score metrics for Bayesian network learning. *IEEE Trans Syst Man Cybernetics A* 32, no. 3 (2002).

Zukiel, R., S. Nowak, A.-M. Barciszewska, I. Gawronska, G. Keith, and M.Z. Barciszewska. A simple epigenetic method for the diagnosis and classification of brain tumors. *Mol Cancer Res* 2 (2004): 196–202.

Chapter 9
Validation and Benchmarking

In Chapter 8, we introduce bias and variance, overfitting, and key classifiers as bioinformatics applications. These techniques have been used successfully with both clustering and classification methods. In this chapter, we describe the evaluation strategies used to test a hypothesis and evaluate the performances of the clustering and classification techniques described in Chapters 6 to 8. For the readers' convenience, we have divided this chapter into two parts. The first part contains an explanation of model selection and evaluation techniques used on classification models. The second part contains an explanation of cluster evaluation techniques.

9.1 Introduction: Performance Evaluation Techniques

With the exponential growth of data and the growing importance of data mining, the roles of clustering and classification techniques have become of an integral part of research in bioinformatics. Despite this importance, the significance of the results and knowledge mined from biological data is formalized using evaluation techniques. A wide range of performance evaluation techniques are available in data mining. These techniques have been derived using well-known statistical principles. This section of the chapter is dedicated toward explaining how these principles can be used for better inference evaluation.

Before we delve into the techniques of model evaluation, we would like to remind the readers that classification techniques are used to generate learning models that are used to classify samples into the most unbiased forms possible using the least possible variance. As discussed in Chapter 8, a trade-off exists between the bias and variance of the learning model, i.e., models with low bias have high variance and vice versa. It is known that classification model bias remains constant as the size of training set D increases, whereas the variance decreases (i.e., as $D \to \infty$, the model's variance $\to 0$) in such cases. Variance is therefore an indicator of the performance of a model (Guyon 2009).

However, while models with low variance seem to be the best logical choice, it is observed that models with the least possible variance tend to be overfitted models. Therefore, though variance is an effective estimate of model performance, it fails to gauge a model's ability to generalize across training sets and compare model performances. To avoid these errors, it is a common practice in data mining to use the generalization error as an alternate means to evaluate and compare learning models.

The generalization error ($\mathcal{G}$) of a model is defined as follows (Nadeau and Bengio 2003). For example, if a large dataset $X_1^n = \{X_1,\ldots,X_n\}$ consists of n samples of the form $X_i = (x_i, y_i) \in \mathbb{R}^{p+q}$, where p and q denote the dimensions of x_i and the class label y_i, then let D represent the training set of $n_1 \leq n$ samples drawn at random from the dataset X_1^n. Furthermore, let f_A represent the supervised learning algorithm trained using the training set D. The generalization error ($\mathcal{G}_A$) is defined as the inaccuracy of a decision $f_A(x)$ when y is the associated class label. The difference between the corresponding generalization errors ($\mathcal{G}_A, \mathcal{G}_B$) of two models, f_A and f_B, is used to compare the two learning algorithms, provided they use the same learning data (Vapnik 1999). The following sections describe the data mining strategies used to estimate the generalization errors of models.

9.2 Classifier Validation

To generate accurate generalization error estimates, various validation strategies can be used in tandem with data mining. These validation techniques are motivated by two factors: model selection and performance estimation.

1. **Model selection:** Almost invariably, classification techniques have one or more parameters that dictate the performance of the model generated. For example, in the case of the SVM and its respective kernel function, the parameters of the kernel function dictate the performance of the classifier, or in the case of the random forest classifier, the determination of the number of trees is necessary to obtain optimal model performance. Therefore, model selection enables the users to choose and optimize the set of parameters of a classifier for optimal model performance.

2. **Performance estimation:** Once a model is selected with an optimal parameter set, performance estimation is used to estimate the performance of the selected model. Performance is typically measured by a model's ability to classify samples to their corresponding classes for a given dataset.

9.2.1 Model Selection

With the various classification models available in data mining, it is a challenge to choose those models that best suit the data in the application domain. Model selection is used to choose that model that best fits the data set being analyzed. Typically, bioinformatics datasets consist of a large number of samples (n), and each sample is described by a fixed number of features (p) (i.e., $n \gg p$). In such situations where $n \gg p$, any learning strategy can be applied. However, many high-dimensional datasets used in bioinformatics consist of a small number of samples, and the features that describe these samples are larger in number ($n \ll p$). In such situations, the choice of model affects the results and inferences that can be derived.

Determining which model to use is driven by a *heuristic of model choice* (Guyon 2009) that encapsulates the conditions described below. The benefits of using a heuristic of a model is that this model reduces the chances of model overfitting, prioritizes learners to be used, and reduces the computational complexity that can be avoided. The following heuristics highlight the importance of choosing appropriate classification models by taking into consideration overfitting, linear models, and nonlinear models.

Overfitting: The naïve Bayes classifier is least prone to overfitting and least computationally expensive of known classification techniques. As described in Chapter 8, the naïve Bayes classifier simplifies the assumption of feature independence (i.e., there is no relation between features), and thus renders the easy implementation of a model that is computationally effective. Moreover, this assumption of feature independence may create models that underfit the data when (1) there are a larger number of features than samples and (2) the number of samples is not sufficient to estimate the classifier performance using cross-validation. Due to these limitations, it is advisable to use the naïve Bayes classifier as a baseline model.

Linear models: Linear models are derived from classifiers, such as the support vector machine (SVM) with a linear kernel, and have low computational complexity. These models are most effective on datasets that have a large number of samples (n) with a lower number of features (p). However, these models are preferred as they can provide a better fit of the data.

Nonlinear models: Nonlinear classifiers such as the J48 or C4.5 decision trees should be considered only if sufficient amounts of training data are available to perform cross-validation. Nonlinear SVM models work well with datasets that have a large number of features. However, computational complexity increases with the number of features, and these nonlinear models could require feature selection in such cases.

9.2.1.1 Challenges Model Selection

Before discussing the various validation strategies used to estimate the performance of a model, we provide an overview of the procedure used to validate a selected model in this section. Performance estimation of a model is dependent on the number of samples in the dataset. Let us consider a dataset consisting of n available samples in a study, of which m number of samples is used for model training. In model training, the different parameters of the selected models are tweaked, and the best model is selected for performance estimation. Performance estimation is then performed on a test set T. It is imperative that the test set $T = n - m$ be reserved solely for testing throughout the study. Therefore, performance estimation poses the following challenges.

9.2.1.1.1 Sufficient Number of Samples in Train and Test Sets

This challenge is based on the characteristics of the dataset. If the dataset consists of a relatively large number of samples (n), as compared to the number of features (p) (i.e., $n \gg p$), then it is believed that a model trained on a training set that consists of randomly chosen samples and tested using a testing set would provide relevant error estimates that reflect the characteristics of the entire dataset. However, if the dataset has a lower number of samples (n) than the number of features (p) (i.e., $n \ll p$), we could face a situation where it is not always possible to reserve a sufficiently large test set without compromising the number of samples used for training the model. This problem could invariably generate inaccurate error estimates that could be misleading. As discussed previously, the performance of the model is closely related to the size of the train set. Therefore, appropriate train and test sets (with a minimum number of samples) should be determined based on a predetermined model performance confidence interval prior to model creation.

9.2.1.1.2 Handling Imbalanced Datasets

Models are affected by an imbalance in the number of samples in each class. Bioinformatics is plagued by imbalanced datasets (Chawla et al. 2004). Classifiers trained on imbalanced datasets create models that classify the test instances to the majority class (i.e., the class that has the most samples) that is least important. These misclassifications of samples that belong to the minority class deteriorate the overall performance of the model. The relationship between the training set and the equal representation of all classes is made worse when there is large overlap between classes or when majority classes can be further divided into smaller subclasses.

Two approaches can be adopted to handle imbalanced datasets: unsupervised (data specific) and supervised (algorithmic) approaches. Unsupervised approaches rely on various resampling strategies, such as the random oversampling of the minority class with replacement of samples (Liu et al. 2009), random undersampling of the majority class, directed oversampling (in which no new samples are created, but the choice of which samples to replace is informed rather than random), and directed undersampling (where the choice of samples to eliminate is informed).

Supervised (algorithmic) (Fu et al. 2002) approaches rely on weighing and thresholding strategies that prioritize minority classes to counter the class imbalance caused by the majority classes (Mease et al. 2007). These strategies include adjusting the decision threshold or one-class learning rather than multiclass learning. Other approaches, such as the ensemble of undersampled SVMs (EUS SVMs), include a mixture of data and algorithmic approaches. These methods use ensembles (Kang and Cho 2006) in which the results of many classifiers are combined after oversampling or undersampling the data using different over/undersampling approaches.

9.2.2 Performance Estimation Strategies

Bioinformatics applications have access to a finite set of samples that are often insufficient for testing a hypothesis using classification models. Because of the small sample sets, overfitting is prominent in several bioinformatics applications, especially those that have a large number of features (p).

Performance estimation strategies are used to avoid overfitting the error estimates of a model to overly optimistic (i.e., lower than the true error rate) results.

This section describes the performance estimation strategies used to effectively test a hypothesis despite the shortage of samples in bioinformatics datasets.

9.2.2.1 Holdout

The holdout method is considered to be the simplest form of performance estimation that partitions the data into two disjoint sets: a train set and a test set. The train set is used to train the chosen classifier for model generation during the training phase. During this phase, the optimal values of the model parameters are determined, and an appropriate performance measure is evaluated. Once the model is generated, the testing set is used to obtain an unbiased estimate of the generalized performance of the models.

Though it is the simplest form of performance estimation with a single training and testing experiment, the holdout estimates of error could be misleading when the testing set is not sufficient to provide good error estimates. These data insufficiencies are brought about by sparse datasets that are common in bioinformatics. Such limitations can be overcome using bootstrapping and cross-validation techniques described in the following sections (Figure 9.1).

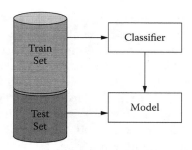

Figure 9.1 A schematic representation of the holdout technique for performance estimation of a model.

9.2.2.2 Three-Way Split

One alternate approach to the holdout technique is the three-way split. In the three-way split, model selection and performance (true error) estimates are computed at the same time. As the name suggests, this technique splits the data into three independent sets: the training set, the validation set, and the testing set.

As with the holdout technique, the three-way split train set is used for model selection and parameter estimation. The difference between the methods is that the three-way split creates an additional split referred to as the validation set, as noted above. The validation set consists of a set of samples that are used to fine-tune the estimated parameters of the model selected using the train set. This additional fine-tuning enables the removal of biases from the true error estimates created during the model training using the train set. Furthermore, all parameter estimations should terminate with the validation set.

Finally, the testing set is used to assess the final performance of the fine-tuned model. It should be noted that just as in the holdout method, the testing set is an independent set of samples that are used to generate the true error estimates of the final model.

The following steps encapsulate the process of performance evaluation using the three-way split method (Figure 9.2):

1. Dataset D is divided into three disjoint (independent) sets: train (t), validation (v), and testing sets (T).
2. Choose an appropriate classifier (F) and determine the parameters that need tuning.
3. Use the training set (t) and the classifier (F) to generate the model (f).
4. Determine optimal parameters of model (f) using the validation set (v).
5. Repeat steps 2 through 4 if there are multiple classifiers or if multiple parameters need to be optimized.

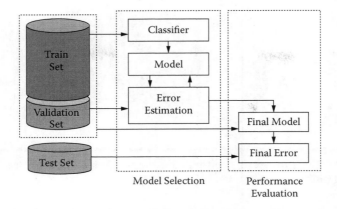

Figure 9.2 A schematic representation of the application of the three-way split approach to performance estimation.

6. Select the best model (f), and train it using the combined training and validation sets.
7. Perform parameter estimation using the final model (f) and the independent test set (T).

9.2.2.3 k-Fold Cross-Validation

The k-fold cross-validation is the most prominently used performance estimation technique in data mining and bioinformatics applications. k-Fold cross-validation divides the data set into k disjointed (independent) subsets consisting of equal (or nearly equal) samples in each subset. Each of the k disjointed data subsets is referred to as a fold, thus the name k-fold. The k-fold cross-validation process is an iterative procedure in which one of the k subsets (chosen at random) is used as a test set for performance estimation at each iteration, while the remaining $k-1$ disjointed subsets are combined to form the training set that is used to train the model.

It should be noted that the number of iterations in the k-fold cross-validation is set to k; i.e., the number of iterations is equal to the number of disjointed subsets used for performance evaluation. Having the number of iterations fixed to k is done such that there is an equal probability of each fold being used as the testing set for performance evaluation. Once all the iterations of the k-fold cross-validation are carried out, the average of the error estimates is computed to provide a generalized performance estimate performed over all k-folds. This generalized performance estimate, though slightly pessimistic, is considered justified as it is carried out over the entire sample space.

Another form of the k-fold validation technique is the leave-one-out cross-validation (LOOCV) (Efron and Tibshirani 1997), in which each subset contains

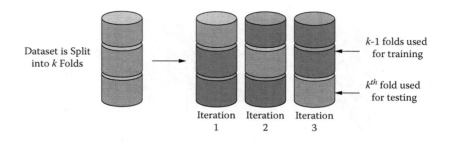

Figure 9.3 The process of splitting the dataset into folds followed in *k*-fold cross-validation.

one sample, i.e., $k = N$, where $N = $ the number of samples in the dataset D (Figure 9.3).

9.2.2.4 Random Subsampling

Random subsampling performs k data splits of the dataset. Unlike *k*-fold cross-validation, the number of splits is not equal to the number of iterations by which the procedure is repeated. Random subsampling is also referred to as Monte Carlo cross-validation (MCCV).

In this approach, each split consists of a fixed number of samples (determined by the user) that are randomly chosen without a replacement from the dataset.

The error estimates (E_i) are carried out on multiple iterations for a given dataset. In every iteration of the algorithm, a new set of samples is chosen from the dataset independently for training and testing. The *true error* estimate is obtained by taking the average of the separate estimate E_i, as shown in Equation 9.1.

$$E = \frac{1}{K}\sum_{i=1}^{K}E_i. \quad (9.1)$$

The error estimates generated using random subsampling are believed to be pessimistic (i.e., worst-case estimates), whereas those generated using the holdout test are overly optimistic.

9.3 Performance Measures

In this section, we discuss the measures proposed in data mining to test the performance of a model. The most fundamental of these measures is the ROC analysis and its application to the binary (or two-class) classification problem. A binary classification algorithm maps a sample (for example, an unannotated sequence)

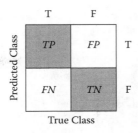

Figure 9.4 A schematic representation of a confusion matrix in the case of a binary classifier. The different performance measures that are derived from the confusion matrix are true positive (TP), false positive (FP), true negative (TN), and false negative (FN).

into one of two classes, denoted as C+ and C−. Building on our discussions in Section 9.3.2, the parameters of any classification algorithm are derived using the train set that consists of samples obtained from the known C+ and C− classes, and then the classifier is tested on the C+ and C− samples that are disjoint from the train set.

Such a binary classifier predicts only the classes to which test samples belong. There are four possible outcomes for this classifier: true positive (TP), true negative (TN), false positive (FP), and false negative (FN). These outcomes are schematically known as a confusion matrix (see Figure 9.4).

If a sample that belongs to the true positive class C+ is correctly classified as positive, then the result is counted as a true positive (TP); however, if the sample is misclassified as negative, it is counted as a false negative (FN). Similarly, if a sample that belongs to the true negative class C− is correctly classified as negative, it is counted as a true negative (TN); if it is misclassified as positive, it is counted as a false positive (FP).

9.3.1 Sensitivity and Specificity

The TP, FN, TN, and FP counts can then be used to derive other measures of classifier performance. The true positive rate (also known as the hit rate or recall) of a classifier is derived from the following relation:

$$TP\ Rate = \frac{Positives\ correctly\ classified}{Total\ number\ of\ positives}. \qquad (9.2)$$

As shown in the confusion matrix (see Figure 9.4), the positives correctly classified refer to the true positive (TP) count, and the total number of positives refers to the sum of both the true positive and false positive counts (i.e., TP + FN).

Similarly, the false positive rate (also known as the false alarm rate) of the classifier is computed using the following relation:

$$FP\ Rate = \frac{Negatives\ incorrectly\ classified}{Total\ number\ of\ negatives}, \qquad (9.3)$$

where negatives incorrectly classified refer to the false positive (FP) count and the total number of negatives refers to the FP + TN.

The TP and FP rates are two of the most import measures of model performance. It is important to know that a model that is effective for discriminating between samples of the C+ and C− classes will have both a high TP rate and a low FP rate. The interplay between the TP rate and FP rate is best captured using the ROC plot described in Section 9.3.3.

The true positive rate (TP rate) is also referred to as the *sensitivity*. Another important measure of model performance is known as *specificity* or *TN rate* and is calculated using the following relation.

$$Sensitivity = 1 - Specificity \qquad (9.4)$$

Typically, sensitivity represents a model's ability to identify samples that belong to the positive class (C+), and specificity represents a model's ability to identify samples of the negative class (C−).

9.3.2 Precision, Recall, and f-Measure

Similar to the measures of sensitivity and specificity, the measures of precision and recall are used to estimate the performance of a model. Precision and recall are measures used to evaluate the retrieval performance of a classifier and are suited to biological applications that deal with information retrieval (Huang and Bader 2009; Abeel et al. 2009). In this section, we provide the formal definition of precision and recall, and their derivative *f*-measure used as a comprehensive measure to gauge the performance of a classifier.

Precision (p) is the ratio of the number of true positives (TP) to the total number of positives (TP + FP) used and is represented by Equation 9.5:

$$p = \frac{TP}{TP + FP}. \qquad (9.5)$$

Precision, therefore, represents the *positive predictive value* of a model. Similarly, we have the measure of recall (r). Sometimes referred to as the TP rate, *sensitivity* is to the ratio between the number of true positives (TP) and the total outcomes (TP + FN) generated by the model. Recall (r) is represented as follows:

$$r = \frac{TP}{TP + FN}. \qquad (9.6)$$

To determine model accuracy using both p and r, we use the *f*-measure. The *f*-measure is the harmonic mean between p and r and is represented as follows:

$$F - measure = 2 \times \frac{p \times r}{p + r}. \tag{9.7}$$

In Equation 9.7, the *f*-measure is believed to be high when both the p and r values are high. The *f*-measure is effective in capturing the compromise between p and r. Therefore, a model that has a higher *f*-measure is unbiased and is an effective classifier.

9.3.3 ROC Curve

The receiver operating characteristics (ROC) curve is a classification evaluation technique that is used to visually compare the performance of classifier. In order to analyze the performance of a model, it is important to compare the interplay between the true positives and the false positives of independent classifiers. The ROC is a graphical plot of the true positive rate and the false positive rate of a classifier in the ROC space. The ROC space is represented by the specificity (FP rate) on the *x*-axis versus sensitivity (TP rate) on the *y*-axis. A point in the ROC space is the representation of a classifier in terms of its FP rate and TP rate as coordinates in the ROC space using a test set. This representation of the ROC space enables the capture of the trade-off between the true positives and the false positives of a classifier so that the result is beneficial for comparing the classifier performance.

An ROC curve is a step function that tracks the performance of a classifier as the number of samples in the test set increases (i.e., as it tends to ∞). Figure 9.5 provides a schematic representation of the performance of a classifier using the ROC curve. If the ROC curve of a classifier is skewed toward the northwest corner of the ROC space, the classifier exhibits a higher TP rate and a lower FP rate as the number of samples in the test set increases. Classifiers that follow this skewed trend are believed to be liberal when the skew identifies positive samples that are true positives with weak evidence.

If, on the contrary, the curve is skewed toward the southeast corner of the ROC space, the classifier exhibits a higher FP rate and a lower TP rate. In such a scenario, the classifier is believed to conservative when it is biased toward false positive classifications along with a lower TP rates. Similarly, if the ROC curve of a classifier falls along the diagonal of the ROC space, it is believed that the classifier has no bias toward the TP rate or the FP rate, and performs like a random guess, as in the case of making a decision by flipping a coin (head or tail). Typically, it is desirable to have a classifier that has a higher TP rate and a lower FP rate.

In order to quantify the performance of a classifier using the ROC curve, we use the measure of area under the curve (AUC). A relative measure that ranges

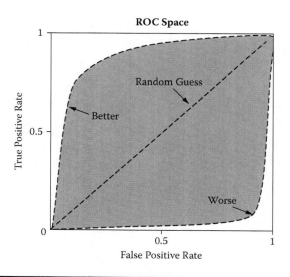

Figure 9.5 ROC curve. (From Fawcett, T., *Pattern Recog Lett* (2006): 861–874. With permission.)

from 0 to 1, the AUC refers to the area under the ROC curve in the ROC space (see Figure 9.5). A classifier is believed to perform well if the AUC is higher and approaches closer to 1, and vice versa.

9.4 Cluster Validation Techniques

With the large volume of unlabeled data being generated in the field of bioinformatics, it is vital to understand the underlying distribution of the data. Unsupervised clustering techniques of data mining aid in the understanding of the inherent properties of data. However, with the gamut of clustering techniques available, it becomes increasingly difficult for users to choose and validate these findings. Refer to Chapters 6 and 7 for a description of clustering techniques and their applications in bioinformatics. In this section, we describe the validation techniques that can be used to quantify the quality of a cluster. The evaluation of the results obtained from a clustering algorithm uses three cluster characteristics to quantify the quality of a cluster. These cluster characteristics include compactness, connectedness, and spatial separation (see Figure 9.6) (Handl et al. 2005; Halkidi et al. 2001).

Compactness: Compactness, the formation of compact clusters, is achieved if the clustering algorithm is effective in keeping the intracluster differences small. Compactness can be achieved with algorithms that enable the formation of spherical and well-separated clusters such as the k-means algorithm.

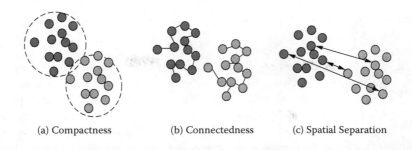

(a) Compactness   (b) Connectedness   (c) Spatial Separation

Figure 9.6 Dataset exhibiting the different properties. (From Handl, J., et al., *Bioinformatics* 21, no. 15 (2005): 3201–3212. With permission.)

While compactness is useful for characterizing clusters with well-formed boundaries, this property is ineffective in characterizing complicated clusters.

Connectedness: As the name suggests, connectedness can be used to characterize arbitrary shaped clusters based on the connectivity between points of a cluster. Compactness is based on the assumption that neighboring data items belong to the same cluster.

Spatial separation: Spatial separation is a criterion that enables the characterization of clusters that are sparse (i.e., data points between two clusters are widely separated). Therefore, spatial separation usually combines with other characteristics, like compactness along with a distance measure. Spatial separation between clusters is measured using three approaches: (1) single linkage, (2) complete linkage, and (3) average linkage.

9.4.1 The Need for Cluster Validation

All clustering methods are driven by the choice of distance measure, and the objective to form clusters with high intracluster similarity and low intercluster similarity. Those bioinformatics applications that use clustering strategies for hypothesis testing are plagued with datasets that are noisy and sparse. These inherent properties of the dataset make it difficult to interpret the results obtained using clustering algorithms. Typically, researchers rely on visual inspections of clusters and use prior biological information to estimate the quality of a cluster, making cluster validation subjective. Moreover, these counterproductive practices of users undermine the clustering algorithms' abilities to discover useful information possessed by the data necessitating the use of stringent validation techniques.

Clustering techniques are primarily used to discover significant groups present in high-dimensional datasets. However, different clustering techniques generate varied results. These discrepancies in results are attributed to factors that govern clustering techniques.

Clustering techniques are biased toward cluster parameters: Clustering algorithms are biased toward the formation of clusters as the creation of clusters is governed by the parameters used by the technique. For example, the k-means algorithm is governed by the predetermined value of k that corresponds to the number of clusters in the data. This is the fundamental problem that leads to observable discrepancies between the solutions produced by different algorithms.

The sensitivity of the clustering technique to the number of features in the dataset: Clustering relies on the existence of distinct naturally occurring clusters of data points within the feature space. As most clustering techniques are governed by the use of a distance measure, it is a challenge to identify naturally occurring clusters in sparse high-dimensional spaces. This inherent problem results in the clustering of data points in the absence of any observed distribution in points, leaving it to the user to detect the significance of the resultant clusters returned.

It is therefore necessary to validate a clustering algorithm to determine that the clustering algorithm is not biased toward particular cluster properties and that the clusters formed are significant. In this section, we describe the cluster validation techniques that are categorized into external and internal measures of cluster quality.

9.4.1.1 External Measures

External validation measures consist of those techniques that use existing information (correct class labels) to evaluate the quality of a cluster. These validation measures are therefore used to evaluate a predefined objective or hypothesis. The measures are also used to validate a cluster with a known set of benchmark data. In situations where no known benchmark is available to evaluate a cluster, we rely on an internal measure of cluster goodness. Internal measures therefore do not rely on class labels, but rather use information intrinsic to the structure of the data (Handl et al. 2005).

External measures are divided into unary measures and binary measures, which are described as follows.

Unary measures: Unary measures are used to validate whether a cluster partition complies with the ground truth. The ground truth typically consists of a dataset with each sample assigned a unique class label. Unary measures are evaluated based on purity and the completeness of the cluster evaluated with respect to the ground truth dataset. Purity denotes the fraction of the cluster taken up by its predominantly occurring class label, whereas completeness denotes the ratio of the number of samples in the predominant class that are classified to the cluster being evaluated to the total number of samples in the class. To obtain an assessment of a cluster, it is important to consider purity and completeness together. For a comprehensive assessment of purity and completeness, we use the *f*-measure as described in Section 9.3.2 (Handl et al. 2005).

Binary measures: Binary measures are used to assess the consensus between a cluster and the ground truth based on the contingency table of the pairwise assignment of data items. Most of these indices are symmetric and are therefore equally well suited for use as binary measures, that is, for assessing the similarity of two clustering results.

The Rand index is a binary measure that is used to determine the similarity between two clusters as a function of positive and negative agreements in pairwise cluster assignments. Other binary measures include the Jaccard coefficient, which, unlike the Rand index, takes into consideration only the positive matches between clusters for evaluation.

9.4.1.2 Internal Measures

Internal measures, unlike external measures, do not rely on a ground truth dataset. All internal measures of a cluster are relative to the dataset from which the cluster is derived and use intrinsic information of the cluster and dataset to assess the quality of the clustering. As discussed in the previous section, measures of compactness, connectedness, and separation are effective internal measures of cluster goodness. Apart from these three internal measures, we describe other measures that are derived from these measures.

Combinations: As the name suggests, combination measures are combinations of the internal measures of compactness and separation. In clustering, it is believed that as intracluster homogeneity increases with the number of clusters, the distance between the clusters decreases. Therefore, the measures that fall into this category measure both intracluster homogeneity and intercluster separation. A final score is computed as a linear or nonlinear combination of the two measures. An example of a linear combination is the SD validity index, and an example of a nonlinear combination is the Dunn index.

Predictive power/stability: Another form of cluster validation techniques that assess the predictive power or stability of a cluster forms a special category of internal validation measures. These techniques rely on repeated resampling or perturbation of the original dataset and reclustering the resulting data. The consistency of the corresponding results provides an estimate of the significance of the clusters formed.

Compliance between partitioning and distance information: An alternative measure of cluster quality is an estimation of the degree of distance information preserved from the original datasets in clusters. This measure uses the cophenetic matrix C that is symmetric of size $N \times N$, and N is the number of samples in the dataset. Each element $C(i,j)$ of the matrix C acts as an indicator if a pair of samples is assigned to a common cluster. For the evaluation of a hierarchical clustering, the cophenetic matrix can also be constructed to reflect the level within the dendrogram. Here, an entry $C(i,j)$ represents the level within the dendrogram at which the two samples i and j are first assigned to the same cluster.

Several methods have been proposed that capture the correlation between the cophenetic matrix and the original dissimilarity matrix to assess the preservation of distances under different distance functions and within different feature spaces or to compute the dendrograms obtained for different algorithms.

9.4.2 Performance Evaluation Using Validity Indices

A great deal of research is focused on finding the correct or optimal number of partitions. Cluster validity indices help address this problem by estimating the correct number of clusters and finding the quality clusters (Halkidi et al. 2001). The most commonly used validity indices have been described below (Azuaje and Bolshakova 2002).

9.4.2.1 Silhouette Index (SI)

The computation of the silhouette index is described by the following steps:

1. For a given cluster, $X_j (j=1,...,c)$, the silhouette technique assigns a silhouette width, $s(i)(i=1,...,m)$, to the ith sample of X_j. This value is defined as

$$s(i) = (b(i) - a(i)) / \max\{a(i), b(i)\},$$

   where $a(i)$ is the average distance between the ith sample and all of the samples included in X_j, and $b(i)$ is the minimum average distance between the ith sample and all of the samples clustered in $X_k (k=1,...,c; k \neq j).s(i)$ lies between −1 and 1.
2. When the value of $s(i)$ is near 1, it can be assumed that the ith sample has been assigned to an appropriate cluster.
3. When $s(i)$ is near zero, it can be assumed that the ith sample can be assigned to the nearest neighboring cluster.
4. When $s(i)$ is near −1, it can be assumed that the ith sample has been misclassified (Rousseeuw 1987).

A global silhouette value or silhouette index, GS_u, can be used as a validity index for a partition U. This measure can be determined using Equation 9.8, as shown below, which helps estimate the "correct" number of clusters for partition U (Rousseeuw 1987). Thus, a high value of silhouette index indicates that partition U is a better or optimal cluster. This method can be represented as

$$GS_u = \frac{1}{c}\sum_{j=1}^{c} S_j \qquad (9.8)$$

9.4.2.2 Davies-Bouldin and Dunn's Index

Unlike the SI, the Davies-Bouldin (DB) index is defined as the ratio of the sum of the within-cluster scatter to the between-cluster scatter (Davies and Bouldin 1979). A small DB value indicates a compact cluster. Mathematically, such a reading can be defined as

$$DB = \frac{1}{n}\sum_{i=1, i\neq j}^{n} \max\left(\frac{\sigma_i + \sigma_j}{\sigma(c_i, c_j)}\right), \qquad (9.9)$$

where n is number of clusters, σ_i is the average distance of all patterns in cluster i to their cluster center c_i, σ_j is the average distance of all patterns in cluster j to their cluster center c_j, and $d(c_i, c_j)$ is the distance of cluster centers c_i and c_j.

Similarly, the Dunn index (D) is defined as the ratio of the minimum intracluster distance to the maximum intercluster distance. The Dunn index lies within the range of 0 to 1, and values approaching 1 correspond to good clusters. The index is given by

$$D = d_{min} / d_{max}, \qquad (9.10)$$

where d_{min} is the minimum distance between two objects from different clusters, and d_{max} is the maximum distance of two objects from the same cluster.

9.4.2.3 Calinski Harabasz (CH) Index

The Calinski Harabasz (CH) index, proposed by Maulik and Bandopadhyay (2002), is computed as

$$(race(b)/(k-1)/(trace(w)/(n-k)), \qquad (9.11)$$

where b and w represent the between- and within-cluster scatter matrices, respectively, and k and n represent the cluster and data points, respectively.

The trace for the between-cluster scatter matrix B can be written as

$$Trace(b) = \sum_{k=1}^{k} nk \, \| zk - z \|^2, \quad (9.12)$$

where nk is the number of points in cluster k and z is the centroid of the entire dataset. The trace of the within-cluster scatter matrix W can be written as $trace(W)$,

$$Trace(w) = \sum_{k=1}^{k} \sum_{i=1}^{nk} (xi - zk)^2, \quad (9.13)$$

9.4.2.4 Rand Index

A Rand index determines the similarity between two partitions with respect to positive and negative agreements and can be used to assess the degree of agreement between two clusters (Rand 1971; Youness and Saporta 2010). The Rand index ranges in value from 0 to 1; a higher Rand index value indicates a higher similarity between two partitions. This index is defined as the ratio of the number of agreements between two partitions divided by the total number of objects (Hubert and Arabie 1985).

9.5 Conclusion

This chapter provides an explanation of computation techniques used to validate and benchmark results obtained using either clustering or classification techniques on datasets. Moreover, it should be noted that these techniques are used for hypothesis testing in bioinformatics.

References

Abeel, T., Y. Van de Peer, and Y. Saeys. Toward a gold standard for promoter prediction evaluation. *Bioinformatics* 25 (2009): i313–i320.

Azuaje, F., and N. Bolshakova. Clustering genome expression data: Design and evaluation principles. In D. Berrar, W. Dubitzky, and M. Granzow (Eds.), *Understanding and using microarray analysis techniques: A practical guide.* London: Springer Verlag, 2002, 230–245.

Chawla, N.V., N. Japkowicz, and A. Kotcz. Editorial: Special issue on learning from imbalanced data sets. *SIGKDD Explor Newsl* 6, no. 1 (2004): 1–6.

Davies, D.L., and D.W. Bouldin. A cluster separation measure. *IEEE Trans Pattern Anal Machine Intell* 1 (1979): 224–227.

Efron, B., and R. Tibshirani. Improvements on cross-validation: The 632+ bootstrap method. *J Am Stat Assoc* 92, no. 438 (1997): 548–560.

Fawcett, T. An introduction to ROC analysis. *Pattern Recog Lett* 27, no. 8 (2006): 861–874.

Fu, X., L. Wang, K.S. Chua, and F. Chu. Training RBF neural networks on unbalanced data. In *Proceedings of the 9th International Conference on Neural Information Processing (ICONIP'02)*. Piscataway, NJ: IEEE, 2002, pp. 1016–1020.

Guyon, I. A practical guide to model selection. In J. Marie (Ed.), *Machine learning summer school*. Springer, to appear.

Halkidi, M., Y. Batistakis, and M. Vazirgiannis. On clustering validation techniques. *J Intell Inf Syst* 17 (2001): 107–145.

Handl, J., J. Knowles, and D.B. Kell. Computational cluster validation in post-denomic data analysis. *Bioinformatics* 21, no. 15 (2005): 3201–3212.

Huang, H., and J.S. Bader. Precision and recall estimates for two hybrid screens. *Bioinformatics* 25, no. 3 (2009): 372–378.

Hubert, L., and Arabie, P. Comparing partitions. *J Classification* 2 (1985): 193–218.

Kang, P., and S. Cho. EUS SVMs: Ensemble of under sampled SVMs for data imbalance problems. *Lecture Notes Artif Intell* 3918 (2006): 107–118.

Liu, X.-Y., J. Wu, and Z.-H. Zhou. Exploratory undersampling for class-imbalance learning. *IEEE Trans Systems Man Cybernetics B* 39, no. 2 (2009): 539–550.

Maulik, U., and S. Bandopadhyay. Performance evaluation odd some clustering algorithms and validity indices. *IEEE Trans Pattern Anal Machine Intell* 24, no. 12 (2002): 1650–1654.

Mease, D., A.J. Wyner, and A. Buja. Boosted classification trees and class probability/quantile estimation. *J Machine Learn Res* 8 (2007): 409–439.

Nadeau, C., and Y. Bengio. Inference for the generalization error. In *Machine learning*. MIT Press, 2003, pp. 239–281.

Rand, W.M. Objective criteria for the evaluation of clustering methods. *J Am Stat Assoc* 66, no. 336 (1971): 846–850.

Rousseeuw, P.J. Silhouettes: A graphical aid to the interpretation and validation of cluster analysis. *J Comput Appl Math* 20 (1987): 53–65.

Vapnik, V.N. An overview of statistical learning theory. *IEEE Trans Neural Networks* 10, no. 5 (1999): 988–999.

Youness, G., and G. Saporta. Comparing partitions of two sets of units based on the same variables. *Adv Data Anal Classification* 4 (2010): 53–64.

Index

A

AdaBoost algorithm, 288
Adjacency matrix, 223
Agencourt/Applied Biosystems, 15
Alleles, 6
Amino acid substitution (AAS), 24
Amino acids, 8
Applied Biosystems, 15
   SOLiD Sequencer, 17
Approximate similarity computation, 194–195
Atomic strings, 101

B

Bagging, 283–286
Basic local alignment search tool (BLAST), 19, 35, 194–195
Bayes classification, naive, 268–269, 301
Bayes' theorem, 268
Bayesian networks, 104, 115
   classifiers, application in bioinformatics, 275–277
   data distribution capture, 272
   equivalence classes, 273
   gene expression analysis; *see under* Gene expression analysis
   learning networks, 273
   likelihood equivalence, 274
   methodology, 270–271
   multiclass classification, 278
   node order, 274
   overview, 270
   scoring metrics, 273, 274–275
Ben-Hur method, 201–202
Biases, 114, 248

Binary features
   defining, 185
   distance measure for binary variables, 185–186
Binning, 159–160
Bioinformatics
   data analysis, 43
   ontologies; *see* Ontologies
   splice variations, cataloguing, role in, 26, 27, 29–30
Biological databases. *See also specific databases*
   annotation, data, 47
   argumentation, 63–65
   broad, 48
   categorization, 63, 253
   data cleaning; *see* Data cleaning
   data reconciliation, 63, 64
   data transformation; *see* Data transformation
   deep, 48
   dynamic nature of, 47
   general solution types, 49
   heterogeneous nature of, 44
   hierarchical structure, 47
   integration, data, 47, 67–71; *see also specific software and approaches*
   knowledge-based framework, 65–67
   management issues, 47
   multisource integration, 62–63
   noisy data, 114
   overview, 44
   physical data transfer, 108
   point solution types, 49
   primary, 48
   quality issues, 61
   schema issues, 61, 63, 169
   secondary, 48
   semantic incompatibility, 168

single-source techniques, 62
standardization issues, 47
volume, data, 44, 47
volume/size of data sample, 253
warehousing of data; *see* Warehousing, data
wrappers, use of; *see* Wrappers
Biological Magnetic Resonance Data Bank (BMRB) group, 55
BioMart, 73, 74
BIRCH clustering algorithm, 96
BLAST. *See* Basic local alignment search tool (BLAST)
Boosting, 287–291
Bootstrap sampling, 283
Borda count, 286
Brookhaven Protein Database, 53

C

C4.5, 115, 165, 279, 280
CAAT box, 31
Cancer, complexity of genetics of, 30
CART. *See* Classification as Regression Trees (CART)
Cell surface receptors, 4
Cells, human
   functions of, 3
   hereditary material in, 3
   membrane of, 3–4
Centroid shrinkage, 163
Chameleon algorithm, 226
Charge-coupled devices (CCDs), 15
Chebyshev inequality theorem, 89–90
Chromatin immunoprecipitation (ChIP), 32
Chromosomes, 6
Cis-regulatory modules (CRMs), 31
Cisternae, 4
Classification as Regression Trees (CART), 104, 115, 165
Classifiers
   flexible, 248
   linear, 248–249
   nonlinear, 248, 249
Cleaning, data. *See* Data cleaning
CLICK algorithms, 226, 227
Cluster trees, 92
Clustering. *See also Specific techniques and features*
   analysis, 182
   average linking, 202
   binary features; *see* Binary features
   ClusterONE; *see* ClusterONE
compactness, as validation technique, 310–311
connectedness, as validation technique, 311
connectivity measures, 224
defining, 181
density measures, 224
distance measure, properties of, 187–188
distance-based techniques, 183–185, 198–199
fuzzy; *see* Fuzzy clustering
graph-based cluster properties; *see* Graph-based clustering
hierarchical; *see* Hierarchical clustering
internal measures, 313, 314
k-means algorithms, 188, 190, 239, 310
k-modes algorithms, 190–191
kernel-based; *see* Kernel-based clustering
mixed variables, 187
model-based, 237, 238
nominal features, 186–187
overview, 181–182
Pearson correlation coefficient; *see* Pearson correlation coefficient
self-organizing maps (SOMs); *see* Self-organizing maps (SOMs)
spatial separation, as validation technique, 311
steps, 182
tight, 239–240
vertices, computing values for, 222–223
Clustering feature tree (CFt), 96
Clustering with local shape-based similarity (CLARITY), 194
ClusterONE, 230–232
Codons, 7, 10
Coexpressed genes, 192–193
Comparative genomics, 32–33
Complementary DNA (cDNA), 25
Computational functional annotation
   overview, 34–35
   sequence homology-based functional annotation, 35, 36
   structure-based functional annotation, 36–37
Contingency tables, 116
Convex peeling algorithm, 91, 92
Cross-validation, 123, 157
Curse of dimensionality, 58
Cytoplasm, 4
Cytoscape, 74, 230

D

Data cleaning. *See also* Warehousing, data
   approaches to, 86
   decomposition, 86

duplication, 59, 86, 99; *see also* Field matching techniques
erroneous data, 59, 60, 99
hybrid system models, 89, 96–97
inconsistencies, 59
input errors, 86
instance level, at, 65–67
integration; *see* Data integration
machine learning models; *see* Machine learning models
manual curation, issues with, 60–61
modeling data, 88, 89
neural network models; *see* Neural networks
nonparametric methods, 93
outlier detection, 87, 88–89, 90, 95
overlapping data, 63
overview, 85–86
parametric methods, 91–93
preprocessing, 66, 264
process of, 59–60, 85–86
processing, 66–67
proximity-based techniques, 90–91
reassembly, 86
semiparametric methods, 93
statistical models, 89–90
validation and verification, 67
Data integration
annotation, and, 47, 52
data linkage, relationship between, 97–98
database categorization, relationship between, 48
field matching techniques, 99
KDD process, importance to, 97
overview, 97
schema integration issues, 98–99
work flows, through, 42–43
Data mining, 83
association rule, for, 240–241
margins of separation, 260
WEKA software, using; *see* WEKA software
Data preparation. *See* Preparation, data
Data smoothing, 115
Data transformation, 61
decimal scaling, normalization by, 119
discretization, 115, 116–118
evaluation criterion definition, 121, 122
evaluation criterion estimation, 121, 122
feature construction, 121, 131–132; *see also* Matrix factorization
feature extraction, 121, 122, 267

feature selection, 128–130, 146
filters, 122, 123, 124–126, 127–128
generalization, 115
information theoretic ranking criteria, 121
matrix factorization; *see* Matrix factorization
max-min normalization, 118
nested subset selection, 130–131
overview, 115
Pearson correlation coefficient; *see* Pearson correlation coefficient
smoothing, 115
strongly relevant features, 119–120
weakly relevant features, 120
wrappers; *see* Wrappers
z-score standardization, 118–119
Data visualization, 81
Data warehousing. *See* Warehousing, data
Databases, biological. *See* Biological databases
Davie-Bouldin index, 206
DBSCAN, 96
Debauches wavelet functions, 161
Decimal scaling, 119
Decision trees, 92, 165
C4.5; *see* C4.5
challenges in construction, 279–280
conditional entropy, 280
guidelines for models, 279
ID3 algorithm; *see* ID3 algorithm
pruning, 280–281
Deoxyribonucleic acid (DNA), 5
bases, 6
chromosomes, 6
conversion of information, 8, 9
fragments, 168
information carried by, 6
methylation analysis, 16
nuclear, 5
sequencing of; *see* Sequencing, DNA
structure of, 5
Description logics (DLs), 169, 171
Deterministic annealing (DA), 92
Diagonal matrix, 134
Dimensionality
curse of, 58
input space, of, 253–254
reduction of through KDD, 83, 161–162, 257
DiscoveryLink, IBM's, 68, 69–70
Discretization, 115, 116–118
Distance metrics, 84–85

DNA. *See* Deoxyribonucleic acid (DNA)
DNA-protein cross-linking (DPC), 31
DNase footprinting assays, 31

E

E-R diagrams, 108
Edinburgh Mouse Atlas Project (EMAP), 64
Electrophoretic mobility shift assay (EMSA), 31
EMAGE, 64, 65
EMAP, 65
Endoplasmic reticulum, 4
Energy, Department of (DOE), 11
Ensembl, 27, 33, 68, 73
Ensemble learning, 281–283, 292–295. *See also*
        Bagging; Boosting; Random forest
EnsMart, 71, 73
Enzyme Commission Classification (EC), 35, 74
Enzyme-linked immunosorbent assay
        (ELISA), 31
Enzymes, 4
Escherichia coli, 17
Euclidian distance, 84, 142, 184, 223, 253
European Bioinformatics Institute, 68
Evolution, 32
Exact similarity computation, 194
Exons, 9
Expectation maximization, 92
Express sequence tags (ESTs), 29
Expression ratios, 148
Extraction transformation loading (ETL),
        98–99
Extrons, 30

F

Factor analysis (FA), 139–140
Fast fuzzy clustering algorithm (FFCA), 214–215
Field matching techniques
    character-based similarity metrics, 99, 100
    data linkage/matching techniques, 102, 103
    importance of, to duplicate detection, 99
    overview, 99
    probabilistic matching models, 103–104
    token-based similarity metrics, 101, 102
Fisher's linear discriminant analysis (LDA), 162
Fluorescent energy resonance transfer
        (FRET), 14
FOCUS algorithm, 124–126
Fourier transform analysis, 167
Frame-dependent k-mers, 167

Frameless k-mers, 167
Functional genomics
    annotation, 33
    methods, 25
    microarray-based, 30
    objectives, 25
    overview, 24
    prediction aspects, 33–34
Fuzzy c-means (FCM). *See under* Fuzzy
        clustering
Fuzzy clustering
    aim of, 210
    algorithms, comparison, 213–215
    fast fuzzy clustering algorithm (FFCA),
        214–215
    fuzzy c-means (FCM), 209, 210, 212, 213
    fuzzy J-means (FJM), 210–212
    fuzzy k-means, 212
    fuzzy probabilistic c-means (FPCM), 214
    gene, clustering of, 210–212
    mountain function, 208
    objective function, 208
    overview, 207–208

G

Gaussian functions
    data distribution, normalization and
        standardizaion, 118
    distribution, Bayesian networks, 277
    elimination via, 133
    Gaussian mixture models (GMMs), 93,
        197
    kernel-based clustering, 232, 265
    means and variances, distributions with, 227
    membership functions, 214
GenBank, 76, 168
Gene expression analysis, 156–157, 199
    Bayesian networks, computational
        challenges with, 278–279
    patterns, 264
    self-organizing maps (SOMs), using, 206
    shortest path, using, 228
    support vector machines (SVMs), use of,
        263–264
Gene Expression Omnibus (GEO), 49, 51,
        52–53
Gene flow, 20
Gene Ontology (GO), 35, 74, 171–172, 174
Gene regulatory network analysis, 30
GenePattern, 195, 202–203

Genes
    annotation of, 27–28
    codes in, 7
    differential expression of, 155–156
    functions of, 8
    heredity, role in, 6
    human body, number of in, 44
    human characteristics, 6
    prediction of, evidence-based techniques, 28
    prediction of, informant approach, 28
    scattered, 238
    splicing; *see* Splicing, gene
    transcription of, 8, 9
    translation of, 8, 10
Genome Sequence Database (GSDB), 168
Genome-wide association studies (GWASs), 42
Genomics, comparative. *See* Comparative genomics
Genomics, functional. *See* Functional genomics
Genotyping, 16
GenScan, 68
GLIMMER SYSTEM, 166
GO Annotation (GOA) project, 35
Golgi apparatus, 5
Gram-Schmidt process, 128–130, 133
Graph-based clustering
    cut in graph, 221, 225
    edge betweenness, 226
    gene expression analysis using shortest path, 228
    genetic linkage maps, construction of, 228–229
    intercluster, 221, 222
    intracluster, 221–222
    properties of, 219–220
    spectral methods, 225
GXD, 64, 65

H

Hamming distance, 229
HapMap, 74
HCOSM clustering algorithm, 200
Helicos Biosystems, 15
HGMD. *See* Human Genome Mutation Database (HGMD)
Hidden Markov model (HMM), 266
Hierarchical clustering
    agglomerative/bottom-up, 196–197
    applications, bioinformatics, 199, 202–203
    cluster merging, 197
    cluster splitting, 197
    irregular/partially overlapping data, with, 200
    overview, 196
Histidine-ammonia lyase, 35
Holdout method, 303
HTTP, 70
Human DNA, 5
Human Genome Database (HGD), 168
Human Genome Mutation Database (HGMD), 21
Human Genome Project (HGP), 8
    base pairs, 11
    completion, 12
    final phase of, 11
    initiation of, 11
    mapping, 12
    Mutation Database; *see* Human Genome Mutation Database (HGMD)
    objectives, 33
    sequences, 11, 12
    SNP data, 22; *see also* Single nucleotide polymorphisms (SNPs)
Hyperplanes, 258–259, 265
Hypotheses
    validity, 42

I

IBM, 69–70
ID3 algorithm, 279
Illumina, 15
    Genome Analyzer, 17
In vitro studies, large-scale, 25
In vivo studies, large-scale, 25
In-frame k-mers, 167
Independent component analysis (ICA), 140–141
Integration, data. *See* Data integration
IntelliClean, 65
Intensity-based normalization, 148, 149
International Union of Crystallography (IUCr), 57
Interpolated Markov model (IMM), 166
InterPro, 36
Introns, 9, 30
IRIS dataset, 200

J

Jaccard distance, 186
Jaro distance metric, 101

Java, 68, 70
JavaR Expert System Shell (JESS), 66
Jordan decomposition of a matrix, 137–138

K

K-means algorithms, 188, 190
K-modes algorithms, 190–191
KEGG pathways, 74
Kernel-based clustering
  algorithms, 233–234
  Gaussian functions, 232
  kernel functions, 232, 261–263
  kernel principal component analysis
    (KPCA) method, 233
  overview, 231–232
  self-organizing maps (SOMs), use in
    conjunction with, 235–237
  support vector clustering, relationship
    between, 234–235
Khiops discretization, 116, 117
KNN algorithm, 90, 96
Knowledge discovery, 42
Knowledge discovery in databases (KDD),
    42–43. *See also* Biological databases
  data handling and analysis, 43
  data integration; *see* Data integration
  data mining; *see* Data mining
  evolution of, 81
  feature extraction; *see under* Data
    transformation
  overview, 81–82
  process, 43, 78, 81–82, 256
  purpose of, 82, 83
  steps in, 82–83
Kohonen network, 214

L

Linear discriminant analysis (LDA), 162
Lipid bilayer, 3
Localizomics, 44
Locally weighted linear regression (LOWESS)
    analysis, 149–151
LOWESS. *See* Locally weighted linear regression
    (LOWESS) analysis

M

Machine learning models, 89, 95–96
Macromolecular Crystallographic Information
    File (mmCIF) schema, 56, 57

Mahalanobis distance, 84–85, 183–184
MALDI-TOF-MS. *See* Matrix-assisted
    laser desorption/ionization
    time-of-flight mass spectrometry
    (MALDI-TOF-MS)
Markov model, 166
MartView, 71
Mass spectrometry, 14
  applications, 16
  baseline subtraction/smoothing of data,
    158–159
  binning, 159–160
  dimensionality reduction, 160–161
  embedded multivariate methods of feature
    1, 65
  key data acquisition platform, role as, 16
  matrix-assisted laser desorption/ionization
    time-of-flight mass spectrometry
    (MALDI-TOF-MS), 16
  multivariate methods of feature selection,
    164–165
  normalization of data; *see* Normalization
    techniques
  overview, 157
  univariate methods of feature selection,
    163–164
Matrix factorization, 131
  diagonal matrix; *see* Diagonal matrix
  eigenvalues, matrix, 133, 134
  eigenvectors, matrix, 133, 134
  Jordan decomposition of a matrix; *see*
    Jordan decomposition of a matrix
  LU decomposition, 132–133
  overview, 132
  principal component analysis (PCA); *see*
    Principal component analysis (PCA)
  QR factorization, 133, 135
  singular vector decomposition (SVD),
    135–135
  spectral theorem, 135
  square matrix; *see* Square matrix
Matrix-assisted laser desorption/ionization
    time-of-flight mass spectrometry
    (MALDI-TOF-MS), 16
Max-min normalization, 118
Maximization algorithm, 215
Messenger ribonucleic acid (mRNA), 8, 9, 10
  gene indices, 29
Meta-data, 108
Micro-RNA (miRNA) expression, 146
Microarray Gene Expression Data (MGED)
    ontology, 174

Microarrays, 146, 154
  global filtering of, 155
  intensity-based filtering of, 153–154
  measurement of, 147
  Microarray Gene Expression Data (MGED) ontology, 174
  normalization of; *see* Normalization techniques
Microelectrophoresis, 13, 16
Min-max cut algorithm, 197
Minimum description length (MDL) principle, 116
Minimum volume ellipsoid (MVE) estimation, 91, 92
Minkowiski distance, 184
Mitochondrial deoxyribonucleic acid (mtDNA), 5
MmCIF. *See Macromolecular Crystallographic Information File (mmCIF) schema*
MOLAP. *See* Multidimensional online analytical processing (MOLAP)
MOLGENIS, 70
Mouse genes, databases for, 64, 65. *See also* specific databases
Multidimensional online analytical processing (MOLAP), 107
Multidimensional scaling (MDS), 141–142
Mutagenesis, 25
Mutations, genetic, 20
  genome-wide, 25
Mutual information measure, 193–194, 197

N

Naive Bayes classification. *See* Bayes classification, naive
Nanopore sequencing, 14, 16–17
National Center for Biotechnology Information (NCBI), 74
National Human Genome Research Institute, 11
National Institutes of Health (NIH), 11
National Library of Medicine, U.S., 76
Natural deoxynucleotides (dNTPs), 14
Neural networks, 89, 93–94
  supervised, 94
  unsupervised, 94–95
Nitrocellulose binding assays, 31
Nitrogen lyase, 35
Noisy data, 114, 250–251, 256
Nominal features, 186–187

Nonsynonymous single nucleotide polymorphisms (nSNPs), 22
Normalization techniques, 146, 147. *See also* specific techniques
  canonical normalization, 159
  direct normalization, 159
  global normalization, 149–151
  intensity-based normalization, 148
  inverse normalization, 159
  local normalization, 152–153
  logarithmic normalization, 159
  process of, 257
Nuclear DNA, 5
Nucleic acid database (NDB), 32

O

Object-oriented database management system (OODBMS), 107
OLAP. *See* Online analytical processing (OLAP)
OLTP. *See Online transaction processing (OLTP)*
OMIM. *See* Online Mendelian Inheritance in Man (OMIM)
Online analytical processing (OLAP), 105, 106–107
Online Mendelian Inheritance in Man (OMIM), 21, 74
Online transaction processing (OLTP), 106
Ontologies. *See also specific ontologies*
  defining, 168
  description logics (DLs); *see* Description logics (DLs)
  development, 169
  overview, 167–168
  role of, 169
OODBMS. *See* Object-oriented database management system (OODBMS)
Open Biomedical Ontologies (OBO), 172, 174
Open database connectivity (ODBC), 108
ORFeomics, 44
Organelles, 4
Overfitting, 114
OWL. *See* Web Ontology Language (OWL)

P

Pandemonium, 285
Parameter joint probability density functions (PJPDFs), 274
Partial least-squares-based dimension reduction (PLS), 138–139, 162

Pearson correlation coefficient, 120–121, 185
Performance bounds techniques, 123
Performance evaluation techniques
    classifier validation, 300, 301
    holdout method, 303
    k-fold cross validation, 305–306
    model selection, 301–303
    overview, 299–300
    random subsampling, 306
    receiver operating characteristics (ROC) curve, 309–310
    sensitivity, 307–308
    specificity, 307–308
    three-way split method, 304–305
Perl, 68
Pfam database, 48
Phage display (PD), 32
Pharmacogenomics, 44
Phenomics, 44
Plasma membrane, 3
Polymerase chain reaction (PCR), 14
Preparation, data, 113–114
    data transformation step; *see* Data transformation
Principal component analysis (PCA), 136–137, 138, 162
PRODOM, 36
Protein Data Bank (PDB), 49, 53, 55–57
    data integration in, 74
    structure mapping, 76
Proteins
    ligand interactions, 25
    amino acid sequences of human, 7–8
    expression of, 25
    modifications, 25
    sequences, 34
    synthesis of, 10
Proteomics, field of, 16, 44
PubMed, 76

Q

Q-grams, 101
QR factorization, 133, 135
Quantitative reverse transcriptase polymerase chain reaction (qPCR), 146

R

Rand index, 316
Random forest, 291–292

RCSB. *See* Research collaboratory for structural Bioinformatics (RCSB) PDB
RDBMS. *See* Relational database management system (RDBMS)
Receiver operating characteristics (ROC) curve, 309–310
Relational database management system (RDBMS), 107
Relational online analytical processing (ROLAP), 107
RELIEF algorithm, 126, 127–128, 164, 165
Research collaboratory for structural Bioinformatics (RCSB) PDB, 55, 76
Rete algorithm, 66
Ribonucleic acid (RNA), 8
    conversion of information, 9
    hybridization to arrays, 147
    sequencing, 16
    untranslated, 25
Roche, 15, 17
ROLAP. *See* Relational online analytical processing (ROLAP)
Rough endoplasmic reticulum (RER), 4

S

Saccharomyces Genome Database (SGD), 48
Sanger Institute, 68
Sarcoplasmic reticulum (SR), 4
Self-Defining Text Archive and Retrieval (STAR) language, 57
Self-organizing maps (SOMs)
    algorithm, 203, 205
    gene expression, distinct, use in identifying, 206
    kernel-based clustering, analyzing gene expression in conjunction with, 235–237
    overview, 203
Self-organizing maps (SOMs), 95
Selfridge's pandemonium, 285
Sequence data, genomic
    content analysis, 166
    overview, 165–166
    sequence features, 167
    signal analysis, 166
Sequence Retrieval System (SRS), 68–69
Sequencing by hybridization (SBH), 13, 15
Sequencing, DNA
    cost of, 13
    cyclic array sequencing, 13, 15

de novo assembly, 18, 19
dideoxy technique, 13, 14
    fragments, from, 17–18
    handling, 18–19
    hash table-based algorithms, 19
    Human Genome Project, as part of; *see under* Human Genome Project (HGP)
    hybridization; *see* Sequencing by hybridization (SBH)
    mass spectrometry; *see* Mass spectrometry
    microelectrophoresis; *see* Microelectrophoresis
    nanopore technique; *see* Nanopore sequencing
    natural selection/variation, 20–21
    next-generation technology, 17–18
    nucleotides, of, 10, 14
    polymorphisms, 20
    suffix-based trees, 19–20
Sex, genetic variations based on, 20
Silhouette index (SI), 314–315
Single nucleotide polymorphisms (SNPs), 21, 71
    abundance of, 21
    characterization, 22, 24
    genetic markers, use as in ensemble approach
    intronic, 22
    mapping, 21
    nonsynonymous; *see* Nonsynonymous single nucleotide polymorphisms (nSNPs)
    random forest analysis with, 294
    screening, for variation, 292
    size, of data, 24
    SNP interaction identification, 293
Singular vector decomposition (SVD), 135–135
SMART, 36
Smith-Waterman alignment, 19
Smith-Waterman distance, 100
Smooth endoplasmic reticulum (SER), 4
Snowflake schemas, 108
SOAP family of DNA alignment tools, 19
Solexa/Illumina, 15
SOTA technique, 206–207
Spectal theorem, 135
Splicing, gene
    ab *initio* programs, 27, 28
    alignments, clustering of, 29–30
    alternative, 28
    alternative, of pre-mRNA, 26–27
    annotations, 29
    cataloguing, 26
    regulation of, 27, 28
SQL. *See* Structured query language (SQL)

Square matrix, 134, 137
Star schemas, 108
START codon, 7
STOP codon, 7, 8
Structural Classification of Proteins (SCOP), 266
Structured query language (SQL), 68, 69
Supervised learning, 104
Support vector clustering, 234–235
Support vector machines (SVMs), 104, 257–258
    Fisher algorithms, 266
    pairwise algorithms, 266
    gene expression analysis, use in, 263–264
    optimization algorithm, 267
    performance of, 301
    training of, 267
Surface antigens, 4
SwissProt, 33, 48
Systemic evolution of ligands by exponential enrichment (SELEX), 32

T

TATA box, 31
Taverna, 74
Term frequency-inverse document frequency (TF-IDF) weighting, 102
Three-way split method, 304–305
Total intensity normalization, 149
Transcription factors (TFs), 10–11, 30
Transcriptomes, 29
Transfer RNA (tRNA), 8
Transformation, data. *See* Data transformation
Transporters, 4

U

UCSC Genome Browser Database, 27
UniProt, 74, 76
Unsupervised learning, 104
Uracil, 9

V

Variances, 248, 250

W

Warehousing, data. *See also* Data cleaning
    defining, 104
    designs of, 70–71
    flexibility, 71

focus of, 105
lifecycle, 107–109
overview, 104–105
purging, 109
query interfaces, 73–74, 76
success of, 70
Web Ontology Language (OWL), 169, 172
WEKA software, 195, 215
WHIRL, 102
Wilcoxon rank test, 163
Within-class scatter matrix, 162
Work flows, scientific
constraints, 42
overview, 41–42

Worldwide Protein Data Bank (wwPDB), 55. *See also* Protein Data Bank (PDB)
Wrappers, 70, 107, 108, 157
overview, 123

X

XQL, 69
XQuery, 69

Z

Z-score standardization, 118–119